Biochemical Evolution

Second Edition

The Pursuit of Perfection

Athel Cornish-Bowden

NEW YORK AND LONDON

Garland Science
Vice President: Denise Schanck
Senior Editor: Summers Scholl
Editorial Assistant: Michael Roberts
Production Editor: Natasha Wolfe
Cover Designer: AM Design
Copyeditor: John Murdzek
Proofreader: Chris Purdon

ISBN 978-0-8153-4552-7

The Library of Congress Cataloging-in-Publication data:

Names: Cornish-Bowden, Athel, author.
Title: Biochemical evolution : the pursuit of perfection / Athel Cornish-Bowden.
Other titles: The pursuit of perfection
Description: 2nd edition. | New York, NY : Garland Science, Taylor & Francis Group, LLC, [2016] | Revised edition of The pursuit of perfection : aspects of biochemical evolution / Athel Cornish-Bowden. 2004.
Identifiers: LCCN 2016013922 | ISBN 9780815345527
Subjects: | MESH: Evolution, Molecular | Biochemical Phenomena
Classification: LCC QH325 | NLM QU 475 | DDC 572.8/38--dc23
LC record available at http://lccn.loc.gov/2016013922

Published by Garland Science, Taylor & Francis Group, LLC, an informa business, 711 Third Avenue, New York NY 10017, USA, and 3 Park Square, Milton Park, Abingdon, OX14 4RN, UK.

Printed in the United Kingdom

15 14 13 12 11 10 9 8 7 6 5 4 3 2 1

Visit our website at http://www.garlandscience.com

To Enrique Meléndez-Hevia,
the originator of many of the ideas discussed in this book

Contents

Foreword

JUST OVER A DECADE AGO, a smallish book with the intriguing title *The Pursuit of Perfection* came out. The subtitle, *Aspects of Biochemical Evolution*, indicated its contents. It was a little gem. In this, its successor, Athel Cornish-Bowden has exchanged the title and subtitle and produced a gem with more facets. It should attract biologists, scientists from diverse areas and, not least, anyone who appreciates good scientific writing. Rather than the second edition it purports to be, this version is a new avatar. Its heart consists of a series of demonstrations of the ways in which the intricacies of metabolism can be unravelled by viewing them as products of evolution by natural selection—that is, as a set of adaptations. At the same time, attention is drawn to two classes of phenomena. Some are outcomes of chance and are best understood as "frozen accidents" of history. Others are apparent maladaptations. To understand their persistence, one has to invoke an evolutionary past when they were beneficial. *Biochemical Evolution* combines information, instruction, and entertainment in equal parts. Dire warnings accompany the risk of arguing from analogy. But its intuitive appeal cannot be denied. When employed with care, as here, its pedagogic benefit cannot be denied.

I can give only the barest outline of the rich contents of this volume. Each of the several themes addressed by Cornish-Bowden has been chosen to illustrate how biochemistry, long neglected by evolutionary theorists on account of its seeming intractability, actually provides illuminating examples of the working of natural selection (Enrique Meléndez-Hevia is repeatedly acknowledged for having pioneered selectionist thinking in biochemistry). In order to recognize this, it is necessary to move away from thermodynamics and its emphasis on potentials—of doubtful utility anyway, if living systems function far from equilibrium—and adopt a kinetic approach, one that focuses on fluxes and concentrations. While doing so, it is important to keep in mind that chemical molecules play two distinct roles in the cell. They can be intermediates in the transformation of one molecule to a second and be utilized for themselves, so to speak. On the other hand they can function as regulators, as signals. Then they play a symbolic role, which is a different thing altogether.

At first sight the thicket of metabolic pathways makes the task appear forbidding, but the feeling derives from a fundamental mistake. The mistake is to assume that something that is complicated, for example metabolism, also has to be complex. The inference is incorrect. By focusing on the pattern of interactions that underlie a pathway, one can ask: Given the constraints within which the system operates, is there an optimal route to the outcome, and if so, what is it? Once the criteria of optimality are decided on, the question can be tackled by logical analysis. The analysis is on the whole qualitative and rests more on topology and geometry than algebra. The strategy turns out to be successful, and it is interesting to see why. It is because often the biochemical network within a cell can be disentangled piecemeal, as it were, by regarding it as a restricted subnetwork that contains the reactions of immediate interest and an environment. The environment contains sources and sinks and also imposes constraints within which the subnetwork must operate.

Once an appropriate separation of variables has been achieved, the subnetwork becomes amenable to analysis. From an evolutionary perspective, one might wonder why the approach works at all. Clearly the answer must have to do with the fact that the subnetworks fulfill distinct needs within the overall economy of the cell. To put it differently, in a certain sense they operate like modules. How different modules came together, to what extent they overlap, and so on, are fascinating questions for the future. The notion of modularity has long permeated the fields of anatomy and physiology, and of late is much discussed in developmental biology. *Biochemical Evolution* shows that it may be meaningful in biochemistry as well.

The quality of Cornish-Bowden's writing enhances the pleasure of going through *Biochemical Evolution*. At the same time, the text requires close attention. The deceptively simple style can easily convince you that you have understood something when you have not. After a concise introduction to biochemistry and molecular biology, the contents encompass biochemical adaptation, nonadaptive (neutral) evolution, the genetic code (might it be an adaptation, too?), DNA-based evolutionary trees (with the interesting aside that Adam and Eve may never have met), Mendelian genetics, inborn errors of metabolism, feedbacks and regulation, how flux is shared between different steps in a biochemical network, and how aneuploidy might lead normal tissue to transform into a cancer. In every case there is a clear statement of the problem, a discussion of what a solution might entail and, finally, a demonstration of the solution itself. The book ends with two chapters that stand apart from the rest. They contain possible chemistry-based approaches to the origin of life problem and a forthright attack on the hypothesis of "intelligent design"; the quotation marks are there for a purpose. The chapters that deal with the sequential steps of the pentose phosphate cycle and the structure of glycogen will evoke the most admiration. The issues behind the two are treated as mathematical diversions in the manner of Martin Gardner (part of its attractiveness is that *Biochemical Evolution* is put together as a series of mathematical diversions). After sketching their essence in formal terms, Cornish-Bowden shows how straightforward the solutions are. He uses careful, step-by-step reasoning to make the point that in both situations evolution has indeed pursued perfection: by optimizing the number of intermediate steps in the cycle in the first, and the size and branching configuration of the glycogen molecule in the second. Who would believe that the solution had anything to do with open key cryptography, or with the childhood puzzle of the farmer who had to transport a fox, a goose and a sack of oats across a river in a boat that could hold only two of them at a time? This is pedagogy at its best.

Along the way, Cornish-Bowden takes spirited swipes at the supercilious-ness towards biochemistry (or biochemists?) that has long been displayed by those who found it difficult to grasp. People have railed against the attitude before, famously Erwin Chargaff when he said molecular biology was the practice of biochemistry without a licence. Cornish-Bowden is equally forthright in expressing his annoyance with the freewheeling ways of those theoretical physicists who have displayed a tendency to "lead ... the reader from sets of more or less plausible but entirely arbitrary assumptions to precisely stated but irrelevant theoretical conclusions", as Wassily Leontief[1] said about practitioners of his own field. The quotations that precede each chapter are worth returning to after finishing the chapter. They are apt and reflective, excellent provokers of thought. Everyone will have their own favorite; mine is the very first sentence of a quote from Frédéric Dardel at the head of the new Preface.

An abundance of good literature is available to anyone who wants to get a feeling for the excitement that pervades the study of evolution. Today it is possible for a lay person to realise why "nothing in biology makes sense except in the light of evolution". However, there is an anomaly. What is central to a living cell, what endows it with life—namely, the gamut of chemical processes that go on within it—has not had the attention that it deserves: Biochemistry is conspicuously missing from the literature of evolution, at least in English. *Biochemical Evolution* amply makes up for the deficit. Readers will get a glimpse of what lies ahead in the little-explored and potentially vast field of biochemical adaptation. A feast awaits them.

Vidyanand Nanjundiah
Centre for Human Genetics, Bangalore, and
Stellenbosch Institute for Advanced Study, Stellenbosch

[1]Wassily Leontief (1906–1999), an American economist of Russian origin, developed methods for analyzing interindustry transactions and won the Nobel Memorial Prize in Economic Sciences in 1973.

Preface

> Don't tell my mother that I'm a biochemist; she thinks I'm a pianist in a brothel. This paraphrase of the title of a book by the publicist Jacques Séguéla illustrates, perhaps in rather an extreme way, some aspects of our subject. In the same vein I could perhaps entitle this editorial "I'm a biochemist, but I'm having treatment." What I want to say is that biochemistry ... has become shameful, with an inferiority complex in comparison with its sister subjects, more cellular, and more appealing because they offer more global and integrated approaches.
>
> Frédéric Dardel, 2010[1]

BOOKS ABOUT EVOLUTION, at all levels from the most popular to the most technical, are widely available, and some of them are excellent. They tend to be books about anatomy or behavior and to ignore biochemistry except insofar as it impinges on these other aspects, but our understanding of evolution has been transformed by biochemical information. All paleontologists agreed 50 years ago that humans had no close relatives in the animal kingdom: The great apes were far more closely related to one another than any of them was to us, and they separated from the human line about 30 million years ago, about the same time as both apes and humans diverged from the monkeys. Nearly all paleontologists now agree that the separation between the African apes and humans is far more recent, after the separation of these species from other apes such as orangutans and gibbons, and long after the separation of the apes and monkeys. The reassessment has not been driven by abundant new fossils, or by a better understanding of anatomy or behavior, but by the deluge of biochemical data that began in the 1960s and has continued ever since.

In a sense this is a revised edition of *The Pursuit of Perfection: Aspects of Biochemical Evolution*,[2] but it contains a large amount of new text and many more illustrations. The earlier book led to graduate courses in several universities: Patrice Soumillion invited me to the Université Catholique de Louvain, in Louvain-la-Neuve, Belgium; afterwards Rafael Vicuña organized a similar course at the Universidad Católica de Chile, in Santiago, and Juan-Carlos Slebe organized one at the Universidad Austral de Chile, in Valdivia

[1]F. Dardel, President of the French Society of Biochemistry and Molecular Biology (2010) *Regard sur la Biochimie*, July 2010, 1. Ne dites pas à ma mère que je suis biochimiste, elle me croit pianiste dans un bordel. Cette paraphrase du titre d'un livre du publicitaire Jacques Séguéla illustre de manière un peu extrême certains aspects de la situation de notre discipline. Dans la même veine, j'aurais aussi pu intituler cet éditorial "je suis biochimiste, mais je me soigne." Ce que je veux dire par là, c'est que d'une certaine manière, la biochimie ... est devenue honteuse, complexée devant ses sœurs, plus cellulaires, plus séduisantes parce que proposant des approches plus globales et plus intégrées.

[2]Oxford University Press, 2004.

(this last enlivened by tear gas and bomb threats, because it occurred during a period of student unrest). These were not only interesting for me, but they also constituted an educational experience, because I learned much from interacting with the participants—in particular how some of the material could be presented more clearly.

Many of the ideas discussed here, especially on the relationships between evolution and biochemistry, were developed originally by Enrique Meléndez-Hevia, and in recent years he has analyzed two major problems that natural selection has failed to solve: those of collagen-related diseases and of obesity in the human population. The first of these is ancient, with its roots in the appearance of the first animals, and the other recent (in evolutionary terms), derived from the spread of agriculture. Both are biochemical in character, and for both we need to understand why natural selection has not managed to solve them. A new Chapter 11 deals with them.

As Franklin Harold remarks at the beginning of Chapter 14, biologists concentrate their efforts on the detailed mechanisms of life, largely ignoring the question of what life is. One can regard that as a serious omission, and in the closing paragraphs of the earlier book I touched on the question of self-organization. Reading it now I am embarrassed at how superficial the treatment was. Here, therefore, I develop this idea more seriously, trying to bring one of the least studied (though far from the least important) themes in modern biology to a broader audience.

At the beginning of the century, the rise of creationism was just a speck on the horizon so far as much of the world was concerned—something for scientists in the USA to worry about, but with little effect on the teaching of biology elsewhere. We were already wrong about that then, but now it is obvious to everyone who does not keep their head in the sand that creationism has become a serious threat to biology teaching everywhere. Chapter 15 deals with this subject.

I have included brief biographical sketches of many of the people who have contributed to our understanding of biochemistry and evolution, some of them well-known, like Gregor Mendel, others far less well-known than they deserve to be, like Elizabeth Fulhame, Nettie Stevens, and Marthe Gautier. (It is not by chance that these are all women: If they had been men they would be far better known than they are.)

The book ends with a quotation from the great evolutionary biologist George Gaylord Simpson, and I end this Preface with two other quotations, the first from the same article:[3]

> I was on the receiving end of high school education in one of the

[3] G. G. Simpson (1961) "One hundred years without Darwin are enough" *Teachers College Record*, **60**, 617–626.

> good public school systems (by the then standards) of the 1910s.... After a bit of a struggle, I achieved a sound routine knowledge of mathematics at the intermediate levels. As for science, that was limited to one course called "physics," which, as far as my memory goes, consisted of measuring things (lengths, weights, times, temperatures) and making the measurements agree with the book. I learned, and later had to unlearn in order to become a scientist myself, that science is simply measurement and the answers are in print.

Much of science education still, unfortunately, consists of measuring things, and much of biochemistry education, especially in metabolism, consists of memorizing things. In this book I take a different approach, that even some of the most unmemorable parts of biochemistry may follow a logic, and understanding the logic is helpful for remembering the facts. As for assuming that "the answers are in print," one of the proudest moments in my teaching life came when I was talking with someone who had attended my lectures on enzyme kinetics some 20 years earlier, and she told me that the most important thing she had learned from these lectures was that she should not believe everything she read in a textbook, but should approach it critically.

The second comes from a speech made in 1911 by Francis Darwin, the third son of Charles Darwin, at the opening of the Darwin Laboratories at Shrewsbury School:

> When science began to flourish at Cambridge in the 'seventies,[4] and the University was asked to supply money for buildings, an eminent person objected and said, "What do they want with their laboratories? — why can't they believe their teachers, who are in most cases clergymen of the Church of England?" This person had no conception of what the word "knowledge" means as understood in science.
>
> Another characteristic of science is that it makes us able to predict.

In the course of the book, I will describe several examples to illustrate these points, especially the last one. The idea that links them all is that biochemistry and evolutionary biology are closely linked. Not only is a biochemical perspective necessary for understanding many of the advances in our knowledge of evolution during the past 50 years, but an evolutionary perspective allows us to rationalize many details of biochemistry that would otherwise be mysterious.[5]

[4] 1870s

[5] A. Cornish-Bowden, J. Peretó, and M. L. Cárdenas (2014) "Biochemistry and evolutionary biology: Two disciplines that need each other" *Journal of Biosciences* **39**, 1–15.

I am very grateful to Joe Felsenstein, Enrique Meléndez-Hevia, Ron Milo, Vidya Nanjundiah and Jeffrey Wong for their useful comments, and I am particularly grateful to Vidya for his Foreword. Special thanks are due to Marilú Cárdenas, who went through the entire book with me in detail, and made many very valuable suggestions. I also thank the Director of the CNRS unit *Bioénergétique et Ingénierie des Protéines*, Marie-Thérèse Giudici-Orticoni, and all of her research group, for making the environment in which the book was written pleasant and welcoming. I am also grateful to the CNRS itself for granting me the emeritus status that made it possible for me to work on it.

Athel Cornish-Bowden, Marseilles, 2016

1. *Some Basic Biochemistry*

From the elephant to the butyric acid bacterium—it is all the same!

Albert Jan Kluyver, 1926[1]

"Best enzyme man in the world," I said.
I heard Dawlish cough.
"Best what?" he said.
"Enzyme man," I said "and Hallam would just love him."
"Good," said Dawlish.

Len Deighton, 1964[2]

I think if people are passionate about something, it could be real estate or biochemistry, and that spark gets turned on in them, everyone's beautiful in that zone.

Cindy Crawford, 2011[3]

POPULAR BOOKS ON EVOLUTION are typically books about behavior, and fascinating they are to read, too. Their authors have an easier task than mine, because they can assume that their readers have at least some idea of the basic terminology of their subject. They do not have to spend their earlier chapters explaining what eating is for, what sex involves, how elephants differ from artichokes, and so forth; instead they can march straight into their subject secure in the knowledge that their readers have at least some idea what they are talking about. Later on, they may want to define some of their terms more precisely, but they can safely leave this task until the reader's attention is well and truly engaged.

Biochemical evolution is more difficult to describe, because biochemistry is a less well-known science that does not feature much in popular literature. When you do find a mention of a biochemist in a novel, it is likely to be as vague as the quotation from *Funeral in Berlin* at the beginning of this chapter, and even reading the whole book will not tell you what an enzyme is, why this may be useful or important to know, or even that enzymes constitute an essential topic in the study of biochemistry. Likewise, we are introduced quite early in *Wilt*[4] to a character described as a biochemist, but a few pages later we

[1]Handwritten lecture notes quoted by H. C. Friedmann (2004) "From 'Butyribacterium' to 'E. coli'—*An essay on unity in biochemistry*" *Perspectives in Biology and Medicine* **47,** 47–66.

[2]Len Deighton (1964) *Funeral in Berlin*, Jonathan Cape, London.

[3]Quoted in *Visionary Artist Magazine*, December 18, 2011

[4]Tom Sharpe (1978) *Wilt,* Secker and Warburg, London.

learn that he wrote his doctoral thesis on "role play in rats," and nothing in the rest of the book suggests that he ever did anything that a biochemist would call biochemistry.[5] Fortunately, most of the details of biochemical knowledge are unimportant for this book, but some of them matter, and so I will begin with a few pages introducing some of the most basic ideas of biochemistry.

Unity of biochemistry

Before getting down to any details, there is one point about biochemistry that is so familiar to biochemists that they do not always remember to mention it, though it appears bizarre in the extreme to other biologists. It is essential in general biology to be clear about which organism or group of organisms you are talking about, because almost everything varies wildly from one sort of organism to another. Biologists are accordingly rather shocked at the vagueness—to the point of sloppiness—of biochemists about species. To take one example, the enzyme hexokinase is necessary for the first step in obtaining energy from glucose, a major dietary component; it is found in organisms as diverse as yeast, wheat, rats, and bacteria. The form of hexokinase that was studied in most detail until recent years comes from brewer's yeast, *Saccharomyces cerevisiae*, and when biochemists referred to "hexokinase" without qualifying it they usually meant yeast hexokinase.

ALBERT JAN KLUYVER (1888–1956) was a Dutch microbiologist and biochemist. He and Hendrick Jean Louis Donker published *Die Einheit in de Biochemie* (*Unity in Biochemistry*) in 1926. Its vision of biochemical unity was strongly influential, and he is regarded as the father of comparative microbiology.

This sort of thing can seem quite absurd to a zoologist: How could you guess some unknown fact about human anatomy by looking at the corresponding feature of a fungus? What would be the "corresponding feature," anyway, for creatures so different from one another? It works amazingly well most of the time in biochemistry, because all organisms are made from proteins, carbohydrates, fats, and nucleic acids, use the same sort of chemistry to make these out of the food they consume and convert them into one another, and use the same method for storing the information they need for reproducing themselves. To a first approximation, therefore, biochemistry is indeed the same in all organisms, the point of Albert Kluyver's remark quoted at the beginning of this chapter.[6] This is nowadays more familiar in various versions used by Jacques Monod, of which "anything that is true of *Escherichia coli* must be true of elephants, only more so" expresses the idea most forcefully. *Escherichia coli*, a bacterium originally isolated from

[5]By contrast, although Cindy Crawford is better known as a model than as an engineer, she did study biochemistry for a brief period as a student in chemical engineering at Northwestern University, in Chicago.

[6]There are important differences as well: *Photosynthesis* is an essential part of plant metabolism, but does not occur in animals; all animals depend on the protein *collagen*, as discussed in Chapter 11, but other organisms do not.

Figure 1.1 A bird's knee. A flamingo appears to bend its "knee" the wrong way. The confusion arises from interpreting its ankle as its knee. Its true knee is far up the leg and hidden by feathers. The same is true of other birds, but it is particularly obvious in flamingos.

human feces, is one of the most intensively studied organisms in biochemistry. Much of what we know about metabolism came from studies of it, and biologists have sometimes contemptuously referred to it as "the only animal known to biochemists."

This idea of the unity of biochemistry is not just a lazy shortcut for second-rate scientists unwilling to spend the time filling in all the details: It is taken quite seriously as a way of rationalizing and integrating the subject, and, as noted a moment ago, was popularized by Jacques Monod, the great bacteriologist and biochemist who won the Nobel Prize in Physiology or Medicine in 1965. Along with Francis Crick and James Watson, he is among the few Nobel laureates whose names remain familiar some years after the award. Much earlier than him, we can find Geoffroy Saint-Hilaire commenting in 1807 that "there is, philosophically speaking, only a single animal," or Thomas Henry Huxley, in 1868, that "all living forms are fundamentally of one character." Much earlier still, as we will see in Chapter 14, Denis Diderot suggested that "there was never anything but a single animal prototype of all the animals." Whatever reservations we may have about such statements in biology as a whole, there is no doubt that they have taken biochemistry a long way. Incidentally, we may be tempted to assume that when Saint-Hilaire referred to animals he meant just mammals, but that would be wrong, as he intended the term much more broadly, including, for example, insects and spiders.[7]

ÉTIENNE GEOFFROY SAINT-HILAIRE (1772–1844) was a French naturalist who established the principle of *unity of composition*. He was a colleague of Jean-Baptiste Lamarck, and he expanded and defended Lamarck's evolutionary theories. He participated in Napoleon's expedition to Egypt, and was subsequently Professor of Zoology in Paris.

Elephants are not quite the same as *Escherichia coli*, but so far as their biochemistry is concerned the similarities are more striking than the differ-

[7]If you look at flamingos or other long-legged birds, it may puzzle you that their "knees" appear to bend the wrong way (Figure 1.1). How can all vertebrates—let alone insects and spiders—be built on the same plan if birds and mammals differ in such a fundamental way? The explanation is that what we interpret as a flamingo's knee is actually its ankle—its true knee is almost hidden by its body.

ences: Both contain protein, their proteins are made with the same 20 basic ingredients, and the choice between these ingredients is made according to the same genetic code, written in the same genetic language; both base their cellular economy on the extraction of energy from glucose; and so on, and so on. The proteins that both contain are, moreover, used for the same sorts of purpose: Protein forms the principal building material for both *Escherichia coli* and elephants, and both use proteins as their preferred material for making the molecules that allow all the chemical reactions in a living organism to take place. Substances that need to be present for a reaction to be possible, even though they are not themselves consumed or produced by it, are called *catalysts*, and these are used not only in living organisms but also in many industrial processes. Natural catalysts in living organisms are usually called *enzymes*, and it is these that are nearly always made of protein.

Later in this book (particularly in Chapters 4 and 5) we discuss *metabolic pathways*, the sequences of chemical reactions that interconvert different cellular components. These are not absolutely identical in all organisms, but again, the similarities are striking. For example, the sequence of reactions that convert sugar into alcohol in yeast fermentation occurs in almost exactly the same form in humans, and it has almost exactly the same function: to extract energy from glucose. Only the end of the process is different, because yeast makes alcohol and we do not. Despite that, the enzyme (alcohol dehydrogenase) that makes the alcohol in yeast exists also in ourselves, but the reaction goes in the opposite direction: We use it to remove alcohol.[8]

ELIZABETH FULHAME (late eighteenth century) studied the chemical nature of the oxidation of metals and the reduction of their oxides. She set out her conclusions in *An Essay on Combustion* (London, 1794), a work of genius. Almost nothing is known of her life other than the fact that her husband Thomas Fulhame had studied medicine at Edinburgh. She not only introduced the idea of *catalysis*, usually attributed to Jöns Jakob Berzelius, who studied it 40 years later, but her work on the effect of sunlight on silver compounds also made her a founding figure in the origins of photography.

Water

It is sometimes said in England that even the Archbishop of Canterbury consists of 99% water, and doubtless similar ideas are current in other countries. This is an exaggeration, because the Archbishop is only about two-thirds water.[9] Nonetheless, the idea is qualitatively right, because water is by far the most abundant component of the human body, and

[8]That is perhaps an oversimplification. It certainly acts to remove alcohol when we consume it in substantial amounts, but it also exists in animals that do not consume alcohol as such. In some cases these are animals that eat food like rotting fruit that contains ethanol, or they receive ethanol from bacterial fermentation in their digestive systems.

[9]The version that was current when I was a child was probably an exaggerated adaptation of a much more accurate statement of the great geneticist J. B. S. Haldane (Chapter 9) that even the Pope was 70% water.

indeed of almost any other living organism. The chemistry of life, therefore, is largely the chemistry of processes in water. The eighteenth-century pioneer of chemistry, Elizabeth Fulhame, argued that *all* oxidation and reduction reactions involved water as an intermediate—another exaggeration, but with a measure of truth. Although we now know that some of the most interesting processes occur in or across the membranes that separate cells or parts of cells from one another, water is usually involved even then.

This may seem obvious, because water is also by far the most familiar fluid in our lives, but it is important to realize that the chemistry of reactions in water under mild conditions of acidity and temperature is not what most chemists study most of the time, and that what a chemist regards as mild conditions may be quite extreme from the point of view of a living cell. On the shelves of any (old-fashioned) chemistry laboratory you are likely to find a bottle labeled "dilute HCl," meaning dilute hydrochloric acid. A substance is called an acid if it is a good source of protons (hydrogen atoms that are positively charged because they have lost their negatively charged electrons). A large part of chemistry, especially the chemistry of processes in water, is concerned with moving protons around. Now, what a chemist calls "dilute acid" contains a concentration of protons more than 10 million times greater than that in a typical living cell.

Enzymes

It follows that the reactions necessary for life must proceed under conditions vastly milder than those that chemists regard as mild. As virtually all the reactions we are concerned with are extremely slow under such mild conditions—for practical purposes most of them do not proceed at all—they require catalysts, which in living systems are, as noted earlier, enzymes. Until fairly recently, all were thought to consist mainly or entirely of protein, and although the discovery in the 1980s of enzymes based on RNA (one of the major classes of nucleic acids, which I will come to shortly) changed this perception somewhat, it remains true of the majority of enzymes known.[10] For most biochemical purposes, and specifically for those discussed in this book, we will not go far wrong if we continue to think of enzymes as proteins.

We are usually impressed that enzymes are highly effective catalysts—that is, they can make reactions that are normally barely detectable proceed rapidly. This misses a crucial point, because making a reaction go faster is one of the easiest things a chemist can do. Virtually any reaction can be made as fast as you like by confining the reactants in a sealed container and heating them to a high temperature. There are no important exceptions to this generalization, so why should we be impressed at the capacity of enzymes to do the same thing?

[10]We will see in Chapter 14 that the usual definition of an enzyme as a *biological catalyst* fails to exclude many molecules that are not usually regarded as enzymes, but for the moment we will ignore that.

The fact that an enzyme can do it at low temperatures is part of the answer, but the really impressive aspect of an enzyme is not that it is a *good* catalyst for a given reaction, but that it is an extremely *bad* catalyst—not a catalyst at all—for virtually every other reaction. Heating the reactants, effective though it is, is just as effective at accelerating large numbers of other reactions that we do not wish to accelerate; in other words, it lacks *specificity*. That is what enzymes abundantly provide, and it is their specificity that ought to impress us, because it is specificity that allows all the reactions of life to proceed in an orderly fashion under the mild conditions that exist in a living cell. These mild conditions are not a luxury, nor are they a chance consequence of living on a planet with abundant water at a moderate temperature; they are an absolute necessity for life. Without mild conditions it would be impossible to prevent a mass of unwanted reactions from proceeding in parallel with the desirable ones. This raises another question: If mild conditions are essential, how is it that certain organisms, the *extremophiles*, manage to live in conditions far less mild than what we take for granted—near to deep ocean vents, for example, at temperatures far above the ordinary boiling point of water? This is an interesting question, but I will not develop it here. From the point of view of some organisms, including all of those at the origin of life, and some that still exist today, we are extremophiles ourselves, because we can tolerate, and indeed depend on, the molecular oxygen in the atmosphere, which is highly toxic to most living cells.

Specificity is expensive, however. If the catalysts are to be highly specific there must be many different kinds of catalyst, roughly one for each reaction that the cell needs to undergo. Even reactions that go quite fast without one may need to be catalyzed in the living organism. For example, removing the water from dissolved bicarbonate so that it can be released in the lungs as a gas, carbon dioxide, is a fast reaction in the absence of any catalyst. It goes almost to completion in a few minutes, but that is not fast *enough*, and many living organisms use the enzyme carbonic anhydrase to accelerate it by many thousandfold. Some bacteria even use two different carbonic anhydrases on the two sides of a membrane: one to convert the negatively charged bicarbonate ion into neutral carbon dioxide so that it can pass through the membrane as an uncharged molecule, the other to convert it back into bicarbonate on the other side. Specificity also means that enzymes have to have complicated (and hence large) structures: A simple structure such as that of the metal platinum may work well as a general catalyst, but there is no way to make a specific catalyst with anything as simple as platinum.

An enzyme is a large molecule that contains at least 1000 atoms (it is normally at least 20,000 times greater in mass than a hydrogen atom), but usually many more than that: Compare this, for example, with a molecule of the sugar glucose, which is a mere 180 times heavier than a hydrogen

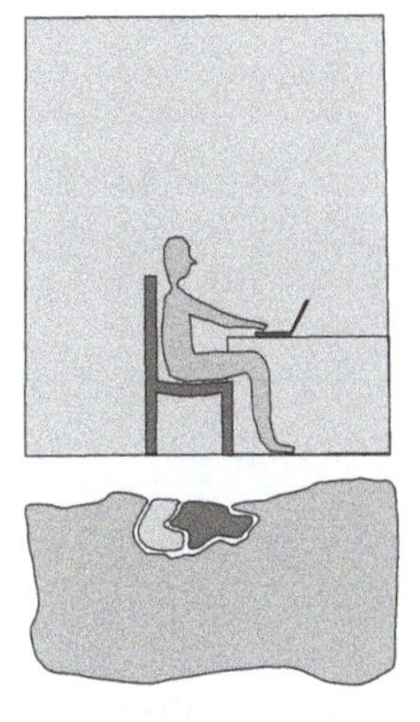

Figure 1.2 Sizes of familiar and unfamiliar objects. A typical small room used as an office for one person has a volume about 200 times the volume of the person. On the other hand, an enzyme is often regarded as a very large molecule even if it occupies no more than 50 times the volume of the molecules it acts on. The objects illustrated have *depth* as well as areas, so, for example, although the area occupied by the enzyme in the lower diagram is only about 13–14 times the area occupied by the substrates, this ratio corresponds to a volume ratio of the order of $13.5^{1.5}$, or about 50. The actual ratio depends on the relative depths of the various objects, which are not indicated.

atom. Proteins (including enzymes) and other large biological molecules are often called *polymers*, which means that their large structures are achieved by assembling large numbers of much smaller building blocks, these small building blocks being chosen from a fairly small number of possibilities. A protein may contain a small number of identical (or nearly identical) *subunits*, which are large themselves. In this book I will largely ignore the structure of multiple subunits of a protein, and when I talk about the structure of a protein I will in most cases (until we consider the differences between the oxygen-binding proteins hemoglobin and myoglobin in Chaper 10) be talking about the structure of one of its subunits.

It is hardly satisfying to say that enzymes are large compared with the molecules they act on and leave it as that. Why do they have to be so big? This is partly a misunderstanding and partly a real question that deserves an answer. Before trying to answer the valid question, I will dispose of the misunderstanding, which comes from the way in which we commonly perceive the sizes of things. When we talk about things we can handle, like beds or bicycles, we normally think in terms of linear dimensions. So, if we say that a bed is a bit bigger than the person sleeping in it, we mean, and are understood to mean, that it is a bit longer. It is also much wider, so that its volume is much larger than that of the person, but that goes if not unnoticed then at least unremarked. On the other hand, if we say that a bicycle is about the same size as its rider, we mean, and are understood to mean, that it has about the same length, even though it occupies a much smaller volume.

When we talk about the sizes of things that we cannot handle and are far outside the range of everyday objects, like proteins or planets, we are usually referring to volume rather than length. So, noting that an enzyme like liver hexokinase is some 70 times larger in volume (and also in mass, as the densities are similar) than the combined volume of the two molecules that it acts on, we tend to be impressed by the large factor, forgetting that its cube root is not much greater than four: The enzyme is only a bit more than four times the "size" (as

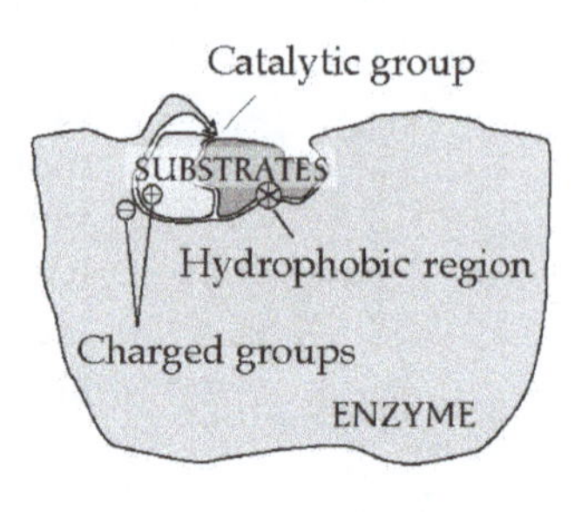

Figure 1.3 Features necessary for an enzyme to do its job. An enzyme molecule needs a cavity into which its substrates will fit (and into which, as far as possible, unwanted molecules will *not* fit), and it needs to have regions that attract the substrate molecules, such as charged groups with the opposite charge from groups on the substrates, and hydrophobic (water-repelling) regions to interact with similar regions on the substrates. When all this is achieved, its reactive groups need to be in the right positions to bring about the required reaction in the substrates.

we would use the word for beds or bicycles) of its substrates. A volume ratio of 70 is not far out of line with what we regard as normal in everyday life, such as the size of a car considered suitable for transporting a single commuter to work, and it is far less than many ratios in the macroscopic world: A room only 70 times bigger in volume than the person who worked in it would be regarded as tiny. People who complain of having to work in "boxes" or "cupboards" are usually referring to volume ratios of the order of 100, but they may be the same people who marvel at how much bigger enzymes are than the molecules they act on. This difference between the way we perceive familiar and unfamiliar objects is illustrated in Figure 1.2.[11]

Yet when all is said, a valid question remains. Could not an enzyme like hexokinase have about half the linear size that it has, thus requiring only one-eighth as many cell resources to synthesize, while still being large enough to make all of its necessary interactions with its substrates? If it could, then why has evolution failed to produce a smaller and more efficient enzyme? But we must bear in mind that an enzyme is not just a lump of amorphous glob of roughly the same size as the molecules it works on. If it were, it would be unlikely to do anything much and certainly nothing precise. An enzyme is a precision instrument, capable of recognizing its substrates, distinguishing them from other molecules of similar size and chemical behavior, and transforming them in precise ways. In general, the more precise any instrument has to be, the bigger it must be, and the relative size difference between an enzyme and the substance that it acts on (commonly called its substrate) is not very different from what you find if you compare a precision drill with the object to be drilled, if you include the clamps and other superstructure as part of the drill.

There is an example in Figure 1.3. It is quite crude, but it does illustrate

[11]We describe a striking example in Chapter 12: Although we tend to think of the Universe as unimaginably large, whereas a bacterial cell is something we can see if we have a powerful enough microscope, the ratio of the mass of the Universe to that of the Earth is much smaller than the ratio of the mass of the Earth to that of a bacterial cell.

some of the things that an enzyme must achieve in order to catalyze a reaction. Notice first that not just one substrate is shown, but two, and the enzyme has to promote a reaction between them. This is typical, but not universal: A few enzymes act on just one substrate at a time, but it is much more common to act on two, and some act on three or more.[12] So the enzyme must be able to provide a cavity into which the substrate or substrates will fit, and if there are multiple substrates they need to fit side by side with the parts that are to be changed in the reaction close to one another. It is not enough just to provide a cavity of the right shape. If a substrate has a charged group (marked ⊕ in the example), it will bind much better to a cavity that has an oppositely charge group lined up with it (marked ⊖). Many substrates have *hydrophobic* (water-repelling) regions (⊛), and these will bind best to similar water-repelling regions on the enzyme, and so on. Binding in a cavity is also not enough, because the enzyme needs to do things to the substrates once they are bound, and for this it needs *catalytic groups*, which, again, need to be positioned exactly where they can interact with appropriate parts of the substrates. Moreover, as these catalytic groups may need to move in order to do their jobs, the enzyme will have to have the superstructure necessary for moving them in precise ways.

It follows, then, that an enzyme cannot be arbitrarily small, and it must be built according to a structural plan that allows for precise positioning of its chemically active groups. Moreover, these groups must be extremely unreactive when not in the presence of the molecules they are intended to transform, to make sure that they do not bring about any unwanted reactions, as we saw in discussing specificity. This is much more difficult for working on a small molecule than it is for a large one, and we sometimes find that the largest enzymes work on the smallest molecules, and vice versa. Enzymes involved in digesting large molecules, such as pepsin in the human stomach, are among the smaller enzymes, whereas catalase, which acts on two molecules of four atoms each, has more than 3000 times their combined weight.

Catalase occurs in blood, and it offers an excellent opportunity to illustrate the tremendous catalytic powers of enzymes, as it acts on a commonly used domestic antiseptic, hydrogen peroxide. Fill a cup about three-quarters full with ordinary hydrogen peroxide of the strength used as an antiseptic (probably labeled as "10 volumes," or as "3%"). Do this in a place where you do not mind if the cup overflows, and then allow a drop of blood to fall into the cup: The result will be instantaneous, as the liquid froths like champagne from

[12]The record, as far as I know, is held by cyclosporin synthetase, which catalyzes a reaction with 29 substrates. The complicated reaction that it catalyzes consists of several clearly distinct stages, however, and differs from a metabolic pathway only in that the various intermediate products remain attached to the enzyme and are not released into solution. Not surprisingly, cyclosporin synthetase is a huge molecule, much larger than the usual run of enzymes.

Figure 1.4 Peptide bond. Alanine and valine (*top*) are two of the 20 amino acids commonly found in proteins. Removal of water allows formation of a *peptide bond* between the two amino acid *residues* (*center*), and a protein contains many residues linked in the same way. The $-NH_3^+$ group present in all three structures can donate a proton (H^+) and is therefore very weakly acidic; the $-CO_2^-$ group can accept a proton and is very weakly basic. The $-NH_2$ and $-CO_2H$ groups shown in the **wrong** structure of alanine at the *bottom* cannot exist simultaneously to significant extents in the same molecule: If they could, they would cause proteins to be far more indiscriminately reactive than they are and incapable of the specificity necessary for fulfilling their functions, but they would convert one another to the much weaker acidic and basic groups that actually exist.

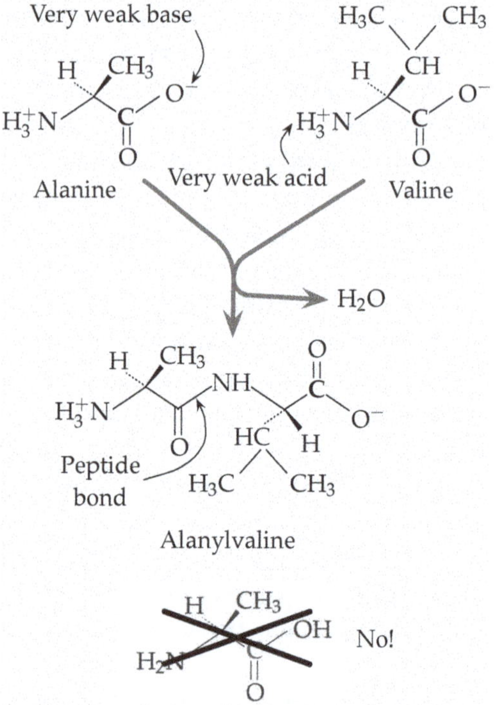

a bottle that has been shaken before being opened, and becomes quite warm to the touch. You may wonder why blood (like other cells from living organisms) needs an enzyme to destroy an ordinary chemical like hydrogen peroxide, and needs it enough to have it available in quite large amounts. The reason is that any living organism needs to be able to destroy harmful chemicals of one kind or another that get into cells. There are a great many potential hazards of this kind, and as the cell cannot prepare a separate solution to every conceivable problem that might arise, it deals with many unwanted chemicals by making them react with ordinary oxygen, in a process known as *oxidation*. This is often a useful step toward making insoluble poisons soluble in water, so that they can be excreted in the urine. The exact nature of the chemical hazard cannot be predicted, and so it is not possible to make a separate enzyme to deal with each one. Instead, general-purpose enzymes have evolved, but unfortunately their general-purpose character makes them unspecific, and they tend to oxidize water instead of the intended substrates, generating hydrogen peroxide. As this is itself a dangerous substance, the cell needs a way of destroying it instantly whenever it appears and that is what catalase is for.

Proteins

It turns out that protein molecules fulfill the requirements for a biological catalyst admirably. A protein is composed of a large number of building blocks known as *amino acids*, arranged head-to-tail in a long line (Figure 1.4).[13] Each amino acid has at least one *amino group* and at least one *carboxyl group*, and the protein structure involves bonds between the carboxyl group of one and the amino group of the next. Contrary to what is said in most books, the amino group exists normally in a form in which it is very weakly acidic (and is more correctly, though in practice almost never, called an ammonio group), whereas the carboxyl group normally exists in a form in which it is a very weak base. There are 20 different kinds of amino acids normally found in proteins, as listed in Table 1.1 together with their standard three-letter codes, not counting a few that are found in only a few proteins and are inserted into them in special ways, but the order of amino acids in any protein is neither haphazard nor repetitive. Although a casual glance at the order of amino acids in any protein might suggest a random or haphazard arrangement, it is not. Some substitutions or rearrangements, especially in the less important parts of a protein, may be tolerable, but simply rearranging exactly the same building blocks into a random order will no more yield a meaningful protein than rearranging the letters and spaces of a sentence will yield a meaningful sentence: "Novolulogense except iny makeg in bios s tthint of ehe lightion."[14] Even though some of the original structure of the sentence is still there, the modest amount of rearrangement (arbitrary blocks of characters moved from where they were to arbitrary positions, repeated four times) has made it unintelligible. Incidentally, the sentence appears in its uncorrupted form elsewhere in this book, and because it is a famous sentence in the development of modern biological thought, you may be able to recognize it immediately.

Table 1.1 Amino acids and their three- and one-letter codes[a]

Ala	A	alanine
Arg	R	arginine
Asn	N	asparagine
Asp	D	aspartate
Cys	C	cysteine
Glu	E	glutamate
Gln	Q	glutamine
Gly	G	glycine
His	H	histidine
Ile	I	isoleucine
Leu	L	leucine
Lys	K	lysine
Met	M	methionine
Phe	F	phenylalanine
Pro	P	proline
Ser	S	serine
Thr	T	threonine
Trp	W	tryptophan
Tyr	Y	tyrosine
Val	V	valine
Xaa	X	*unspecified*

[a]It is intuitively obvious which three-letter symbol refers to which amino acid. The much more opaque one-letter code, popular with authors who value conciseness over clarity, is not used in this book.

This example may be misleading, because our eyes and brains are much

[13]Strictly speaking that is a description of a *polypeptide chain*, and many proteins have more than one of these. We will not worry about that here.

[14]Nonetheless, as we will see in Chapter 12 (Figure 12.3), proteins can stand a much greater degree of arbitrary rearrangement than most of us would have guessed, until Willem Stemmer and his colleagues put it to the test.

better at resolving confusion than any machine can be expected to be. Consider, for example, the following sentence, which also appears in an uncorrupted form elsewhere in the book: "As Mark Tawin oerbvsed, while maielnvlrg at our aniamzg aitdtopaan: Our legs are jsut long eoungh to reach the gunord." Here each word has the correct first and last letters, but all the others are in completely random orders. It is striking how much of it can be read almost without effort (though not "aitdtopaan"). This is not a good guide to the capabilities of mechanical devices, however, which lack our capacity to extract some order from disordered information.

The reason why we need as many as 20 different kinds of amino acids is that a protein needs to have various different kinds of properties, which need to be available at specific locations in space, and the 20 amino acids differ among one another in the properties they can provide: Some are large, some are small; some interact well with water, others repel water; some interact strongly with metals like magnesium, others with metals like zinc; some behave as acids, some as bases; one (histidine) is versatile and can act either as an acid or as a base. We should not be misled by the word "acid" in "amino acid": Most amino acids are no more acidic than they are basic. A free amino acid is not exactly the same thing as the component derived from it in a protein structure, because it has been chemically changed while making it into a part of a protein, and biochemists often refer to the components of proteins as amino acid "residues"—that is as the bits of amino acids that are left after they have been joined up. The distinction is not always made, however, and the word "residue" is sometimes omitted.

DNA and RNA

It follows that to produce functional proteins a living organism needs a way of specifying the order in which amino acids are to be arranged, and this order is stored in *deoxyribonucleic acid*, usually shortened to DNA. Certain viruses store the information in the same way in a somewhat different structure, *ribonucleic acid*, usually shortened to RNA. Organisms that store the information in DNA also use RNA, as an interface between DNA and proteins, and for other functions.

Both DNA and RNA are long sequences of other kinds of building blocks, quite different from amino acids, called *bases*, and each protein is specified by a particular stretch of DNA called a *gene*. There are four kinds of bases in DNA, usually known by the first letters of their names, A, C, G, and T,[15] and so a gene can be regarded as a long message written with a four-letter alphabet. Their order is what specifies the order in which amino acids are to be assembled into a protein. A group of three bases is called a *codon*, and because there are four different kinds of bases, there are $4 \times 4 \times 4 = 64$ different kinds of codons. This is more than enough to specify 20 kinds of amino acid, and most amino

[15]In RNA the base T is replaced by a similar base, known by its first letter as U.

UUU Phe	UCU Ser	UAU Tyr	UGU Cys
UUC Phe	UCC Ser	UAC Tyr	UGC Cys
UUA Leu	UCA Ser	UAA *stop*	UGA *stop*
UUG Leu	UCG Ser	UAG *stop*	UGG Trp
CUU Leu	CCU Pro	CAU His	CGU Arg
CUC Leu	CCC Pro	CAC His	CGC Arg
CUA Leu	CCA Pro	CAA Gln	CGA Arg
CUG Leu	CCG Pro	CAG Gln	CGG Arg
AUU Ile	ACU Thr	AAU Asn	AGU Ser
AUC Ile	ACC Thr	AAC Asn	AGC Ser
AUA Ile	ACA Thr	AAA Lys	AGA Arg
AUG Met	ACG Thr	AAG Lys	AGG Arg
GUU Val	GCU Ala	GAU Asp	GGU Gly
GUC Val	GCC Ala	GAC Asp	GGC Gly
GUA Val	GCA Ala	GAA Glu	GGA Gly
GUG Val	GCG Ala	GAG Glu	GGG Gly

Figure 1.5 The genetic code. With four exceptions, each group of three bases encodes a particular amino acid, symbolized here with the standard three-letter codes listed in Table 1.1. Three of the exceptions are the *stop* codons, but the fourth is more complicated: AUG at the start of a gene acts as a *start* signal, but within a gene it codes for methionine (Met).

acids can be specified in more than one way, but it makes no difference to the resulting protein which particular codon has been used to specify which particular amino acid. The set of relationships between codons and amino acids is called the *genetic code* (Figure 1.5), and it turns out that 61 of the total of 64 codons do code for amino acids. The other three are *stop* codons, messages telling the protein-synthesizing machinery that it has come to the end of the sequence of a particular protein. The genetic message also needs to include information about where to begin making a particular protein, and how much of it to make, but these are far more complicated aspects that I will not consider here.

What about *start* codons, you may ask: How does the translation machinery know where to start making a protein? In a sense the codon AUG is a start codon: When it occurs in the middle of a gene, it codes for the amino acid methionine, but initiating translation is more complicated than that, because the *initiation sequence* that precedes the AUG codon needs to encode other information, such as whether the particular protein is needed in a particular condition. By contrast, the only thing the system needs to know once translation has started is where to stop. After translation the polypeptide sequence undergoes various modifications before being folded into a mature protein, and the initial methionine residue does not usually remain in this mature protein.

Naively you might expect each gene to be a separate molecule, but it is not so. Instead, genes are strung together one after another in a relatively small number of enormously long DNA molecules. In the human, for example, all of the genes taken together are organized as a mere 46 different *chromosomes*, each of which contains just one stretch of DNA, of total length when uncoiled of about 3 meters. The whole collection of genes in an individual is called the *genome*, but in popular accounts, for example in news reports of DNA testing in court cases, you will often hear the genome called the "genetic code," which is by no means the same thing: Your genetic code is exactly the same as mine, and would be useless as evidence in court, though our genomes are different.

As always when discussing living systems, there is a complication, that the DNA consists not only of genes but also long stretches of DNA that look at first sight like genes, but which do not code for proteins (or for RNA molecules with regulatory roles). In humans this noncoding DNA accounts for around 90% of all the DNA. Some of the noncoding DNA may have a simpler function, such as acting as spacers, but much of it is almost certainly just an evolutionary relic with no function. This has often been called "junk DNA," a term that has always offended people with religious objections to the idea that 90% of human DNA could have no function, but in 2012 it became controversial, even among some biochemists, on account of a project called ENCODE that was claimed to show that as much as 80% of DNA could be functional. At present, most people not involved in the project and who have studied the results believe that the conclusion depends on an absurdly loose definition of the word *function*, and that 90% is still a reasonable estimate of the amount of nonfunctional DNA in the human genome. Skepticism about the ENCODE project does not mean thinking that *all* of the DNA in the genome that does not code for proteins is junk: There are certainly functions yet to be discovered. I will return to this question in Chapter 3.

Sydney Brenner[16] made a useful distinction between "junk" and "garbage" when discussing why noncoding DNA remains in the genome even if it has no function:

> Some years ago I noticed that there are two kinds of rubbish in the world and that most languages have different words to distinguish them. There is the rubbish we keep, which is junk, and the rubbish we throw away, which is garbage. The excess DNA in our genomes is junk, and it is there because it is harmless, as well as being useless, and because the molecular processes generating extra DNA outpace those getting rid of it. Were the extra DNA to become disadvantageous, it would become subject to selection, just as junk that takes up too much space, or is beginning to smell, is instantly converted to garbage by one's wife, that excellent Darwinian instrument.

As we will see in the next chapter, some organisms, such as the alga *Gonyaulax polyedra*, have far larger genomes than the human, despite having no obvious need for far more coding capacity.

[16]S. Brenner (1998) "Refuge of spandrels" *Current Biology* **8**, R669.

H4A1	AUG	UCU	GGC	CGC	GGC	AAA	GGC	GGG	AAG	GGC	CUU	GGC	AAA	GGC	GGC ...
H4	AUG	UCC	GGC	CGU	GGA	AAG	GGC	GGA	AAG	GGC	UUA	GGC	AAA	GGU	GGC ...
Fetal H4	AUG	UCC	GGC	AGA	GGA	AAG	GGC	GGA	AAA	GGC	UUA	GGC	AAA	GGG	GGC ...
Protein	Met	Ser	Gly	Arg	Gly	Lys	Gly	Gly	Lys	Gly	Leu	Gly	Lys	Gly	Gly ...
					5					10					15

H4A1	GCU	AAG	CGC	CAC	CGU	AAA	GUA	CUG	CGC	GAC	AAU	AUC	CAU	GGC	AUC ...
H4	GCU	AAG	CGC	CAC	CGU	AAA	GUC	UUA	AGA	GAC	AAU	AUU	CAG	GGC	AUC ...
Fetal H4	GCU	AAG	CGC	CAC	CGU	AAA	GUC	UUA	AGA	GAC	AAU	AUU	CAG	GGC	AUC ...
Protein	Ala	Lys	Arg	His	Arg	Lys	Val	Leu	Arg	Asp	Asn	Ile	Gln	Gly	Ile ...
					20					25					30

Figure 1.6 Genes for histone H4. The figure shows the messenger RNA sequences of three of the 14 genes for histone H4 in humans encoding the first 30 amino acid residues of the protein before the initiating methionine residue is removed. (In the corresponding DNA sequences every U would be T.) Although the base sequences are different, they all specify the same amino acid sequence. Notice that although many of the substitutions affect only the third base in a codon, not all of them do: Arginine at position 4, for example, is encoded by three codons that differ in both the first and the third codons.

Each DNA molecule carries a substantial negative charge, and because negative charges repel one another you might expect the molecule to be extended and wonder how it is possible to fit a molecule several centimeters in length inside a cell. Association between the DNA and *histones*, positively charged proteins that we will meet again in Chapter 12, allows a compact structure for the chromosome. Packing all this not just into each cell, but into the nucleus of each cell, constitutes a folding problem of tremendous complexity, but we can largely ignore it here: For our purposes, it is sufficient to consider a gene as a straight sequence of bases. It is not quite as sufficient to think of proteins as straight sequences of amino acids, because understanding their function requires some understanding of how they are folded up into compact globular structures, but it will do for now.

FRANCIS HARRY COMPTON CRICK (1916–2004) made many major contributions beyond his work on the structure of DNA and the early development of molecular biology. He was born at Northampton, England, and studied physics at University College London. His work on magnetic and acoustic mines during World War II led to his first important discovery, the invention of "magnetic tweezers," now a major tool for experimenting with single molecules. In his later life he worked principally on neuroscience at the Salk Institute in La Jolla, California.

The traditional idea of enzymologists was that life was basically about enzymes and that DNA existed to store the information about how to make them. Much of modern molecular biology has turned this on its head, so that proteins are often regarded as no more than "gene products," as if what they do is of little or no interest. Indeed, much of evolution, at many different levels,

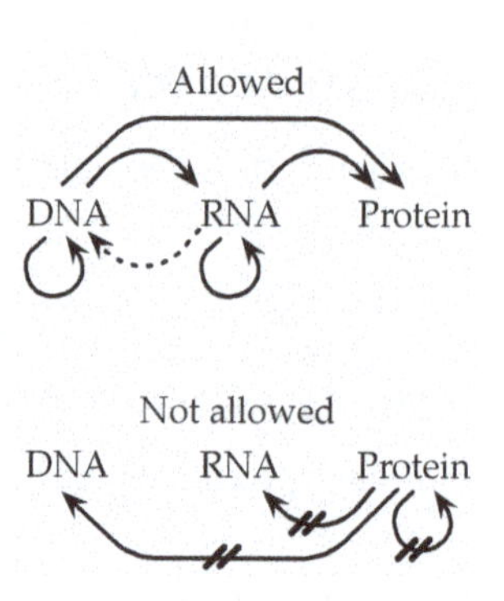

Figure 1.7 The central dogma of molecular biology. This is based on a sketch made by Crick in 1956, and it conveys the essential idea that *once information has passed into protein, it cannot get out again.* The diagram does not imply (and Crick did not say) anything resembling the catchphrase "DNA makes RNA makes protein." The arrow from RNA to DNA is shown as a broken line because at that time information transfer in that direction had not been discovered, but even in 1956 Crick envisaged it as a possibility. The situation today is exactly as Crick foresaw it: All of the arrows in the top diagram have been observed, but those in the bottom diagram have not.

can best be understood by accepting the view of Richard Dawkins in his books that a body is just the genes' way of perpetuating themselves. This may be useful for understanding why living organisms evolve as they do, but it is of little help for understanding the detailed machinery of how living systems work, which is still mainly about enzymes.

Thinking of proteins as no more than gene products also encourages confusion about biochemical evolution. Although there may be a *slight* advantage in coding an amino acid sequence in one way rather than another, this is overwhelmed by the important differences in functionality between different protein sequences, and it is these latter instances that decide whether a change in a gene can be tolerated, and if so, whether it can be accepted. The *protein* histone H4 varies hardly at all across all animals and plants, but its gene varies very greatly, because there are many ways of coding the same sequence.[17]

The central dogma

When the relationship between nucleic acids and protein sequences was first becoming understood, Francis Crick proposed what he called the *central dogma of molecular biology* (Figure 1.7). As this term has been and continues to be widely misunderstood and misused, it is worth pausing a moment to consider what it means: "Once (sequential) information has passed into protein it cannot get out again." Crick wrote these words to clarify what he really meant,[18] essentially that sequence information passes only from nucleic acids to proteins, never from proteins to nucleic acids or to other proteins. Most claims that the central dogma has been violated refer either to RNA viruses, which store the information as RNA rather than DNA, or retroviruses, which can convert RNA information into DNA sequences. These do not contradict the central dogma as Crick formulated it.

[17]There are about 1.8×10^{55} ways of encoding the sequence of histone H4. Not all of these are known, but a surprising number occur even in one species: Humans have 14 different genes coding for the same histone H4 sequence, of which three are illustrated in Figure 1.6.

[18]F. Crick (1970) "Central dogma of molecular biology" *Nature* **227**, 561–563.

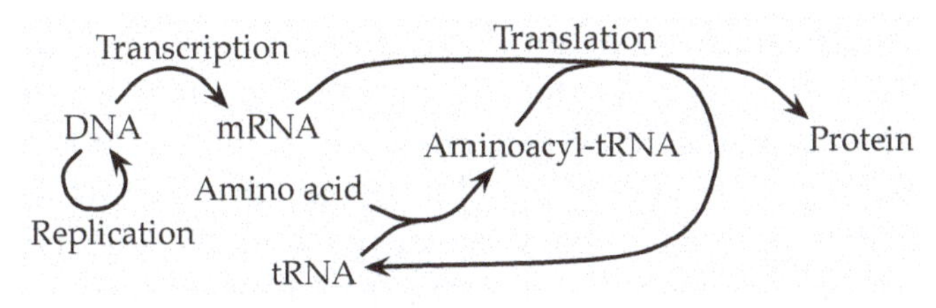

Figure 1.8 Translation in higher organisms. Not all of the steps shown as "allowed" in Figure 1.7 occur in every organism. Those found in higher organisms are illustrated here and can be classified as *replication, transcription,* and *translation.* Translation is more complicated than implied by Figure 1.7 and requires a pool of different molecules of *transfer RNA,* or *tRNA,* each "charged" with a different amino acid by an enzyme, aminoacyl-tRNA synthetase, specific for each amino acid and for a corresponding tRNA. It follows that there must be many different aminoacyl-tRNA synthetases.[19] Each tRNA, and thus each aminoacyl-tRNA, has a three-base sequence called the *anticodon,* which pairs with the corresponding codon in the mRNA, allowing the appropriate amino acid to be added to the growing chain of protein and releasing the uncharged tRNA. There are additional complications that are not shown.

Not all of the steps shown as "allowed" occur in all organisms. The process as it occurs in higher organisms, including humans, is illustrated in a simplified way in Figure 1.8 and involves three distinct steps: Copying an entire DNA molecule to produce two identical molecules is *replication*; copying the part of the DNA sequence that constitutes a gene into a molecule of *messenger RNA,* or *mRNA,* is *transcription*; and reading of the sequence three bases at a time by molecules of *transfer RNA,* or *tRNA,* is *translation.* The enzyme, an aminoacyl-tRNA synthetase, that charges the tRNA with an amino acid needs to select both the correct tRNA and the correct amino acid from a pool containing many different tRNA molecules and all of the possible amino acid molecules.

Carbohydrates and lipids

Enzymes and nucleic acids are not the only important classes of molecules in living systems. Cells are not just made of proteins and water; and plant cells are not even mainly made of protein and water. There are many kinds of small molecules needed in cells, and in addition to proteins and nucleic acids (the class that contains DNA and RNA), together with their building blocks, there are at least two other major classes, the *carbohydrates* and the *lipids.* Carbohydrates include well-known sugars such as sucrose (the every-day "sugar" found on the breakfast table) and glucose, and other sugars that are less well-known in everyday life but just as important, such as ribose and

[19]There must be at least one different kind of tRNA for each amino acid, and thus a minimum of 20. However, each tRNA must also recognize at least one codon for that amino acid, and because there are 61 codons that code for amino acids, one could suppose that there would be 61 different tRNA molecules. In fact, there are fewer than that (but many more than 20) because some tRNA molecules can bind to more than one of the codons for the same amino acid.

deoxyribose, essential components of the genetic material. Glucose itself is a versatile molecule: Not only does it provide the main way in which fuel for doing work is available to all the cells of the body, but it is also the major building block for polymers such as cellulose, the primary structural material in plants, and other quite different polymers used for storing energy—namely, starch in plants and glycogen in animals. I will return to glycogen in more detail in Chapter 6. Lipids include the fats and oils familiar in the diet, as well as other water-repelling substances such as the phospholipids, which constitute the major structural component of the membranes that separate cells from one another, and different compartments within cells from one another.

Adenosine phosphates

In addition to these, living systems use a small group of metabolites that have structures similar those of the bases of nucleic acids, but which have major metabolic functions quite separate from their relationship to nucleic acids. These are the *adenosine phosphates*: Two of them, adenosine 5′-triphosphate and adenosine 5′-diphosphate, participate in many metabolic reactions (more, indeed, than any other substance, apart from water); a third, adenosine 5′-monophosphate, participates in relatively few reactions but affects many enzymes as an inhibitor or as an activator, and in Chapter 8 we will see why it was selected for this role. These names are cumbersome for everyday use, and biochemists refer to them nearly all of the time as ATP, ADP, and AMP, respectively. In animals, the ATP needed for driving all the functions of the cell is generated in small compartments of cells called *mitochondria*. For the purposes of this book, we will not need to know any details of how mitochondria fulfill their functions, but we do need to know that they exist, because we will meet them again in a quite different context: They contain small amounts of their own DNA, and this allows the investigation of "mitochondrial Eve," discussed in Chapter 2. Adenosine, the skeleton from which ATP, ADP, and AMP are built, has a separate importance as one of the four bases that define the sequence of DNA.

Thermodynamics

Before discussing the adenosine phosphates further, we need to pause a moment to consider *thermodynamics*. The name suggests something to do with heat and power, especially the power to move heavy objects, and this is indeed what thermodynamics is principally about when engineers use it to design power stations or engines. However, thermodynamics has become as important in chemistry, including biochemistry, as it is in engineering, but the reason for its importance in chemistry may be less obvious. When engineering was being developed in the nineteenth century, early physicists were trying to understand why some designs of engines worked well whereas others did not. It gradually became clear to them that the principles that decided how energy could be converted into work or different sorts of energy transformed into one another were the same principles regardless of what sorts of energy were under discussion. An electric battery is just as much a machine for converting chemical energy into work as a steam engine is, even though in the case of a battery the conversion is direct: There is no initial conversion of chemical energy into heat followed by the conversion of the heat into work. In living organisms, muscles behave like motors driven by electric batteries, as they convert chemical energy directly into mechanical work.

Everyone finds thermodynamics difficult, not so much because the underlying mathematics is difficult, but because the concepts are not easy to understand. As Keith Laidler explained,[20] even the great scientists who developed the subject in the nineteenth century were confused about many points, and the following comment of the physicist Arnold Sommerfeld is quite appropriate:

> Thermodynamics is a funny subject. The first time you go through it, you don't understand it at all. The second time you go through it, you think you understand it, except for one or two small points. The third time you go through it, you know you don't understand it, but by that time you are so used to it, so it doesn't bother you any more.

In Chapter 7 we will see that it still leads even competent scientists into absurdity.

In biochemistry, thermodynamics is almost exclusively the study of equilibria. When we say that something is thermodynamically possible we mean that the equilibria are such that a process can happen without requiring an external source of energy. This does not mean that it *will* happen, however, because thermodynamics is not the only factor that decides what will happen. It decides what is allowed and what is not allowed, but no more than that. We see the same thing in the everyday world, in which gravity provides enough energy to bring a large rock from the top of a hill to the bottom, but does not ensure that it will happen: There needs to be a push to get the process started.[21] In biochemistry, many reactions are thermodynamically possible but need a catalyst to bring them about.

ATP and ADP

The adenosine portions of ATP and ADP are more or less irrelevant from the chemical and thermodynamic point of view, and their job could just as well be done by inorganic diphosphate (two phosphate ions joined together to make a single ion) and inorganic phosphate. Indeed, it is still done by these simpler inorganic ions in some organisms that exist today.[22] The important point is that ATP has three phosphate groups, one of which can be passed to some other molecule such

[20] K. J. Laidler (1995) *The World of Physical Chemistry*, Oxford University Press, Oxford.

[21] The winter of 2013–2014 was extremely wet in the south of France, and in February a rock of about 10 tons detached itself from a mountain and fell onto a passing train. From the thermodynamic point of view, this had always been possible, but it did not happen until weeks of rain had weakened its support.

[22] In higher plants, for example, various reactions involved in the mobilization of sucrose

as glucose, the glucose becoming glucose 6-phosphate and the ATP becoming ADP. Exchanging the diphosphate linkage in ATP for the sugar–phosphate linkage in glucose 6-phosphate is thermodynamically favored, so the process as a whole is thermodynamically favored, but it does not occur without a catalyst. The qualification "thermodynamically" may seem cumbersome in this sentence, but it is absolutely necessary: In organic chemistry as a whole and the chemistry of life in particular, there is a vast array of chemical reactions that are thermodynamically perfectly possible, but which occur extremely slowly in the absence of a suitable catalyst.

ATP is often seen as the universal "cell currency." An energy-harvesting machinery couples the conversion of glucose and other foods to, ultimately, carbon dioxide and water to the resynthesis of ATP from ADP. Biochemists often refer to this process as *combustion*, in an echo of nineteenth-century ideas of what is involved, but remember that a muscle is not like a steam engine, and the heat produced by muscular activity is a side effect, not because chemical energy needs to be converted to heat before it can produce mechanical work. The ATP thus produced is simultaneously reconverted to ADP in a host of metabolic processes that use it to convert metabolites into one another. Although ATP and ADP are themselves metabolites, it is often convenient to put them in a separate class, the *coenzymes*, because instead of following a long chain of interconversions, they are just converted backwards and forwards into one another at many different points in the metabolic network. They are not alone in this category, and several other pairs or small groups of metabolites fulfill similar roles as metabolites: For example, oxidized and reduced NAD (nicotinamide adenosine dinucleotide, but the name is unimportant) are similarly converted back and forth in many different metabolic oxidation and reduction reactions. Contrasted with them are the "real" metabolites like glucose, which occur in relatively few distinct steps and form parts of long chains of reactions, or pathways.

"Real" metabolites

The names of some of these metabolites, such as glucose itself, are familiar in everyday life because they are mentioned in the lists of ingredients of breakfast cereals and other processed foods, and in leaflets about diet or drugs. Others, like dihydroxyacetone phosphate or 3-phosphoglycerate, may seem rather obscure. Because it will often be unnecessary to define their structures at all in order to understand their role in the discussion, I will not do so, or I will confine the definition to a statement about how many carbon atoms they contain. In general, you may assume that if I refer to a series of three or four reactions that transforms the familiar molecule glucose into the unfamiliar dihydroxyacetone

(ordinary sugar) are driven by the conversion of inorganic diphosphate (often called "pyrophosphate") to phosphate: M. Stitt (1998) "Pyrophosphate as an energy donor in the cytosol of plant cells; an enigmatic alternative to ATP" *Botanica Acta* **111**, 167–175.

phosphate, then dihydroxyacetone phosphate must retain enough of the structure of glucose to be made from it in a few steps, and this is all you will need to know about its structure. This is because metabolic reactions always proceed in simple steps, so that you can never find a single reaction that rearranges dozens of atoms all at once, any more than evolutionary biologists propose the transformation of a dog into a cat, or a baboon into a human, in one generation.

Dihydroxyacetone phosphate is a much simpler molecule than glucose, one of the few important molecules in biochemistry that looks exactly the same as its image in a mirror, so do not be put off by the long name. Shorter does not necessarily mean simpler in biochemical names, any more than it does with names of places. Los Angeles is said to have started its existence as El Pueblo de Nuestra Señora la Reina de los Ángeles de Porciúncula: Maybe, but it needed a simpler name when it became an important city. Only places you do not often need to mention can afford the luxury of names like Lake Chargoggagoggmanchauggagoggchaubunagungamaugg (Massachusetts). So, because dihydroxyacetone phosphate does not feature in lists of cereal ingredients, and is not (yet) considered by dietary advisers to be the secret of a long and healthy life, it continues to be known by its chemical name, whereas with more fashionable molecules like folic acid and riboflavin, the simple names conceal much more complicated structures.

The presence of all of these components in even the simplest cell implies many different reactions to connect them all and correspondingly many enzymes to catalyze them all. The whole network of reactions is called *metabolism*, and the core of this book is about metabolism. It has been a major component of teaching biochemistry to medical and life-science students, and has typically been taught as if the whole complicated organization was arbitrary or haphazard, the result of a whole series of accidents—chance solutions to problems, no better or worse than other solutions that happen not to have been adopted—over the long period of evolution since the origin of life some thousands of millions of years ago. I will try to convince you that this view is wrong, because although the amount of metabolism that has been analyzed from the point of view advocated here is only a tiny proportion of the whole of metabolism, it leads to the conclusion that the solution adopted is not normally just one of a series of equally good or bad solutions that could have been chosen, but it is often the *best* solution.

First, however, I need to examine what is involved in protein evolution, and Chapter 2 is devoted to this. In Chapter 3 I will describe a point of view that is called *Panglossian*, so that you can approach the rest of the book in a proper spirit of skepticism and you can judge whether I am seeing optimal solutions to metabolic problems because they really are optimal, or because that is what I want to see.

2. *The Nuts and Bolts of Evolution*

> I returned, and saw under the sun, that the race is not to the swift, nor the battle to the strong, neither yet bread to the wise, nor yet riches to men of understanding, nor yet favour to men of skill; but time and chance happeneth to them all.
>
> *Ecclesiastes* **9:** 11 (Authorized Version)

> It is not enough to succeed. Others must fail.
>
> Iris Murdoch, 1973[1]

> Mechanisms and molecules have been preserved in bacteria, fungi, plants, and animals, essentially intact through billions of years of Darwinian evolution.
>
> Arthur Kornberg, 2000[2]

AS WE HAVE SEEN, a protein is, to the first approximation, a string of amino acids coded by the sequence of bases in a gene. After the essential relationship between DNA sequence and protein sequence was recognized from the famous work of James Watson and Francis Crick in the 1950s, and the code itself was worked out in the 1960s and 1970s, it appeared for a while to be absolutely identical in all organisms and all cells, and it was called the *universal code*. As I will discuss in the next chapter, this was an oversimplification but in this chapter I will ignore the small variations in the genetic code and treat it as if it really were universal. I will also ignore other complications in the way the code is read, in particular the noncoding DNA, which can in some organisms, including ourselves, represent a high proportion of the total DNA.

Charles Darwin devoted many pages of *The Origin of Species* to discussing the variations between the individual members of any species. His aim in the book was to convince his readers that evolution was a natural process, which he called *natural selection*, similar to the way in which horse breeders or pigeon fanciers breed each new generation by choosing as parents those individuals that come closest to displaying whatever characteristics they consider desirable. Before he could convince them of this, Darwin first needed to show that a sufficient amount of variation already exists in a population to allow the necessary selection. Although he knew nothing of genes or their role in coding for proteins, we can now say that the variation he was discussing

[1] I. Murdoch (1973) *The Black Prince*, Chatto & Windus, London. Other versions have been attributed to other authors.

[2] A. Kornberg (2000) "Ten commandments: Lessons from the enzymology of DNA replication" *Journal of Bacteriology* **182,** 3613–3618.

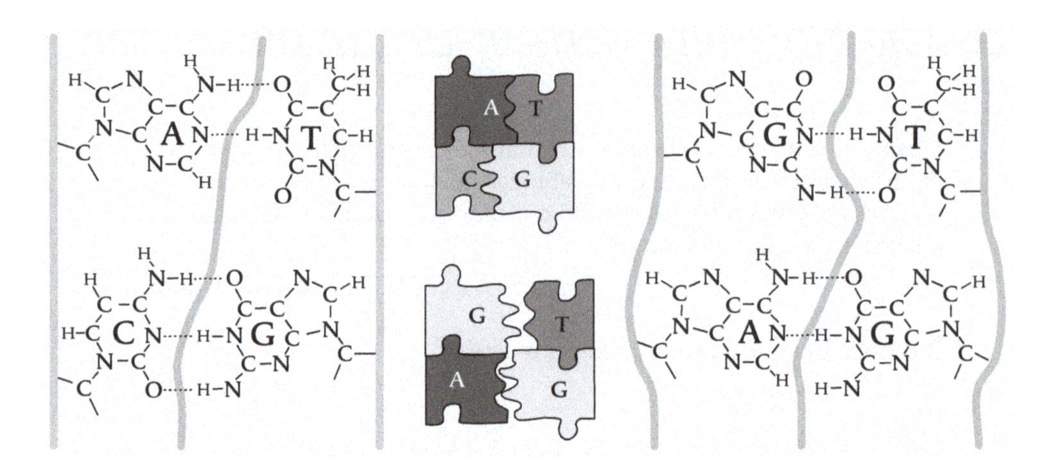

Figure 2.1 Base pairs in DNA. When the bases are paired correctly (*left*), they form strong interactions, and the width of the combination is almost exactly the same for each of the allowed combinations (T–A and G–C as well as A–T and C–G illustrated), so there is no distortion of the double helix. Nonetheless, it is not utterly impossible (*right*) to form "wrong" base pairs, held together by weaker interactions than the correct pairs (though not infinitely weaker), and when these produce a pair of about the right width, as with the G–T pair illustrated, they can occur fairly easily. On the other hand, A–G and several other pairs do not have the right widths and introduce obvious distortions in the whole structure if they occur. The naive idea that the bases fit together like pieces of a jigsaw (*center*) and that only an exact fit is possible is thus misleading.

was due to the existence of multiple variants of genes, with the result that different individuals have different collections of proteins. In biochemical terms, therefore, evolution is the result of changes in genes.

Base pairing

Perhaps the first question we should ask is why any gene changes at all. Would it not be better from the point of view of the gene itself to be copied from one generation to another without any changes whatsoever? Yes, it would, but we are dealing with mechanisms that depend for their efficiency on the laws of chemistry. The genetic code "works" because the double-stranded structure of DNA is more stable if every A is lined up with a T in the other strand, every C is lined up with a G, every G is lined up with a C, and every T is lined up with an A. This allows an exact replication when the two strands are separated and each strand is then allowed to build a new partner strand by making the correct pairings from a pool of bases. This ability to recognize base sequences also applies to transcription and translation, but in the last case the tRNA also needs to be able to select the right amino acid from a pool that contains all of them.

Even though it is true that a G lines up better with a C than with a T, and makes stronger interactions with a C, the wrong pairing is by no means impossible (Figure 2.1). Pairing an A with another A or with a G is more difficult, because there is not enough room for the larger bases without distorting the whole structure, but again, it is not utterly impossible. The four bases are not infinitely large and the interactions between them do not involve infinite numbers of chemical groups. Not only are they not infinite, they are not even particularly large: The chemical energy of a whole pairing can be obtained by adding up the contributions from the individual groups involved; when this is done, the differences between the energy of the correct pairing and that of the wrong ones are insufficient to exclude all possibility of a wrong pairing.

Moreover, additional steps are involved in going from DNA to protein: First a transfer RNA molecule is produced by reading and transliterating the corresponding gene in the DNA; then this is translated into protein by an enzyme that not only reads the code in the RNA but also selects the appropriate amino acid from a pool in which all the other 19 are present. All of these steps are subject to mistakes. For example, the appearance of GU pairs in the RNA, corresponding to GT pairs in DNA, are a common cause of infidelity.

CHARLES ROBERT DARWIN (1809–1882) was the creator of modern biology, his theory of natural selection providing the first convincing mechanism for evolution. He was born in Shrewsbury (on the same day as Abraham Lincoln), and he intended to follow his father into medicine. However, his five-year voyage as naturalist on HMS *Beagle* led to his reflections on evolution, and to *The Origin of Species*.

It is like putting a left-handed glove on the right hand: not comfortable, but not so difficult as to be impossible to do, accidentally or deliberately. By calculating the energies involved from chemical considerations and then using different chemical principles to assess how frequent wrong pairings are likely to be in DNA, you can calculate a frequency of about one in 100,000 base pairs. This may seem very good—it is certainly much less likely than accidentally putting a left-handed glove on the right hand—but it is not nearly good *enough*, because the huge number of base pairs in all the genes in an organism means that even an error rate as low as this would add up to an intolerable number of errors. For example, in humans, with about 3 billion base pairs in the whole genome, an error rate of one in 100,000 implies about 30,000 mistakes every time the DNA is replicated. To prevent this from happening, the machinery for replicating DNA includes sophisticated mechanisms for detecting and correcting mistakes. In the *germ-line DNA*, the DNA that is passed on to the next generation, these mechanisms achieve a final error rate of the order of one in a billion per generation (taking into account the number of times DNA is replicated, even in germ-line cells, before a sperm or ovum is produced[3]), so the average number of mistakes passed on to the new generation is of the order

[3]This number is much higher for sperms than for ova, because a man continues producing sperms throughout his reproductive life, whereas all the ova that a woman will ever have are produced before she is born.

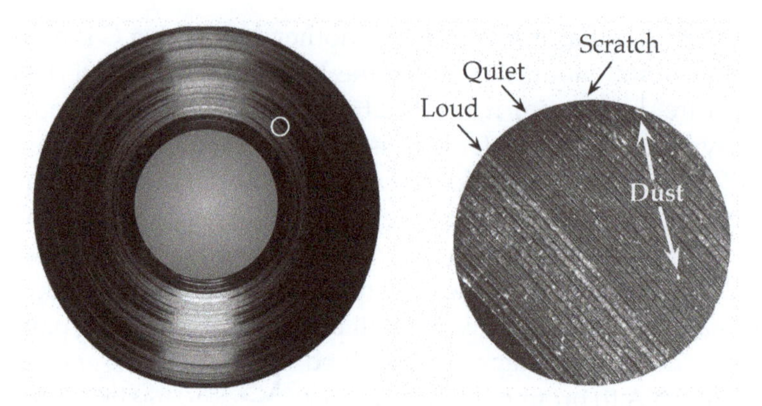

Figure 2.2 An old phonograph record. The area inside the circle on the main image is shown at greater magnification at the right. The relationship between what you see and what you hear is reasonably clear. Just by looking you can distinguish between loud and quiet passages, and you can see the scratches and dust that degrade the sound quality.

of three or four. In the cells that are used in the lifetime of the individual but are not passed on to descendants, which are called *somatic cells*, the error rate is higher, but still impressively small.

Somatic cells tolerate a higher error rate than that found in germ-line DNA because error detection and correction come at a price. As we have seen, the simple interaction energies in base pairs are not enough to ensure that errors never occur and detecting them after they have occurred consumes energy: Ultimately every base pair checked implies the consumption of a certain quantity of glucose, a small quantity, certainly, but glucose that could otherwise have been used for something else.

Quality control

The body has to make the same compromises that a factory making a specialized and complicated instrument has to make: The master copy of the design needs to be carefully protected from damage, and if copied for use in another factory, it needs to be copied with meticulous accuracy to avoid errors. The more routine copies used for the actual manufacturing need to be accurate as well, because it is not good to produce large quantities of defective instruments, but a certain low proportion can be tolerated.

Now that the original designs for most everyday objects are stored in computers, it is no longer easy to find an example from everyday life to illustrate this idea. Before vinyl phonograph records disappeared from the market in the face of the onslaught from compact disks (and now memory cards), they made an excellent example (Figure 2.2), but I am writing this book too late. The old principles still probably applied to commercial films for the cinema until quite recently,[4] however. Somewhere there was a master copy of every film as it was completed in the studio and approved by the producer and director. This copy was much too precious to be used for showing the film in a cinema, and instead a number of sub-masters were made from the master copy, as accurately as possible, and sent to various distribution centers. The master copy was stored away in a vault, protected as far as possible from wear and tear, and if possible never looked at again. The sub-masters were not used in cinemas either, but used to make distribution copies. The distribution copies were then used, and the sub-masters carefully stored. No errors could be tolerated in the master copy, because any error in the master copy would be copied everywhere else. Errors in the sub-masters were marginally less harmful, because they only affected a subset of copies made for use in cinemas, but a high degree of accuracy was still desirable. Faults such as scratches and even missing frames could often be tolerated in the distribution copies, because unless they were severe and frequent they were likely to pass unnoticed.

The same sort of considerations apply to all industrial processes, and whatever the final product, an economic decision has to be made as to the point at which greater expenditure on error avoidance costs more than the losses incurred by sending a few defective items to market. The place where a manufacturer decides to put this point is a major factor in determining whether a product is destined for the mass market or for users who accept only the best: A top-of-the-range car may cost 10 times more to buy than a mass-market car of similar dimensions, but it certainly does not use raw materials that cost 10 times more, and the manufacturing costs are unlikely to be 10 times greater either; on the other hand, the amount of time, effort, and money devoted to checking for errors and discarding substandard items may be much more than 10 times greater.[5] In Chapter 11, we will meet an example where this principle appears not to hold, and we will try to understand why not: The protein collagen is by far the most abundant protein in the human body and is thus clearly a mass-produced item, but it undergoes a rigorous quality control in which a high proportion of molecules are rejected.

Now you may object that the analogy of vinyl phonograph records or celluloid film is invalid, because these are examples of analog recording techniques, whereas the genetic code is, like modern recording, digital. This objection is

[4]Celluloid film is rapidly on its way out, and by the time this book is in print it will probably have been entirely supplanted by digital recording, at least in the western world.

[5]In human economies, the prices of expensive items also depend on what people are willing to pay, regardless of their intrinsic worth. As we will see in Chapter 10, this sort of consideration is not important in the metabolic economy: No cell decides to use platinum instead of zinc in the active site of its alcohol dehydrogenase in order to show its neighbors that it can afford to do so. Partly as a result of this absence of "vanity" choices, the law of supply and demand works well in metabolism.

more apparent than real, because, as we have seen, copying mistakes can and do occur in gene replication and no digital recording technique is perfect, as anyone who has tried to watch a film on a faulty video disk can attest, and indeed the problems may be worse: A damaged vinyl record may produce the sounds of scratches and dust, but a faulty video disk may be completely unreadable. So there is still a need for a master copy of an electronic medium that has undergone a far more rigorous quality control than is applied to distribution copies.

As mentioned above, there are other points where accuracy is needed, but these are less important from the evolutionary point of view. Occasional protein molecules that cannot fulfill their functions because they have incorrect amino acids inserted in some positions represent a waste of resources for the individuals that possess them, but they do not get passed on to the next generation, so they do not accumulate over evolutionary time. Uncorrected errors in DNA replication do get passed on, and do accumulate during evolution. Ultimately, they are what evolution consists of. Every single genetic difference that exists between one present-day organism and another—whether we are talking about two different members of the same species, or about two individuals as different as a rhododendron and a tiger—has its origin in *mutations*—that is, errors in the replication of the genetic information.

Fixation

There are two stages in the process of fixing a modified gene in the population, of which the second is often forgotten in elementary accounts. A mutation is a necessary beginning, but it is insufficient. If it arises in an organism that is never born, whether because it is lethal itself, or for any other reason, it will never be manifested in any living creature. Even if the individual is born and reaches maturity, the mutation can only affect evolution if there are descendants. In many cases even then, the line containing the mutation may die out in a few generations, so although it may affect a few individuals it never becomes part of what distinguishes one species from another. To do this, the mutation must spread through the entire population, a process known as *fixation*.

It is tempting to think that a favorable mutation will inevitably become fixed, but this is far from being correct, because in the first few generations, when it can only be present in a few individuals, there are many other factors that may cause those individuals to leave no descendants in the long term. Even a highly favorable mutation, say one that increases fitness (loosely the probability of having offspring) by 1%, is almost exactly as likely to be eliminated in the first few generations, and lost forever, as a mutation that has no effect at all on fitness, or a mildly unfavorable one that decreases that fitness by 1%. In these early stages, the role of chance is too great for success to be guaranteed. Consider the simplest case, a man with a highly favorable mutation in a gene on his Y chromosome. As we have seen, human genes are

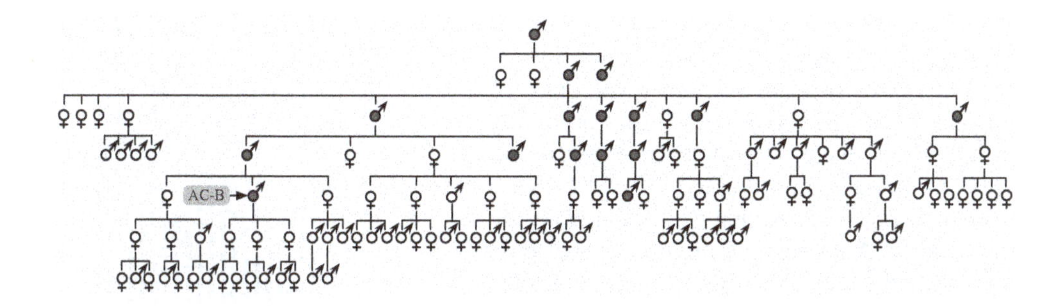

Figure 2.3 Extinction of a Y chromosome. The individual at the top of the tree is my great-great-grandfather in the direct male line (and I am labeled with an arrow). Male and female descendants are indicated with the usual biological symbols ♂ and ♀, with filled symbols ♂ representing people with the original Y chromosome.

organized into 46 chromosomes, but one of these, the Y chromosome, occurs only in males (females have two X chromosomes instead), so any mutation in the Y chromosome is inherited only by sons, not by daughters. If the man with such a favorable mutation has no sons, then no matter how many daughters he may have, the mutation will be lost. Even if the favorable mutation occurs on another chromosome, there is a chance of one-half that it will *not* be passed on to any given child. If there are two children, there is a chance of one-quarter that neither will receive it, and so on. Even if there are 10 children, there is about one chance in 1000 that none of them receive the favorable mutation. This chance may be small, but it is not zero.

Figure 2.3 illustrates this. My great-great-grandfather in the paternal line had two daughters and two sons, of whom one daughter and one son died in infancy. The surviving daughter left no children, so one might guess that there might be few if any descendants, regardless of whether they inherited the original Y chromosome. However, all that changed in the next generation, because the surviving son had 12 children (this was the late nineteenth century, after all!), including six sons, all of whom had children, with five sons between them. One might think that his Y chromosome was well on its way to fixation (ignoring anyone else who inherited it from earlier male ancestors), but in my own generation I am the only male left (I had a second cousin who died without leaving descendants), and I have no sons, so his Y chromosome will die with me. Notice that this does *not* imply anything about the other genes passed on by my great-great-grandfather: He has more than 100 known descendants, essentially all of

NETTIE MARIA STEVENS (1861–1912) was a geneticist at Bryn Mawr College, where she showed that what had been known as the "accessory chromosome" was present in males of various insects and some other species, but not in females, thus establishing the link between the sex chromosomes and sex. Already in her thirties when she started her research, she died from breast cancer at the age of 50.

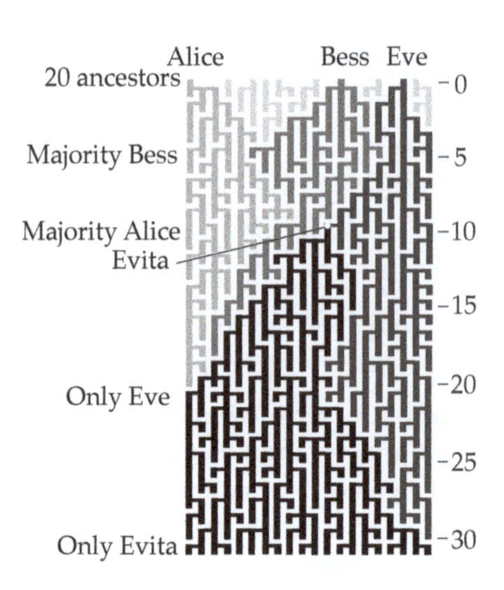

Figure 2.4 Fixation of genes. The top line of the tree represents 20 women, three of them labeled Alice, Bess, and Eve. These three, and seven others, have daughters, and 10 do not. Any unique genes possessed by the 10 who leave no descendants must disappear from the population. Although the descendants of Bess constitute the majority after a few generations, they eventually become extinct, and in generation 21 only Eve's descendants are left. Of these, only one, labeled Evita, still has descendants in generation 31. Evita's descendants are shown in black, and Eve's other descendants in dark gray.

whom will have inherited some genes from him, but I am his only remaining descendant in the direct father-to-son line.

The idea of gene fixation is illustrated in the trees of descent in Figure 2.4. To avoid various complications, we consider only people of the same sex who survive to become adults, we treat the generations as completely separate, all members of the same generation reaching maturity at the same time, and we assume a constant population size. With these conditions, a woman has an average of exactly one daughter, though any particular woman may have no daughters, or more than one. Out of 20 women present in generation 0, only 10 had daughters that survived to maturity, and only seven had surviving granddaughters. Thus, 13 out of 20 lines have disappeared in just two generations. This does not imply that the 13 potential ancestors were inferior in any way; they were just unlucky. In constructing the tree I did not assume that any women were any more or less likely to have surviving daughters than others. I just allowed each the same probabilities of having zero, one, two, or three daughters. But if the population size remains constant, some women can have two or more daughters only if others have none. (I did not consider the possibility of four or more daughters, because in a population of constant size, it would be unusual for a woman to have four daughters.)

If you follow the dark gray part of the tree down, you can see that in generation 21 only the dark gray line continues; all of the other 19 have disappeared. After this generation, the subsequent ones are all shown in dark gray (with shading for some of the descendants, as explained in the legend to the figure), and the genetic composition of the individual ancestor labeled Eve at the top has become *fixed* in the population. No further change is possible unless there are mutations. Notice that even with a small population size, fixation requires a considerable number of generations. With a larger population size, the number of generations required is correspondingly larger and is in general of the same order of magnitude as the population size. For our own species, as well as for many others in nature, the population size (some billions for humans) is much larger than the number of generations since it separated from other species (hundreds of thousands for humans). This means that there has not been nearly enough time for a significant proportion of genes that appeared after the separation to become fixed. The human population was much smaller only a few generations ago, but there have been many generations since it was of the order of hundreds of thousands.

Mitochondrial Eve

The illustration can be applied to a single gene, or to the descent from a single individual, or for anything in between. It can thus be applied to the hunt for *mitochondrial Eve*, a term introduced by Allan Wilson. She is sometimes regarded as the first woman, though that is incorrect, as we will see. Both males and females have mitochondria, the ATP-generating compartments in cells that were briefly mentioned in the previous chapter, but those in males are not passed on to the next generation. Moreover, although most of the enzymes that a mitochondrion needs are coded in the same DNA as all of the other enzymes in the organism, the mitochondrion contains a small amount of DNA of its own, which codes for a few of its enzymes. All people thus inherit their mitochondria exclusively from their mothers, and comparing the mitochondrial genes of present-day individuals allows us to deduce something about the mitochondrion that was the most recent common ancestor of all modern ones. Mitochondrial Eve was thus the woman who was the most recent female common ancestor of all living humans.

ALLAN CHARLES WILSON (1934–1991) introduced the concept of mitochondrial Eve. He was born in New Zealand, but spent most of his career at the University of California at Berkeley, where he used biochemical methods to revolutionize the study of human evolution. He and his student, Vincent Sarich, showed by immunological methods that the separation between humans, chimpanzees, and gorillas was much more recent than had been thought on the basis of traditional methods.

It is sometimes assumed that mitochondrial Eve was an especially successful individual, probably one with many children. It should be clear from Figure 2.4 that she was just lucky, however, and may have had no more than two daughters. (She must have had at least two, because if she had just one, then that one would be mitochondrial Eve herself). In the example, the descendants of the ancestor labeled Bess initially multiplied faster than any other, constituting the majority of all descendants at generation 10, so Bess seemed a likely candidate to become mitochondrial Eve, but her line disappeared in generation 20, and in generation 21 only descendants of Eve were left.

As Figure 2.4 shows, mitochondrial Eve should not be confused with the Eve of the Bible, who is normally regarded as the only woman in existence at the time she was created. On the contrary, mitochondrial Eve was just one of many women alive at the same time. The human population was certainly smaller then than it is now, but it was already large. Because we live in an age of rapid population growth, it is easy to forget that it has not always been so, and that it will not remain so in the future. On the contrary, during most of human evolution, the population size has grown so slowly that a model that treats it as constant is not grossly inaccurate.

Moreover, mitochondrial Eve does not remain always the same person. If the tree represents the descent from a group of 20 women, then the ancestor labeled Eve would be mitochondrial Eve so far as the people at generation 21 are concerned, but working backwards through all the lines from the generation at the bottom line shows a much more recent mitochondrial Eve, labeled Evita. The title of mitochondrial Eve thus advances, and a woman living today or one of her descendants will one day be the mitochondrial Eve for our descendants. She is impossible to identify today, but she must exist. Mitochondrial Eve was not a particularly special person, and did not live at a special moment in human evolution.

Y-chromosome Adam

Some of the same things can be said of *Y-chromosome Adam*, the most recent common male ancestor of all men living today. Just as mitochondria descend through the female line, Y chromosomes descend only through the male line. It is sometimes supposed that Y-chromosome Adam and mitochondrial Eve were a couple living at the same time, but not only is there no reason for expecting this to be true, there are good reasons for thinking it not to be true. Although the average number of children that a man has is the same as the average that a woman has (given that every child has exactly one father and exactly one mother), the variation about the average is not the same for the two sexes. In most mammalian species (including humans), the number of daughters females have is fairly uniform, especially if we just count the daughters who survive to become adults. Except in rapidly growing populations (which can only exist for periods that are short on an evolutionary scale), a typical woman will have zero, one, two, or three surviving daughters, rarely more. A prolific man may have many more than three surviving sons, however, and large numbers of men (me, for example) have no sons at all. In some species, like elephant seals, the variations in fecundity between males are greater than they are in humans; in others, such as gibbons, they may be less, but in most mammalian species males are less uniform than females in the numbers of offspring they have.

This may be illustrated by redrawing the tree according to the same principles as Figure 2.4, except that now most individuals have zero, three, or four sons, as on the right in Figure 2.5 (the left-hand part being the same as in Figure 2.4, with generations after fixation omitted). Notice that fixation occurs much more quickly, with two distinct fixation sequences in only 19 generations. From

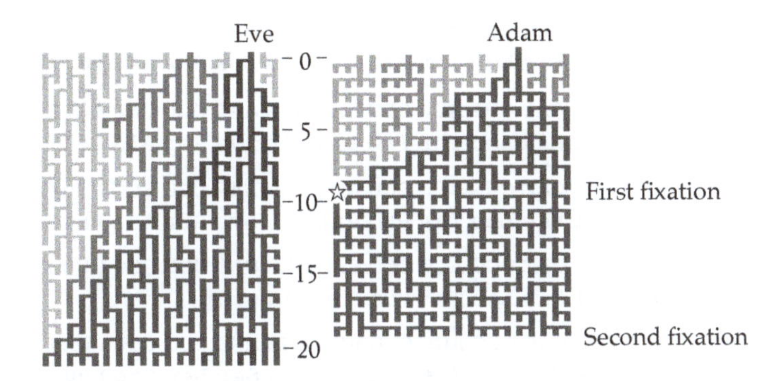

Figure 2.5 Fixation of genes with large variations in fecundity. The left-hand tree is the same as the tree of Figure 2.4, except that the generations after fixation are omitted, and no distinction is made for descendants of Evita. The right-hand tree shows that when individuals vary greatly in fecundity, some lines become extinct almost immediately, and fixation requires fewer generations. The common ancestor of the individuals in the bottom line is marked with a star.

this sort of argument we may expect that Y-chromosome Adam lived much more recently than mitochondrial Eve, so it is not at all likely that they were a couple.

Point mutations

The sort of mutations that I have been mainly discussing, resulting from wrong pairings at specific points in the DNA, and producing individual wrong bases in the daughter DNA, are examples of *point mutations*. They account for most of the changes that occur when protein sequences change over the course of evolution with little or no change in function. For example, hexokinase D exists in the livers of both humans and rats, where it helps to control the concentration of glucose in the blood, as we will consider in Chapter 10. The human and rat enzymes have more than 460 amino acids each, but are almost exactly the same, with no more than 20 differences between them. Most of these differences are of the simple wrong-substitution kind we have considered, but there are also a couple of *deletions*, because the rat enzyme is a little shorter than the human. In properties, the two enzymes are more or less indistinguishable,[6] and it is unlikely that a rat would be any the worse off if it had human hexokinase D, or that a human would be any the worse off with rat hexokinase D. We cannot be certain of this, but I do not know of any biochemist who would claim that all variations between protein sequences in different organisms are adaptive.

Some proteins, such as hemoglobin, which carries oxygen in our blood,

[6]We need to be careful about what we means by "properties": This is true if we are referring to kinetic properties, or properties that can be distinguished by physical measurements, but the immune system can distinguish between proteins that seem almost identical by other criteria, so in this case "properties" does not include immunological properties.

Native	AUG GUG CAC CUG ACU CCU GAG GAG AAG UCU GCC GUU ACU ... Met Val His Leu Thr Pro Glu Glu Lys Ser Ala Val Thr ...
Silent mutation	AUG GUG CA**U** CUG ACU CCU GAG GAG AAG UCU GCC GUU ACU ... Met Val **His** Leu Thr Pro Glu Glu Lys Ser Ala Val Thr ...
Missense mutation	AUG GUG CAC CUG ACU CCU G**T**G GAG AAG UCU GCC GUU ACU ... Met Val His Leu Thr Pro **Val** Glu Lys Ser Ala Val Thr ...
Nonsense mutation	AUG GUG CAC CUG ACU CCU GAG GAG **U**AG UCU GCC GUU ACU ... Met Val His Leu Thr Pro Glu Glu **stop**
Deletion of one base	AUG GUG CAC CUG AUC CUG AGG AGA AGU CUG CCG UUA CU ... Met Val His Leu **Ile Leu Arg Arg Ser Leu Pro Leu** ...
Deletion of two bases	AUG GUG CAC CUG AUC CUG GGA GAA GUC UGC CGU UAC U ... Met Val His Leu **Ile Leu Gly Glu Val Cys Arg Tyr** ...
Deletion of three bases	AUG GUG CAC CGA UCC UGG GAG AAG UCU GCC GUU ACU ... Met Val His Arg **Ser Trp** Glu Lys Ser Ala Val Thr ...

Figure 2.6 Point mutations. The top line shows the beginning of the gene for the β-chain of normal hemoglobin. As in Figure 1.6, the methionine residue at the beginning is not present in the mature protein. Substitution of C by U in the third codon is a *silent mutation*, because it has no effect on the resulting protein. More serious is *missense mutation*, in the third line, which produces valine in the final protein instead of glutamate: The resulting hemoglobin is still functional, but has harmful aggregation properties, and is responsible for *sickle-cell disease*. More serious than either these are *nonsense mutations*, which cause translation to stop prematurely. Nearly always (unless it occurs very near the end of the gene) this will produce an inactive protein. The last three lines show the effects of *deleting* one, two, or three bases. If one or two bases are deleted, the protein sequence is completely scrambled after the first deletion, and the protein is unable to function. These are called *frameshift mutants*. If three bases are deleted, the consequences are much less serious, especially if they occur close together, because the sequence is corrected by the third deletion, and the remainder of the protein is normal. The effects of *insertion* of bases are not illustrated, but follow what one should expect from consideration of deletions: Insertion of one or two bases has more drastic consequences than insertion of three.

vary much more between species than hexokinase D does; others, such as cytochrome c, a protein involved in energy management, vary much less. All of this is useful for constructing trees of relationship from protein sequences, and cytochrome c, for example, was used many years ago to relate many species spanning the plant, animal, and fungus kingdoms—something that would be hard to do with the use of morphology and fossils alone! However, it is not much help in explaining how species become so different from one another during evolution.

Not all point mutations are due to substitution of one base by another. Some can result from deletion of a base or insertion of an anomalous base,

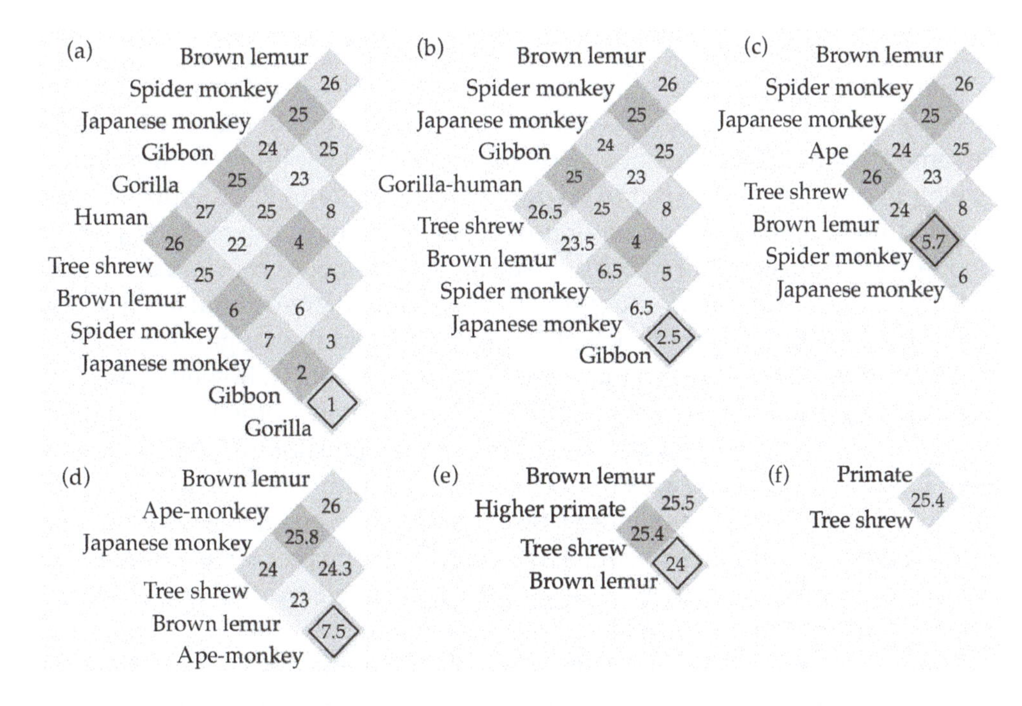

Figure 2.7 Clustering protein sequence data. The figure illustrates the Unweighted Pair Group Method with Arithmetic Mean for clustering amino acid sequences of β-chains of hemoglobin for seven species. (a) Table showing the number of sequence differences for all pairwise comparisons. The smallest number in the table, the 1 between human and gorilla, is shown in a diamond, and the gorilla and human are averaged into a single taxon to produce the next table. (b) Table with gorilla and human data averaged. (c) Table with gibbon, gorilla, and human averaged. The averages are weighted to take account of the two species in gorilla-human and just one in gibbon. (d–f) Continuation of the process until all the data are merged. Each value in a diamond yields a fragment of a tree connecting the seven sequences, and the complete tree is shown in Figure 2.8.

and several cases are illustrated in Figure 2.6. These include the substitution of a different amino acid in the sequence of the β-chain of human hemoglobin, which results in *sickle-cell disease,* a serious form of anemia affecting many people of African descent.

Constructing a phylogenetic tree

Figures 2.7 and 2.8 illustrate how to use amino acid sequence data to construct a tree of relationships. The simplest thing one can do is to line up the different sequences with one another and count the number of differences between each pair. If the sequences are not all of the same length, then complications arise for deciding how to line them up, but we will not take account of such complications now. Instead, we will just consider a set of sequences where they do not arise, as in the data in Figure 2.7 for comparisons between the β-chains of hemoglobin for seven animals.

The first thing to notice is that the tree shrew sequence differs from each of the other species to essentially the same extent. However, although "essentially the same," the numbers are not exactly the same. The variation, with numbers varying in an apparently arbitrary way in the range 24 to 27, is entirely to be expected for a gradual accumulation of random changes over a long period. If you take a sample of a radioactive substance and count the actual number of distintegrations in a period expected on average to give a total of 25.5, you should not be surprised if the actual number varies in the range 24 to 27 (and it will certainly not be 25.5, an impossible result). Curiously, however, Richard Dawkins[7] regarded that sort of variation as "an anomaly, something that in an ideal world would not be there," and spent about a page of his book worrying about it. However, if I saw a set of exactly identical numbers in an example like this, I would not regard it as an indication that I lived in an ideal world; I would regard it as evidence that someone had cheated. In Chapter 9 we will meet an example of an analysis by one great scientist that suggested that another great scientist's results fitted theoretical expectation too well.

There are many ways of constructing phylogenetic trees on the basis of numbers such as those in the table, and there are many arguments about which are best. I will ignore these arguments here, but just describe the method that is easiest to use, and easiest to understand, which has the unfortunate name of the *Unweighted Pair Group Method with Arithmetic Mean*, unfortunate both because it is a cumbersome name for a simple method and because "unweighted" here actually means "weighted." (In contrast, in the related *Weighted Pair Group Method with Arithmetic Mean*, "weighted" means "unweighted"!)

We start by looking for the smallest number, which is the 1 between the sequences for gorilla and human. This suggests, reasonably enough, that the gorilla and the human are the most closely related of the species shown. We then recalculate all the numbers, treating gorilla and human as a single entry. We now again search for the smallest number, the 2.5 separating the gibbon from the great apes, and continue in the same way until all the sequences are united. A complication is that when we recalculate the averages, we take account of the numbers of values in each of the averages used, so in averaging 25 with 26.5 on going from Figure 2.7b to Figure 2.7c, for example, we give double weight to the 26.5, so the average is $78/3 = 26$, not $51.5/2 = 25.75$.

Another point to be clear about is that although we may hope that the numbers are telling us something about the relationship between the seven species, they are actually telling us about the relationships between the seven

[7]R. Dawkins (1995) *River Out of Eden*, Weidenfeld and Nicolson, London.

[8]Motoo Kimura (1983) *The Neutral Theory of Molecular Evolution*, Cambridge University Press, Cambridge.

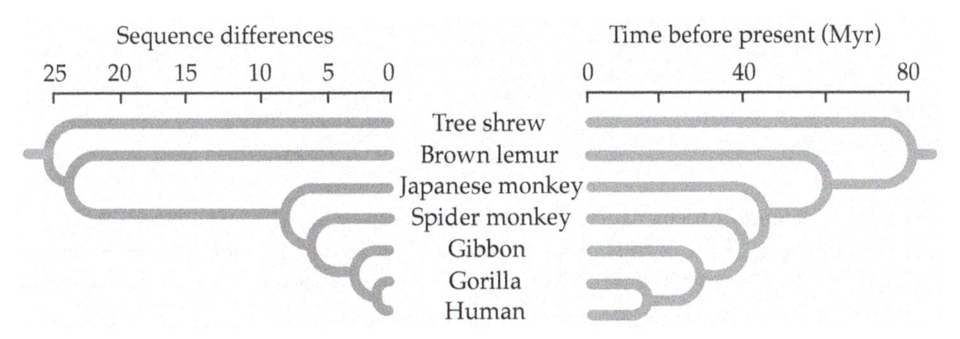

Figure 2.8 Constructing a tree from protein sequence data. *Left*: The tree that results from combining all the data in Figure 2.7; *right*: A tree based on classical ideas of the phylogeny and chronology, as drawn by Kimura in his book.[8]

protein sequences. We may hope that that is the same thing, but we should remember that it may not be. Anomalies can arise if an ancestor had two (or more) different proteins with the same function, and not all modern species have inherited the same one. When the enzyme lysozyme was first studied in various birds, the enzymes from duck and chicken showed the expected similarity, but the one from goose was very different, though ducks and geese are more closely related than ducks and chickens. The anomaly was understood when it turned out that the black swan has two different lysozymes, one gooselike and the other ducklike.[9]

Combining all the numbers leads to Figure 2.8, which has the same topology as the tree obtained from traditional comparison of anatomy, though the scaling is different. There is an important difference, however: Comparisons based on anatomy or fossils are inevitably limited by the small amount of data, but for sequence data the amount of information is already huge and is increasing all the time. Today it is more common to use gene sequences rather than amino acid sequences, but no matter, the principles are the same. One of the most noticeable discrepancies between the two trees in Figure 2.8 relates to the single sequence difference between human and gorilla, which is difficult to reconcile with separation 15 million years ago (as assumed by Kimura). However, this time of separation is now believed to be much more recent than that, of the order of 5 million years ago.

A table of numbers such as in Figure 2.7 contains a little more information than the tree, but the tree is far easier to visualize. Today hundreds or even thousands of species are clustered from sequence data that would be impossible to grasp if presented just as numbers or as aligned sequences. Construction of a tree illustrates Herbert Spencer's characterization of science

[9]N. Arnheim and R. Steller (1970) "Multiple genes for lysozyme in birds" *Archives of Biochemistry and Biophysics* **141**, 656–661.

Figure 2.9 Dimensions of biological entities. Even when they are drawn to look the same size, animals of the same family remain easily distinguishable from one another: A domestic cat is not just a scaled-down lion; a mouse is not just a scaled-down rat.

as "organized knowledge."[10]

Neutral evolution

The approximate constancy of the number of changes seen in each lineage after separation gave rise to the idea of a *molecular clock*, and, more important, to the conclusion that most of the changes seen in comparing proteins with the same function in different organisms are probably *neutral*, with no effect on the fitness of the individual. The realization that most of the mutations detected in such comparisons are neutral was the insight of the great population geneticist Motoo Kimura, and, although there is still some resistance from biologists determined to see adaptation wherever they look, there is increasing acceptance of his view. Despite a sentence in Kimura's book emphasizing the importance of natural selection, Richard Dawkins reported[11] the "rather endearing story told by John Maynard Smith" that Kimura "could not bear to write the sentence himself but asked his friend ... James Crow to write it for him."

This means, therefore, that we need to look elsewhere for an explanation of where new functions come from that allow one species to be different from another. All of the error-correcting machinery works in the direction of forbidding any changes at all, and hence any evolution at all. Such changes as occur are the result of mistakes. Moreover, if Kimura is right, then most point mutations will either be lethal, so the individual will not survive, or they will be neutral, so they will have no gross effect. Yet evolution does produce gross effects: A domestic cat is not just a small lion (Figure 2.9); a mouse is not just a small rat; still less is a bacterial cell a small human. So we need to look beyond point mutations to understand what sort of changes allowed our common ancestor to have descendants as different from

MOTOO KIMURA (1924–1994) was one of the architects of mathematical population genetics, and his book, *The Neutral Theory of Molecular Evolution*, one of the great classics of twentieth century biology, underlies much of our present understanding of evolution. Notice the word *molecular* in its title. Kimura made it clear in the book that he accepted natural selection as the main source of adaptive changes in evolution.

[10]H. Spencer (1854) "The art of education" *The North British Review* **31**, 137–171.

[11]R. Dawkins (2004) *A Devil's Chaplain: Reflections on Hope, Lies, and Love*, Houghton Mifflin, Boston.

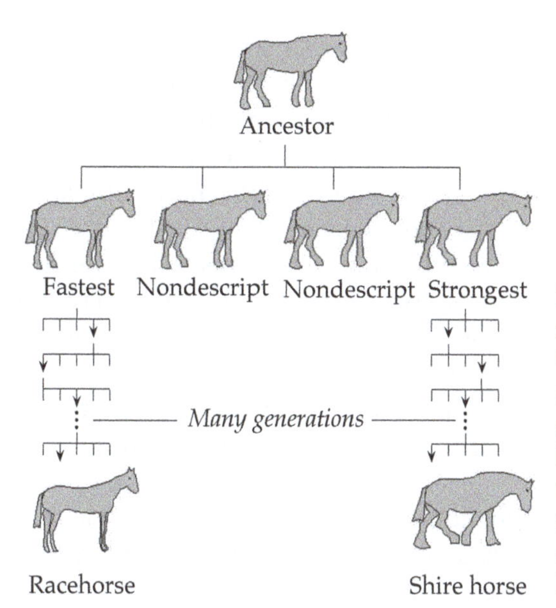

Figure 2.10 Artificial selection. The illustration summarizes how a horse breeder can proceed from a single strain of average horses in the first generation to two distinct types in later generations. In each generation the fastest horses are used as progenitors of the left-hand line, and the strongest are used for the right-hand line.

one another as humans and bacteria.

We find the same sort of results with many proteins. An idea of the problem—and a hint at a possible solution—comes from consideration of what is involved in evolving a new protein with a new function. The new function will nearly always be related in some way to the function of a protein that already exists and is already being fulfilled. For example, hemoglobin and myoglobin are both proteins found in mammals and other vertebrates. We will consider them later in this book from various points of view, but for the present it is sufficient to say that both have a capacity to bind oxygen, myoglobin rather tightly according to a simple mechanism, and without a property known as *cooperativity*, and hemoglobin less tightly and in a more complicated way, with cooperativity. The details of what cooperativity is can wait until Chapter 8; for the moment, it is enough to say that it is important for allowing hemoglobin to do its job well and would not be a desirable property for myoglobin to have.

The two proteins are quite similar in structure and have similar amino acid sequences, so it seems almost certain that they are derived from a common ancestral protein that was also able to bind oxygen. Presumably it resembled myoglobin more than hemoglobin, because myoglobin is somewhat simpler in structure and properties than hemoglobin, and it was certainly needed for binding oxygen in the ancestral organism. The problem is now how to evolve a new protein with new properties, without losing the old properties that continued to be needed.

Consider a horse breeder who wants to breed racehorses from a generic breed of shire horses, but who still needs shire horses for heavy work on his farm. This is not a realistic project for a modern horse breeder, and probably never would have been realistic to expect to achieve within the lifetime of a single breeder, but in past centuries it would have been quite feasible as something to attempt over several generations. Indeed, something of this sort almost certainly did happen before the whole process was accelerated by crossing locally bred horses with imported ones that had desirable characteristics.

In such a case the breeder cannot just breed from the fastest horses in each generation, discarding the others, because this would mean that the still necessary breed of strong horses would be lost. Instead, he would have to maintain separate lines of at least two kinds of horse: fast ones for winning races, and strong ones for doing the heavy work. In effect, therefore, the breeder needs to duplicate the stock in an early generation, and then maintain two (or more) separate stocks thereafter. The process of selection is outlined in Figure 2.10.

In the same way, a new protein function can only appear without losing an essential existing one if the process begins with a gene duplication.[12] If the part of the DNA that coded for the ancestral globin became repeated, so that two separate genes existed to do the same thing, then one of them could be modified in various ways while keeping the other one unchanged; in this way new properties could emerge without loss of the old ones. Alternatively, as with a horse breeder who started by separating an original breed of general-purpose horse—reasonably strong and fast, but neither very strong nor very fast—into two arbitrary groups and then trying to breed a very strong strain from one and a very fast one from the other, one globin line could become more like myoglobin over the same period as the other became more like hemoglobin.

Gross replication errors

Just as errors creep into the basic DNA-replicating system and result in point mutations, more severe errors also can result in long stretches of DNA being deleted or duplicated. Even entire chromosomes may be deleted or produced in extra copies as a result of a mistake in the replication process. In both cases the result may be disastrous, producing an offspring incapable of living, but occasionally it may be less damaging, and very rarely it may have no harmful effects at all. It is these very rare cases that probably supply the capacity of organisms to acquire new functions in the course of evolution.

We can find evidence for such gross replication errors by comparing the DNA of a primitive organism with that of its more complex relatives. Although there is no necessary relationship between the amount of DNA that an organism has and its complexity, there are examples that fit with naive

[12]*Duplication* and *replication* of DNA are sometimes confused, but they are not the same. Replication occurs at every cell division when a new copy of the entire genome is produced, in principle identical to the parent genome. Duplication occurs as a result of an error, when part of a DNA sequence appears twice in the daughter genome.

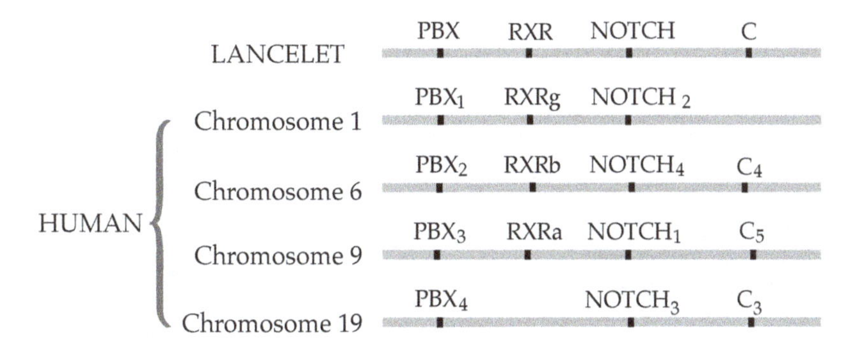

Figure 2.11 Evidence for gene duplication in human DNA. Four human chromosomes contain three or four each of a series of four genes similar to a series of four genes in the DNA of the lancelet, a primitive chordate.

expectations. One is provided by the lancelet, a small animal found on beaches around the world. This looks superficially like a small fish, but it is not a fish, or even a vertebrate, because it has no backbone. It is a primitive *chordate*, which means that it is an invertebrate member of the phylum that includes the vertebrates as its overwhelmingly preponderant members. It has much less DNA than a human, and this DNA includes a series of genes known as PBX, RXR, NOTCH, and C, as illustrated at the top of Figure 2.11. If, now, we look at human chromosome 9, we find that it contains a series of genes called PBX_3, RXRa, $NOTCH_1$, and C_5, similar enough to make it certain that these are the cousins of the similarly named genes of the lancelet. So far this provides no evidence of large-scale duplication, but this comes from examining three other human chromosomes: Chromosome 6 contains PBX_2, RXRb, $NOTCH_4$, and C_4; chromosome 19 contains PBX_4, $NOTCH_3$, and C_3 (RXR is missing); and chromosome 1 contains PBX_1, RXRg, and $NOTCH_2$ (C is missing). More detailed study indicates that chromosome 9 is the one that preserves most of the original structure, but the other three provide clear evidence of ancient replication errors.

As a brief digression, I want to comment on the obscure names of these genes: One (NOTCH) is vaguely suggestive of something or other; the other three appear completely meaningless. At the beginning, such names reflected a lack of knowledge of their functions, but there is an unfortunate habit in modern biochemistry of continuing to use obscure names long after the functions are known. It follows from the conquest of much of biochemistry by molecular biology and the tendency to talk about "gene products" rather than proteins, as if proteins existed only to express what is written in the DNA. This may seem a trivial point (and indeed is considered a trivial point by all those who defend meaningless names on the grounds that "everybody knows" what they

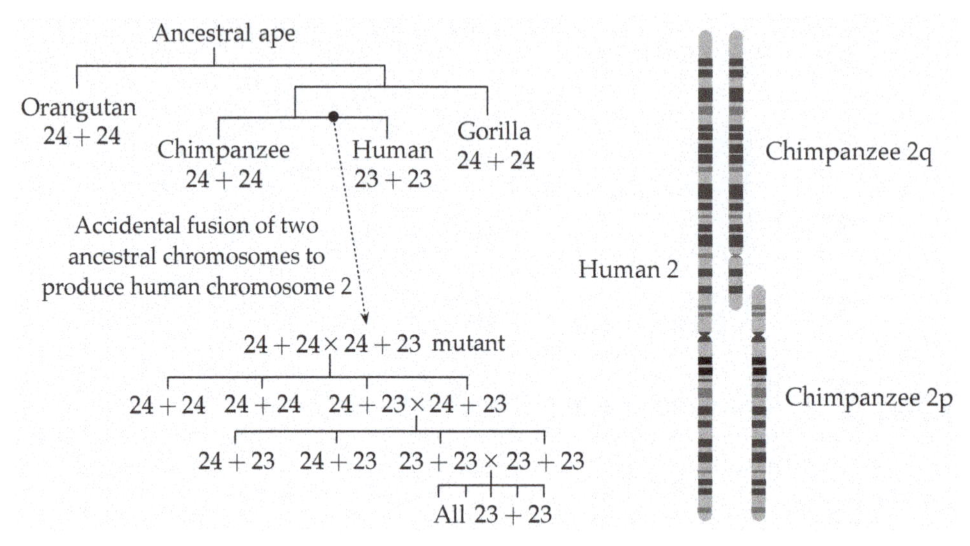

Figure 2.12 A hypothetical scenario for the separation of humans (with 23 + 23 chromosomes) from an ancestor with 24 + 24. Suppose that an accidental gene fusion of two ancestral chromosomes produced what is now chromosome 2 in the human (notice the exact agreement between the pattern of bands in the human chromosome with that in the corresponding two chimpanzee chromosomes). If the mutant 24 + 23 offspring was fertile with an individual with a normal 24 + 24 chromosome count and could produce fertile offspring, then 50% of these offspring would be 24 + 23. Inbreeding between them would produce 23 + 23 individuals who would have only 23 + 23 descendants.

refer to), but it seriously interferes with efforts to understand how biological systems are organized, because if there is selection at all it is the function that is selected, and this is a property of the protein, not of the gene.

Chromosomal abnormalities

Returning to the main theme, we can observe large-scale errors in DNA replication happening even today, in humans as well as in animals and plants. Chromosomal abnormalities are quite common, and by no means all are lethal. The best known (though not the most frequent) in humans is *Down syndrome* (in popular accounts it is often called *mongolism*, an unsatisfactory name that encourages an incorrect perception of the nature of the condition). In its most common form it is also called *trisomy-21*, and it results from having three instead of the normal two examples of chromosome 21. This discovery by Marthe Gautier in the 1950s led to increased understanding of the condition and increased interest in chromosomal abnormalities in general.

People with Down syndrome have disabilities, but frequently live into adulthood, so the condition is hardly lethal, and in Chapter 13 I will discuss why it should have any biochemical effects at all. Certain other trisomies involving the sex chromosomes are mild enough to pass unnoticed in some patients. The existence of these and other abnormalities makes it clear that errors involving large stretches of DNA not only occur quite frequently in all species, but also that even if they are nearly always harmful, their effects may be so mild that they can pass undetected. It requires no great leap of imagination, therefore, to propose that accidental duplication of genes in the course of evolution has been the motor that has allowed new functions to evolve without loss of existing ones.

The fact that chromosomal abnormalities are by no means always fatal, and may pass unnoticed, helps us to understand how humans can have a common ancestor with chimpanzees, given that chimpanzees, like orangutans and gorillas, have 24 pairs of chromosomes, whereas we, their closest relatives, have 23 pairs. Figure 2.12 shows a possible scenario that could explain how an ancestor with 24 pairs of chromosomes has descendants with 23 pairs.

JOHN LANGDON DOWN (1828–1896) was the British physician who described the clinical condition now known as *Down syndrome*. It was Down himself who used the description "mongoloid type of idiot" and thus initiated the now deprecated term of *mongolism*.

MARTHE GAUTIER (1925–) was the young French researcher who discovered in 1958 that the cause of what was then known as mongolism was an extra copy of chromosome 21, a discovery that revolutionized the field of chromosomal abnormalities, but she received almost no credit for it until many years afterwards. Even as late as 2014, when she was 88 years old, efforts were still being made to deny her role in the discovery.

Two Asian species of deer, the Chinese muntjac (*Muntiacus reevesi*) and the Indian muntjac (*Muntiacus muntjac*), provide an extreme illustration of the danger of regarding chromosome numbers as absolutely fixed. The two species are quite similar in appearance (much more similar than chimpanzees and humans) and can form viable hybrids, but whereas the Chinese muntjac has 46 chromosomes, the Indian muntjac has six in females and seven in males.[13]

The six races of brown mice found on the island of Madeira provide a less extreme example, but one that tells us more about the amount of time needed for an ancestral population to evolve into distinct forms with different chromosome counts. Janice Britton-Davidian and colleagues have found that although all of these are thought to descend from ordinary European house mice (*Mus musculus domesticus*), which have 40 chromosomes, the six Madeira races have from 22 to 30 chromosomes, and do not hybridize with one another (in ordinary circumstances they do not meet one another, because the different populations are geographically isolated across a rugged island with deep valleys). The island species have not lost any genes: They have simply reorganized their distribution among the chromosomes, as Figure 2.12 suggested may have happened in humans. The mice in Madeira were initially thought

[13]Females are XX, like humans; males are XXY, a chromosome combination that also occurs, rarely, in human males, where it is known as *Klinefelter syndrome*.

to have arrived with the first Portuguese settlers in the fifteenth century, but more recent evidence indicates that they came from Northern Europe much earlier, in the ninth century. This is anyway recent on an evolutionary scale, and suggests that the appearance of six distinct varieties has only required some thousands of generations.

It is tempting to speculate that the appearance of six new species in a short period of time is related to an apparently quite different characteristic of mice (and of rodents in general): That they are far more prone to cancer than humans and many other animals. If we accept the view discussed in Chapter 13 that cancer is driven by harmful chromosomal rearrangements, then it is reasonable to guess that speciation accompanied by changes in chromosome count may occur more readily in species susceptible to cancer. As far as I know, the incidence of cancer in the mice of Madeira has not been studied and compared with its incidence in other mice.

What is a species?

All of this brings us to a more general question: How do we define a *species*? How can we say that two populations of similar organisms belong to different species? The standard answer, implicit in Darwin's work but rigorously defined by Ernst Mayr, is that they constitute a single species if they can produce fertile hybrid offspring. Thus, horses (32 pairs of chromosomes) and donkeys (31 pairs) are considered to be different species, because although they can hybridize to produce mules, the hybrids are sterile. The entire human population constitutes one species, because all combinations that have ever been studied have normal fertility.[14]

As with any attempt to introduce a rigorous answer to a biological question, difficulties arise, both in time and in space. The spatial problem is easier to observe and concerns what are called *ring species*, well illustrated by seagulls. In Western Europe there are two types of seagull, the herring gull and the lesser black-backed gull, and these clearly constitute two separate species, because they cannot interbreed. If we circle the North Pole, however, we find that the herring gull can interbreed with the American herring gull, which can interbreed with the Vega gull, which can interbreed with the Birula gull, which can interbreed with Heuglin's gull, which can interbreed with the lesser black-

[14]As humans we are sensitive to slight differences between other humans that in other species could pass unnoticed (a point that I will return to at the end of Chapter 9, in relation to the suggestion that mutant genes are more often reported to be dominant in humans than in other species). Moreover, the differences we *see* are the ones on the surface, and thus exposed to differences in climate. There is far greater uniformity in the characters that we do not see with our eyes. Genetic studies (R. Bowden, T. S. MacFie, S. Myers, G. Hellenthal, E. Nerrienet, R. E. Bontrop, C. Freeman, P. Donnelly, and N. I. Mundy (2012) "Genomic tools for evolution and conservation in the chimpanzee: *Pan troglodytes ellioti* is a genetically distinct population" *PLOS Genetics* e1002504) show far less variation over the entire human population than can be found in populations of the common chimpanzee (not including bonobos) in West Africa.

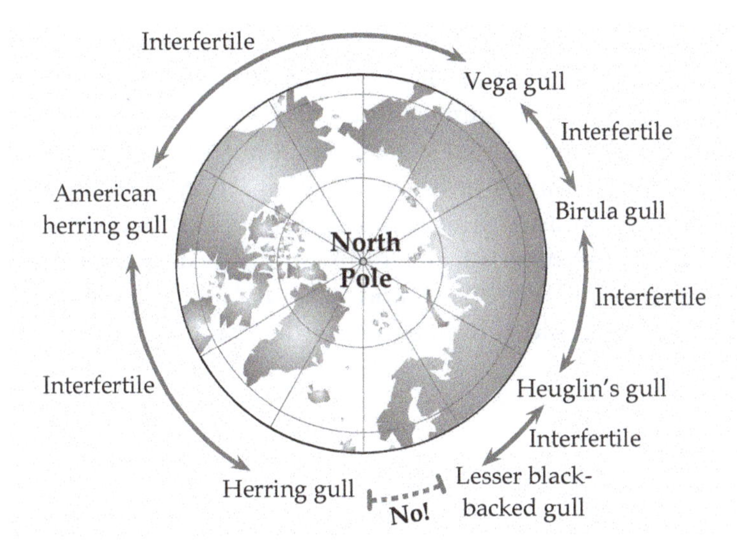

Figure 2.13 Ring species. The gulls that circle the North Pole (but cannot cross it) form a *ring species*. The two varieties found in Western Europe, the herring gull and the lesser black-backed gull, cannot interbreed, but otherwise the neighboring varieties around the circle are fertile with one another.

backed gull, bringing us back to where we started through a continuous series of mutually fertile gulls. Salamanders in California offer a more complicated example, more complicated, because, whereas gulls cannot fly across the Arctic Ocean at all, salamanders can occasionally cross the central valley, albeit with difficulty.

In the strictest sense, a ring species is not merely one in which the neighbors can interbreed, whereas the varieties at the ends of the ring cannot, but also there should be a continuous variation in character along the ring, with no sharp breaks. Now that genetic methods allow a much finer analysis of the different varieties, it appears that none of the supposed examples of ring species, including the seagulls, satisfy this stricter definition.[15]

ERNST WALTER MAYR (1904–2005) was born in Germany, but after 1931 he spent much of his life in the USA, where he died at the age of 100. One of the great evolutionary biologists of the twentieth century, he continued publishing into the twenty-first. He had a "classical" view of evolution and disliked the ideas of population genetics that were developed in the 1930s. He later also disliked molecular evolution and the idea that the gene is the unit of selection.

The difficulty of defining a species rigorously in space, producing relatively few anomalies like the gulls and the salamanders, pales into insignificance beside the problem of defining it in time. We can determine experimentally whether two varieties of gull or salamander can produce fertile offspring, but

[15]D. Liebers, P. de Knijff, and A. J. Helbig (2004) "The herring gull complex is not a ring species" *Proceedings of the Royal Society of London, series B* **271**, 893–901.

we cannot determine whether "Lucy," for example, could have fertile children with a modern father. If you compare yourself with your mother, everyone will agree that you belong to the same species, and no one will seriously question whether you belong to the same species as an ancestor from 100 generations earlier. What about an ancestor from 1000 generations earlier, or 10,000? We should not expect a sudden cut-off point, but a gradual decrease in fertility until after a certain number of generations fertility becomes negligible, but how many generations? We do not know, but it must be at least 800, because Tasmanian islanders could still breed with other humans after 10,000 years of isolation.[16,17]

[16] J. Clark (1988) "Smith, Fanny Cochrane (1834–1905)" *Australian Dictionary of Biography*: `http://adb.anu.edu.au/biography/smith-fanny-cochrane-8466`

[17] $10,000/800 = 12.5$: Am I suggesting that a human generation lasts only 12.5 years? No, with 25 years for each generation, 400 generations will bring you to the ancestor of the Tasmanians, and 400 more are needed to follow the other side of the tree to a descendant on the continent.

3. *Adaptation and Accidents*

> It has been proved, he said, that things could not be otherwise, for everything being made for a purpose, everything must necessarily be made for the best purpose. Notice that noses have been made to support spectacles, and thus we have spectacles. Legs are obviously intended to be trousered, and we have trousers.
>
> Voltaire, *Candide*[1]

> First a simple question. Which of these hands are you more likely to pick up?
>
> ♠ none ♥ A K Q J 10 9 8 7 6 5 4 3 2 ♦ none ♣ none
> ♠ 9 2 ♥ Q 10 9 4 3 ♦ 4 2 ♣ 8 7 6 2
>
> Zia Mahmood, *Guardian* bridge column, May 28, 1996

IN VOLTAIRE'S CLASSIC NOVEL the philosopher Dr. Pangloss taught that "all is for the best in this best of all possible worlds," a belief he stuck to through every disaster that befell him and his companions. He has had many followers in the history of evolutionary biology, who have interpreted every variation seen in nature as an adaptation of some kind. If the left brain controls the right half of the vertebrate body, this must be because that is the best way of arranging matters; if the structure of the hormone insulin is the same in pigs and dogs, but slightly different in humans, it is because humans have different needs from pigs and dogs; if sperm whales have more of the protein myoglobin in their muscles than horses do, it is because they need more of it; if polar bear liver contains so much vitamin A that it is toxic to explorers who eat polar bears, it is an adaptation to protect polar bears from predators; if the amino acid arginine can be represented in six different ways in the genetic code, whereas the more common amino acid aspartate can only be represented in two ways, it is because efficient operation of the protein synthesizing machinery of the cell has more need of redundancy for arginine than for aspartate; if horse liver contains a large amount of alcohol dehydrogenase (the enzyme that makes alcohol in yeast fermentation), it is to allow horses to make as much alcohol as they want without resorting to the methods that humans employ; if 90% of some genes consist of long stretches of bases that appear to code for nothing, the explanation is not that they have no function but that we have so far been too stupid to find it.

And so on, and so on. I have deliberately mixed my examples here: There are genuine cases of adaptation (sperm whale myoglobin), sometimes with a

[1]François-Marie Arouet, *known as* Voltaire (1759) *Candide, ou l'Optimiste*, Sirène, Paris.

wrong explanation attached (horse liver alcohol dehydrogenase), accidents of diet (vitamin A in the polar bear), frozen accidents of evolution (left and right brains, genetic code for arginine and aspartate, and, probably, the structure of insulin), and opportunistic or "selfish" behavior of DNA (noncoding sections of genes). Let us look at these in more detail, because otherwise I may appear to be just expressing personal opinions that others are free to disagree with.

Sperm whale myoglobin

Sometimes, as in the case of the sperm whale, we can see a clear and obvious relationship between the lifestyle of the animal and the biological observation. A sperm whale spends a large amount of its life swimming under water, but as an air-breathing mammal, it cannot obtain its oxygen like a fish from the water. Swimming under water requires a large amount of oxygen: Few people can swim as far as 25 meters under water,[2] and even trained swimmers can often do little better than that. But a large whale can swim for as much as an hour (more than 100 times longer than most humans can manage) without breathing, while expending large amounts of muscular energy. Moreover, we cannot easily imagine a whale adopting a different style of life where it spent most of its time on the surface and breathed more frequently. It has an absolute need, therefore, for a reliable supply of oxygen.

One solution might be to become more like a fish by evolving gills or even a wholly original system for extracting oxygen from water. But evolving gills is not something to be done overnight, or even in the millions of years since the whales returned to the sea: It would require extensive modification of many structures, and is, in short, the difficult solution. An obvious alternative would be to evolve blood with a greater capacity for storing oxygen. It sounds plausible, but it is not as easy as it sounds. Even in humans, with our relatively modest needs for holding our breath, the blood is already so full of red cells, and the red cells so full of hemoglobin, that there is scarcely room for any more. Significantly increasing the oxygen capacity of human blood is not an option, unless we find an alternative to hemoglobin, and it is not an option for whales either.

What the whale has done, therefore, is to increase the oxygen storage capacity of its muscles, which contain a far higher concentration of myoglobin than the muscles of land mammals. Myoglobin is not the same as hemoglobin, but it is similar, and is also capable of binding oxygen reversibly. (The two molecules played an important role in the development of understanding of how enzymes are regulated, which I will discuss in Chapter 10.) In land mammals, the primary function of myoglobin appears to be to allow rapid

[2]This takes of the order of half a minute, and people are sometimes surprised that the amount of time they can swim under water is much shorter than the time they can hold their breath while stationary. The reason is simple: The rate of consuming oxygen while you are engaged in work is far greater than when you are still.

diffusion of oxygen through muscle tissue, but it also helps to increase the total amount of oxygen that can be stored, and by means of large amounts of myoglobin whales can store much larger amounts of oxygen than land mammals can manage. Incidentally, although the name of myoglobin is less familiar to most people than that of hemoglobin, everyone has seen it, and its red color, much like that of hemoglobin, is an everyday sight. If you cut a piece of fresh meat, or you start to cook a hamburger, a red liquid flows out: This red liquid is not, as you might easily suppose, blood, but cell water colored by myoglobin.[3] Traces of blood, and hence traces of hemoglobin, may be present, but myoglobin, not hemoglobin, is the main source of the red color.

If this interpretation of the function of myoglobin is correct, we should expect land mammals to be less crucially dependent on it than diving mammals, and we should also expect small mammals to need it less than large ones do, because the distances that oxygen needs to move across the muscles is much smaller. In this connection, therefore, it is interesting to note that modern genetic techniques have allowed the study of mice that lack myoglobin completely, and they turn out to be just as healthy as ordinary mice, even when exercised—that is, even in conditions where the need for myoglobin ought to be greatest. I will be surprised if humans who have no myoglobin prove to be healthy, however, and amazed if whales do.

Horse liver alcohol dehydrogenase

What about horse liver alcohol dehydrogenase? Drunk wasps are a common sight when the ripe apples fall to the ground in late summer, but who has seen a drunk horse? Horses do not need alcohol dehydrogenase to make alcohol, but to destroy it. It is a *detoxification enzyme,* and horses need it in large quantities because their bacterial flora produce alcohol as an end product of their metabolism (just as it is in yeast fermentation). It is a detoxification enzyme in humans also, even though we get most of our alcohol in a different way and we prefer not to think of it as a poison. So it is correct to regard the high level of alcohol dehydrogenase in horse liver as an adaptation, but wrong to explain it in the silly way that I suggested at the beginning. This example illustrates, incidentally, an important point about enzymes that is sometimes forgotten. Like any other catalyst, an enzyme does not determine the direction in which a reaction proceeds, nor even how far it will proceed given enough time. These are questions that are decided by energetic considerations that are independent of whether a catalyst is present or not; the catalyst only determines how fast the reaction proceeds toward equilibrium. Adding a

[3]When I presented this material in a series of lectures in Louvain-la-Neuve, a distinguished biochemist in the audience objected that the red color is due more to cytochromes, another group of heme-containing proteins, than to myoglobin. I am not sufficiently expert in the matter to argue, but note that other sources that I have consulted agree that myoglobin is the principal one.

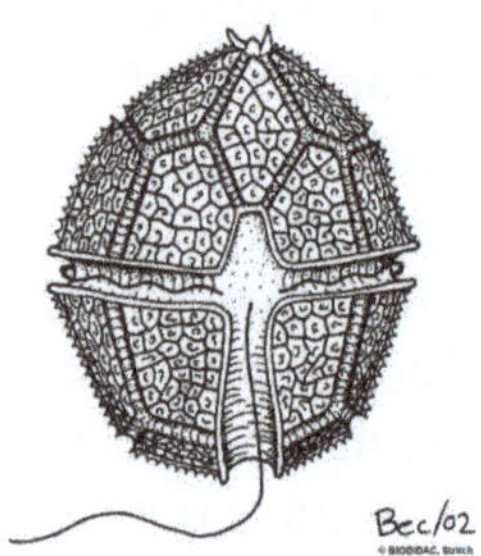

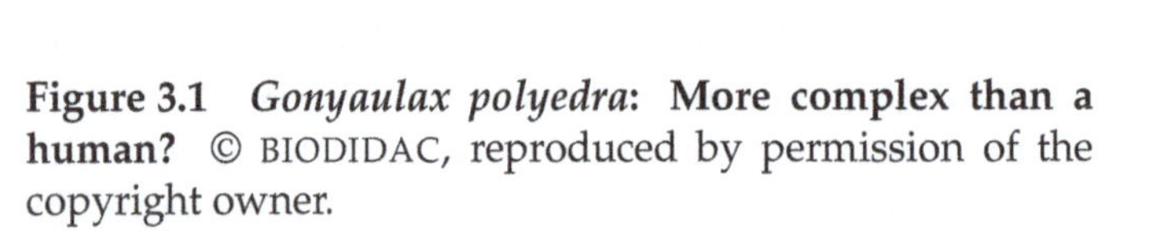

Figure 3.1 ***Gonyaulax polyedra*: More complex than a human?** © BIODIDAC, reproduced by permission of the copyright owner.

catalyst to a chemical system has much the same effect as lubricating an engine: It does not suddenly allow you to roll uphill; it just allows you to go more easily in the direction you would take anyway.

Polar bear liver

By no means everything we observe in a living organism is an adaptation, or if it is, it may be a different kind of adaptation from what it appears to be at first sight. The vitamin A in polar bear liver is not a mechanism to protect the animal from predators: An animal as large as a polar bear has no predators in the normal course of its life (if we ignore occasional unhappy encounters with killer whales), and the unfortunate explorers who fulfilled this role on one occasion did not know about the vitamin A until it was too late.[4] The vitamin A in the bears' livers is a consequence of eating large amounts of fish, of which the liver, as well as being an excellent source of vitamin D, as every child who has been made to consume cod liver oil knows, is also rich in vitamin A. In a sense we can still call it an adaptation, but it is an adaptation in the sense of increased tolerance to vitamin A beyond that of other mammals, not an increased need for the vitamin. It seems better to think of it as an accidental consequence of diet.

Such an accident can fairly easily be reversed. If some descendants of today's polar bears migrate permanently to a different habitat and adopt a different diet (as they may be forced to do by climate change), it will not be surprising if they lose their tolerance to vitamin A. We can see this kind of effect in the human tolerance to lactose: Europeans, and others like the cattle herders of Africa who consume large amounts of milk after infancy, retain the enzymes necessary for digesting lactose, or milk sugar, into adulthood; in other parts of the world, where milk is regarded as a food only for infants, the capacity to digest lactose is largely absent from adults.

[4]As with many such reports, several different stories have probably been confused. The Swiss explorer Xavier Mertz died after (though not necessarily because of) eating liver to avoid starvation, but this was in the Antarctic, and it was liver from dogs, not polar bears. On the other hand, Gerrit de Veer, a participant in the Dutch exploration of Novaya Zemlya in the sixteenth century, reported that eating polar bear liver produced severe symptoms.

Non-coding DNA The existence of sections of noncoding DNA within genes, or *introns*, is hardly an adaptation, at least as long as we think of an organism being organized primarily for its own good. (Introns are just one of various kinds of noncoding DNA, already discussed in Chapter 1.) We must not exclude the possibility that some of it has a function that has so far eluded detection, and certain new functions are coming to light. For example, some DNA codes small molecules of *small nuclear RNA*, which, as the name suggests, are short pieces of RNA located in the cell nucleus. These appear to fulfill important regulatory roles, affecting which protein-coding genes are expressed and in what amounts. There is nothing at present to suggest that these and other functional RNA molecules account for more than a tiny proportion of the DNA that does not code for protein. In an interview[5] after these were discovered, Sydney Brenner reported that he had been asked if he was ready to make "a public confession" that he was wrong about junk, and that he wrote back to say "I am prepared to reduce it from 96% to 95.8%."

If we think of an organism primarily as a vehicle to ensure the survival of the DNA that it contains, then it is not particularly surprising that the genome contains junk. After all, if a stretch of DNA can ensure its survival just by inserting itself in a gene that is ensuring its survival by doing something useful, why go to all the trouble of specifying a useful protein? Perhaps you will find it difficult to believe that all this DNA is no more than junk and prefer to think that it has some function that we have not discovered yet. If so, you may like to reflect on the remarkable fact that the amount of DNA (calculated as the number of base pairs) that different organisms contain bears little relation to their complexity. The lungfish, for example, has about 40 times as much DNA in each of its cells as humans do. Which is easier to believe, that a lungfish is 40 times as complex as a human, or that it has 40 times as much junk? Richard Lewontin has questioned whether a dog is more complex than a fish,[6] but even if we cannot assert with too much confidence that a human is more complex than a lungfish, it is probably not too self-satisfied to claim that we have at least a fortieth of the complexity of a lungfish. Would anyone seriously argue that the alga *Gonyaulax polyedra* (Figure 3.1), which also has much more DNA than a human, is much more complex than a human? Another species with a large genome, the domestic onion, *Allium cepa*, has given rise to the "onion test":[7] If most of the DNA is functional, why does an onion need five times as much as a human?

[5] S. de Chadarevian (2009) "Interview with Sydney Brenner" *Studies in History and Philosophy of Biological and Biomedical Sciences* **40**, 65–71.

[6] R. C. Lewontin (1990) "How much did the brain have to change for speech?" *Behavioral and Brain Sciences* **13**, 740–741.

[7] A. F. Palazzo and T. R. Gregory (2014) "The case for junk DNA" *PLOS Genetics* **5**, e1004351.

```
        4                                  6
Gly-...- Cys-Cys-Thr-Ser-Ile-Cys-Ser-...-Tyr-Cys-Asn

        4                        5                     9         Ala  Pig, dog
Phe-...- Leu-Cys-Gly-Ser-His-Leu-...-Leu-Val-Cys-Gly-... -Lys-{
                                                                 Thr  Human
```

Figure 3.2 Insulin sequences. Pigs and dogs have exactly the same sequences, but human insulin differs in the last amino acid residue, which is threonine (Thr) instead of alanine (Ala). The lines connecting cysteine (Cys) residues represent *disulfide bridges*: Each cysteine residue has an -SH group, but in proteins that exist outside cells, these are usually oxidized to give -S-S- connections between pairs of cysteines. The numbers above the ellipses show the numbers of residues omitted from the figure.

Genetic load

Fitness is a genetic measure of the likelihood that an individual will have offspring. If the population is not completely homogeneous, then not everyone can have the ideal fitness, and so the average fitness must be less than the ideal. The *genetic load*, or more specifically the *mutational genetic load*, is then defined as the reduction in the average fitness of the population due to deleterious mutations. An average human newborn infant has between about 55 and 100 new point mutations that do not occur in either parent. If the neutral theory is correct, then only about one of these is deleterious, a load that the population can easily support: As long as an average individual has somewhat more than one child, there is no danger that the population will collapse from the uncontrolled accumulation of harmful mutants. However, if the claim of 80% functional DNA in the original ENCODE project is correct, the average number of children each parent needs to have to maintain a viable population becomes impossibly large.[8] The fact that new mutations are usually recessive (Chapter 9) complicates the calculation, but it does not alter the qualitative conclusion: Even if mutant children are viable and can reach reproductive age, the mutations will certainly cause serious problems for future generations.

Insulin in pigs, dogs, and humans

Other accidental changes may have been preserved simply because it did not much matter one way or another. The majority of the differences that exist between the proteins of different organisms, including the insulins of pigs, dogs, and humans shown in Figure 3.2, are probably of this kind. Although in the early days of studies of protein evolution 40 years ago, many people tried to apply the Panglossian pan-adaptationist logic to proteins, nearly everyone has since become converted to the neutralist view that I mentioned in the

[8]The argument in this section is based on the calculations by Dan Graur described on pages 514–517 of his recent book: D. Graur (2016) *Molecular and Genome Evolution*, Sinauer Associates, Sunderland, Massachusetts.

previous chapter, that most replacements of one protein structure by another have no functional effects one way or the other. This is almost impossible to test, because calculations indicate that the advantage needed to ensure that a gene is not replaced by an almost equivalent mutant is so small that few experiments are sensitive enough to detect it.

Left and right brains

Other accidents may result from events so long ago that altering them now would entail such a complicated series of changes that it is effectively impossible and will never happen. We can easily imagine that a nervous system in which the left and right brains controlled the left and right halves of the body, respectively, might be marginally more efficient than the present arrangement, but what we cannot easily imagine is the process, with all the rewiring that would be needed in many parts of the nervous system that would allow the change to be made. Presumably the accident arose in a primitive ancestor in which it did not greatly matter how the connections were made: If it had mattered, it ought not to have been too difficult to correct it, but, the more complex the descendants of that early ancestor became, the more difficult it became to change the wiring diagram in such a fundamental way. Here we do not know with certainty that there is no adaptive advantage in the wiring system that exists, but we do know that switching to a different one would be difficult or impossible, and we have trouble, moreover, in imagining any adaptive advantage. It seems most reasonable, therefore, to regard such cases as frozen accidents, and refrain from searching for adaptive advantages.

Arginine codons

Sometimes the case for a frozen accident appears stronger because we can more easily reconstruct both the mechanism that gave rise to the accident and the reasons why it has become frozen. The genetic code is such a case. It makes little adaptationist sense for arginine to have three times as many codons as aspartate, but Jeffrey Wong has shown that it makes good evolutionary sense nonetheless. His *coevolution hypothesis*, illustrated in Figure 3.3, is that at an early stage in the origin of life relatively few amino acids could be separately coded in the primitive DNA. Each of these few, which included arginine and aspartate, could be coded in several different ways: The third letter of the three-letter code was probably little more than a spacer or punctuation mark (as indeed it still is to some extent[9]), and even the first two letters included

[9]Out of 16 permutations of the first two bases in DNA, eight fully specify the amino acid without regard to the third base: For example, TCT, TCC, TCA, and TCG all code for the amino acid serine, so if the first two bases are TC, the third one can be anything. Of the other eight, six result in a two-way selection: For example, CAT and CAC code for histidine, but CAA and CAG code for glutamine. Putting the third base into two classes in this way makes chemical sense, because T and C, *pyrimidine bases*, have similar structures, whereas A and G, *purine bases*, have a different kind of structure. These structures can be seen in Figure 2.1, above.

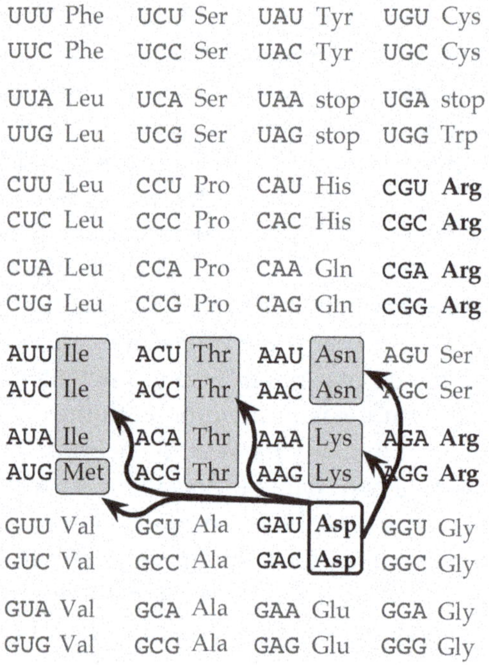

Figure 3.3 Wong's *coevolution hypothesis.* The whole genetic code is illustrated (see Figure 1.5), but we are concerned with the codons for arginine (Arg), of which there are six in the present code, and for aspartate (Asp), of which there are only two, even though most proteins contain more aspartate than arginine. However, aspartate is the metabolic precursor of five other amino acids—isoleucine (Ile), methionine (Met), threonine (Thr), asparagine (Asn) and lysine (Lys)—but arginine is not the metabolic precursor of any other amino acids. Wong's hypothesis is that aspartate had more codons in the primitive code than it has today, but in the course of evolution it lost some of these to satisfy the needs of its metabolic descendants.

UUU Phe	UCU Ser	UAU Tyr	UGU Cys
UUC Phe	UCC Ser	UAC Tyr	UGC Cys
UUA Leu	UCA Ser	UAA stop	UGA stop
UUG Leu	UCG Ser	UAG stop	UGG Trp
CUU Leu	CCU Pro	CAU His	**CGU Arg**
CUC Leu	CCC Pro	CAC His	**CGC Arg**
CUA Leu	CCA Pro	CAA Gln	**CGA Arg**
CUG Leu	CCG Pro	CAG Gln	**CGG Arg**
AUU Ile	**ACU** Thr	**AAU** Asn	AGU Ser
AUC Ile	**ACC** Thr	**AAC** Asn	AGC Ser
AUA Ile	**ACA** Thr	**AAA** Lys	**AGA Arg**
AUG Met	**ACG** Thr	**AAG** Lys	**AGG Arg**
GUU Val	GCU Ala	**GAU Asp**	GGU Gly
GUC Val	GCC Ala	**GAC Asp**	GGC Gly
GUA Val	GCA Ala	GAA Glu	GGA Gly
GUG Val	GCG Ala	GAG Glu	GGG Gly

considerable redundancy.

As life became more complicated, this system became inadequate, because manufacture of more efficient enzymes and other proteins required a larger number of different amino acids to be specified—maybe not all of the 20 that we know today, but more than the primitive organisms needed. How was this greater precision to be achieved without scrapping the whole system and starting again? Wong suggested that evolution of the genetic code occurred concomitantly with the evolution of new metabolic pathways to synthesize the extra amino acids that were needed, and that the amino acids that were coded for in the primitive code would have to "give away" some of their codons to their metabolic descendants. His full hypothesis is more complicated than this, but it explains the observation I started with: We expect an amino acid like aspartate, the metabolic precursor of several other amino acids, to have given away so many of its original codons that it now has rather few. On the other hand, arginine is not the metabolic precursor of any other amino acid, and so it should retain the full stock of its original codons: Even if it had fewer than aspartate to start off with, it has more now. More generally, Wong's hypothesis leads us to expect only a weak correlation between the number of codons that an amino acid has and its frequency in protein structures.

Why should an accident of this kind in the development of the genetic code become frozen so that now it is impossible to modify? Even the simplest of present day organisms needs to make a few thousands of different proteins, and more complex ones need far more. The sizes of these proteins vary, but if we take an average protein to consist of 250 amino acids arranged in a definite order we will not be grossly far from reality. Of these, perhaps 8%, or 20 per protein, will be aspartates. Suppose now we ask what would be the result of altering the genetic code so that one of the two codons that currently codes for aspartate became ambiguous: Instead of coding aspartate all of the time, it started to code for aspartate only half of the time, the other half of the time being read as another amino acid, glutamate. I have deliberately chosen one of the smallest modifications I could think of: I have not assumed a complete unambiguous switch of one amino acid for another, and the particular switch is a modest one, because glutamate is chemically quite similar to aspartate, but a little larger in size. (Together they are responsible for the basic properties of proteins, but in a frozen accident in the evolution of biochemical knowledge, they are known to most biochemists as "acidic amino acids"[10].) So, we may expect that substituting glutamate for aspartate may have a negligible effect at many sites, and that the new protein may be just as good as the old.

This, indeed, is what the study of protein sequences across different organisms leads us to expect. There are many examples of proteins where the sequence in one organism has an aspartate residue and that in another has glutamate. In fact many observed substitutions are far more radical than that, and we can find quite different amino acids substituting for one another in perfectly functional proteins. Contrary to what creationists may believe, there is a great deal of redundancy in protein structures, and it is far from the case that a protein sequence has to be exactly right. On the contrary, most protein functions can be fulfilled by a large number of known different sequences, not to mention all of the possible ways that are not yet known.[11]

Likelihood of surviving a change in the genetic code

Let us return to our example of a conservative change to the genetic code that would allow an aspartate codon to be read as a glutamate codon about half of the time. Because there are two aspartate codons in the present-day code, and we are assuming one of them to remain unambiguous, this implies that about one-quarter of the aspartates in the proteins of the unmodified organism will be replaced by glutamates in the mutant. Insulin, the protein used as an illustration in Figure 3.2, contains no aspartate residues, so it would be completely unaffected by such ambiguity.

[10]This is an important misconception that seriously complicates the efforts of students to understand the properties of proteins, but it is not as crazy as it may appear. If the free amino acids are prepared in their pure forms, they are indeed acids, aspartic acid and glutamic acid. When they are incorporated into proteins, they lose two of their ionizable groups completely, and the third, the one that justifies calling these free forms acids, loses its proton in neutral solution: It then no longer has a proton to donate, so it is no longer an acid; but it can accept one, so it can act as a base.

[11]In Chapter 12 I will mention some evidence to support this statement.

It is a small protein, however, and anyway it is unusual for a protein to lack aspartate completely. Chicken cytochrome c, a protein of 105 residues, has five aspartates, and the effect of allowing the four of these that are encoded by GAT to be changed to glutamate by the ambiguity is explored in Figure 3.4. Notice that two of the allowed *aspartate* $\rightarrow$ *glutamate* substitutions correspond to *aspartate* $\rightarrow$ *alanine* and *aspartate* $\rightarrow$ *lysine* substitutions if the chicken protein is compared with the human—both of these are less conservative than *aspartate* $\rightarrow$ *glutamate*, suggesting that in these positions glutamate should be just as acceptable as aspartate.

Cytochrome c is also a small protein, so we ought to consider what would happen with a larger one. If there are 10 aspartate residues, and a 50% chance that any one of them may be changed to glutamate by the ambiguity, there is about a 0.1% chance that any individual protein molecule will be completely unchanged, and a 99.9% chance that it will contain one or more glutamates where aspartates ought to be.[12]

How likely is it that an organism could survive this? As I have said, most proteins tolerate some variation in the amino acids that occupy most positions, so there is a good chance that some or all of the mutant molecules will be perfectly functional. So, let us guess that there is a 99% chance that the organism will not even notice that a particular protein is modified in 99.9% of its molecules, but a 1% chance that only the completely normal molecule will do. Let us further suppose that for the 1% of proteins that require a perfectly correct sequence to function, the 0.1% of such perfect molecules are not enough for the needs of the organism. So, at the end of all this, the chance that an organism can survive is 0.99 raised to the power of the number of different kinds of protein that it needs to make. Even for a bacterium, the number of different proteins is by no means small: For those that have been most studied, such as *Escherichia coli* and *Bacillus subtilis*, it is believed to be around 4000, but other free-living organisms exist with only about one-third as much genetic material as these, and the parasitic bacterium *Mycoplasma genitalium* has fewer than 500 genes. So, let us suppose we are dealing with a free-living organism that requires a minimum of 1300 different kinds of functional proteins in order to live. Then 0.99 to the power of 1300 gives us 0.000,002, or odds against survival of about half a million to one.

Because I have made a number of assumptions in this calculation, I should perhaps pause a moment to examine whether there are any facts that would enable me to estimate whether the calculation is grossly in error. After all, there may be little reason to care if the final result is wrong by a factor of 10, or even 100, but it would be useful to have an idea if it is likely to be wrong by a factor

[12]If there is a 0.5 chance that any one site is unchanged, then there is a chance of 0.5×0.5 that any two will be unchanged, and so on until a $0.5 \times 0.5 \times 0.5 \times 0.5 \times 0.5 \times 0.5 \times 0.5 \times 0.5 \times 0.5 \times 0.5 = 0.001$ chance that all 10 will be unchanged.

GAT GAT GAT GAC GAT

Chicken Gly-**Asp**-Xaa$_{47}$-**Asp**-Xaa$_{11}$-**Asp**-Xaa$_{30}$-**Asp**-Xaa$_{6}$-**Asp**-Ala-Thr-Ser-Lys

50% Glu | 50% Glu | 50% Glu | No change | 50% Glu

6.25% Gly-**Asp**-Xaa$_{47}$-**Asp**-Xaa$_{11}$-**Asp**-Xaa$_{30}$-**Asp**-Xaa$_{6}$-**Asp**-Ala-Thr-Ser-Lys
6.25% Gly-**Asp**-Xaa$_{47}$-**Asp**-Xaa$_{11}$-**Asp**-Xaa$_{30}$-**Asp**-Xaa$_{6}$-**Glu**-Ala-Thr-Ser-Lys
6.25% Gly-**Asp**-Xaa$_{47}$-**Asp**-Xaa$_{11}$-**Glu**-Xaa$_{30}$-**Asp**-Xaa$_{6}$-**Asp**-Ala-Thr-Ser-Lys
6.25% Gly-**Asp**-Xaa$_{47}$-**Glu**-Xaa$_{11}$-**Asp**-Xaa$_{30}$-**Asp**-Xaa$_{6}$-**Asp**-Ala-Thr-Ser-Lys
6.25% Gly-**Glu**-Xaa$_{47}$-**Asp**-Xaa$_{11}$-**Asp**-Xaa$_{30}$-**Asp**-Xaa$_{6}$-**Asp**-Ala-Thr-Ser-Lys
6.25% Gly-**Asp**-Xaa$_{47}$-**Asp**-Xaa$_{11}$-**Glu**-Xaa$_{30}$-**Asp**-Xaa$_{6}$-**Glu**-Ala-Thr-Ser-Lys
6.25% Gly-**Asp**-Xaa$_{47}$-**Glu**-Xaa$_{11}$-**Asp**-Xaa$_{30}$-**Asp**-Xaa$_{6}$-**Glu**-Ala-Thr-Ser-Lys
6.25% Gly-**Glu**-Xaa$_{47}$-**Asp**-Xaa$_{11}$-**Asp**-Xaa$_{30}$-**Asp**-Xaa$_{6}$-**Glu**-Ala-Thr-Ser-Lys
6.25% Gly-**Asp**-Xaa$_{47}$-**Glu**-Xaa$_{11}$-**Glu**-Xaa$_{30}$-**Asp**-Xaa$_{6}$-**Asp**-Ala-Thr-Ser-Lys
6.25% Gly-**Glu**-Xaa$_{47}$-**Asp**-Xaa$_{11}$-**Glu**-Xaa$_{30}$-**Asp**-Xaa$_{6}$-**Asp**-Ala-Thr-Ser-Lys
6.25% Gly-**Glu**-Xaa$_{47}$-**Glu**-Xaa$_{11}$-**Asp**-Xaa$_{30}$-**Asp**-Xaa$_{6}$-**Asp**-Ala-Thr-Ser-Lys
6.25% Gly-**Asp**-Xaa$_{47}$-**Glu**-Xaa$_{11}$-**Glu**-Xaa$_{30}$-**Asp**-Xaa$_{6}$-**Glu**-Ala-Thr-Ser-Lys
6.25% Gly-**Glu**-Xaa$_{47}$-**Asp**-Xaa$_{11}$-**Glu**-Xaa$_{30}$-**Asp**-Xaa$_{6}$-**Glu**-Ala-Thr-Ser-Lys
6.25% Gly-**Glu**-Xaa$_{47}$-**Glu**-Xaa$_{11}$-**Asp**-Xaa$_{30}$-**Asp**-Xaa$_{6}$-**Glu**-Ala-Thr-Ser-Lys
6.25% Gly-**Glu**-Xaa$_{47}$-**Glu**-Xaa$_{11}$-**Glu**-Xaa$_{30}$-**Asp**-Xaa$_{6}$-**Asp**-Ala-Thr-Ser-Lys
6.25% Gly-**Glu**-Xaa$_{47}$-**Glu**-Xaa$_{11}$-**Glu**-Xaa$_{30}$-**Asp**-Xaa$_{6}$-**Glu**-Ala-Thr-Ser-Lys

Human Gly-**Asp**-Xaa$_{47}$-**Ala**-Xaa$_{11}$-**Asp**-Xaa$_{30}$-**Asp**-Xaa$_{6}$-**Lys**-Ala-Thr-Ser-Lys

Figure 3.4 Effect of an ambiguous code. The figure explores the effect of changing the code so that GAT, one of the two codons for aspartate, becomes ambiguous and has a 50% probability of coding for glutamate (Glu) instead of aspartate (Asp), whereas the other codon, GAC, is unaffected. In the sequence of chicken cytochrome c, there are five aspartate residues, of which one is coded by GAC and the other four by GAT. The code Xaa refers to any amino acid that is not aspartate, and Xaa$_{47}$, for example, means a sequence of 47 such residues. Only one-sixteenth (6.25%) of the protein molecules made by the ambiguous system have exactly the sequence of chicken cytochrome c. Notice that of the four aspartates that can change, human cytochrome c has alanine (Ala) for one and lysine (Lys) for the other.

of a million or more. For this we may consider the ram1 series of mutants of *Escherichia coli*. These mutants contain faulty machinery for translating genes into proteins, and although they are much more accurate than most human keyboard operators, they make far more errors than the normal bacteria make: Out of every 100 molecules that they make of the 1100-amino-acid protein β-galactosidase (much larger than what I have taken as average), no two are identical. These bacteria grow poorly, and in a natural environment would rapidly be overwhelmed in numbers by their healthier cousins, but they do grow, and in the artificial conditions of the laboratory they can survive. We are far from the creationists' fantasy of proteins that have to be exactly correct in every detail for the organism to live and reproduce. This example was given by Jacques Ninio,[13] and he went on to discuss a more technical example that

[13]Jacques Ninio (1982) *Molecular Approaches to Evolution,* Pitman, London.

is even more relevant to the problem of surviving a change in the code. This involves the use of *missense suppressor tRNA* molecules, which cause certain codons to be misread with a frequency of around 10%. Bacteria can survive the presence of one such suppressor tRNA, but not two. Again, this implies that although an appreciable error rate can be tolerated, there are limits that cannot be exceeded.

In general, we can conclude that for obtaining a rough idea of the probability that an organism could survive a change in the genetic code, the earlier calculation is not unreasonable. Thus, even a simple organism has a low probability of being able to tolerate a minimal change in the genetic code, and for a complex organism, the chance would be much smaller. Nonetheless, it is not so small as to be absolutely negligible, and so we might expect to see one or two vestiges of alternative codes in the simplest of organisms.

As mentioned in Chapter 2, the so-called "universal" code is now known to be not quite universal. In bacteria, and in the nuclei of all the cells of nearly all the nucleated organisms that have ever been studied, the genetic code is exactly the same in all its details.[14] Not all genes are in the nucleus, however. As mentioned in Chapter 2, a few proteins are coded in the small bodies within cells known as *mitochondria*, and in these the code is different from the "universal" code. In all cases, the differences from the standard code involve only a few codons, as illustrated for the mitochondria of vertebrates in Figure 3.5. Mitochondria are believed to be remnants of formerly free-living organisms, probably primitive bacteria, that have become so fully adapted to the habit of living within cells of other organisms that they are now part of them. Whether their slightly different codes result from experimenting with alternative codes, or they are so ancient that they came into existence before the "universal" code became frozen is a question I will return to shortly, but it does not matter for the immediate point: They confirm that even though the chance of surviving a change of code may be small, it is not zero.

If Wong's hypothesis is right, the genetic code is not absolutely immutable, and it should be possible to modify it experimentally. He showed that this could be done,[15] albeit with a much more modest change than the introduction of ambiguity into an aspartate codon that I considered earlier. Tryptophan has the most elaborate structure of the 20 amino acids needed for proteins and is one of those that is used least frequently. It has only one codon in the universal code and was almost certainly one of the last amino acids to be recognized in the code. The amino acid 4-fluorotryptophan is chemically almost the same as tryptophan, but it has one of the hydrogen atoms of tryptophan replaced by a

[14]Exceptions may be found in the list maintained at `http://tinyurl.com/kmm7zlo`. Of the 24 variant codes listed at present, nearly all are for mitochondria.

[15]J. T.-F. Wong (1983) "Membership mutation of the genetic code: Loss of fitness by tryptophan" *Proceedings of the National Academy of Sciences of the USA* **80**, 6303–6306.

UUU Phe	UCU Ser	UAU Tyr	UGU Cys
UUC Phe	UCC Ser	UAC Tyr	UGC Cys
UUA Leu	UCA Ser	UAA stop	UGA Trp (stop)
UUG Leu	UCG Ser	UAG stop	UGG Trp
CUU Leu	CCU Pro	CAU His	CGU Arg
CUC Leu	CCC Pro	CAC His	CGC Arg
CUA Leu	CCA Pro	CAA Gln	CGA Arg
CUG Leu	CCG Pro	CAG Gln	CGG Arg
AUU Ile	ACU Thr	AAU Asn	AGU Ser
AUC Ile	ACC Thr	AAC Asn	AGC Ser
AUA Met (Ile)	ACA Thr	AAA Lys	AGA stop (Arg)
AUG Met	ACG Thr	AAG Lys	AGG stop (Arg)
GUU Val	GCU Ala	GAU Asp	GGU Gly
GUC Val	GCC Ala	GAC Asp	GGC Gly
GUA Val	GCA Ala	GAA Glu	GGA Gly
GUG Val	GCG Ala	GAG Glu	GGG Gly

Figure 3.5 A variant code. The genetic code in vertebrate mitochondria differs from the standard code (Figure 1.5) with respect to four codons, the standard values being shown in parentheses. For example, AUA codes for isoleucine (Ile) in the standard code, but for methionine (Met) in the mitochondrial code.

fluorine atom. Because of this high degree of similarity, we may expect that a protein with some or all of the tryptophans replaced with 4-fluorotryptophan will nearly always be functionally indistinguishable from the native protein. Nonetheless, the machinery of protein synthesis distinguishes between them with good accuracy, and in normal bacteria such as *Bacillus subtilis*, the enzyme that starts the incorporation of tryptophan into growing proteins does not readily recognize 4-fluorotryptophan; the fluorinated amino acid is only rarely incorporated in protein, even if the bacteria are grown in a medium rich in 4-fluorotryptophan but lacking tryptophan. Such a medium is not found in nature, but it can readily be produced artificially in the laboratory, and we might expect that bacteria growing in it would find it advantageous to recognize 4-fluorotryptophan, thereby avoiding the need for the metabolically expensive synthesis of the natural amino acid. Wong found that this was indeed the case, and in careful selection experiments he was able to go from a natural strain of *Bacillus subtilis* with a 700-fold preference for tryptophan to a mutant with a 30-fold preference for 4-fluorotryptophan. Wong used classical selection methods to achieve this result more than 30 years ago, but today the techniques of genetic manipulation allow one to proceed much faster.

Is there any evidence that the same sort of evolution may have occurred under natural conditions? The rare amino acid selenocysteine may provide an answer. This is the same as the "standard" amino acid cysteine except that it has a selenium atom in place of the usual sulfur atom. Most organisms do not incorporate it into their proteins, and their genetic codes do not recognize it. The few organisms that use it do incorporate it into proteins, however, and they do so by the normal genetic machinery,[16] having transferred to selenocysteine one of the two codons normally used for cysteine. We can imagine that the original switch may have been brought about by pressure similar to the artificial pressure used by Wong to change the codon for tryptophan into a codon for 4-fluorotryptophan.

Bias in probability calculations

Before continuing, I want to look at the apparently trivial bridge problem that I quoted at the beginning of the chapter: Which is more likely, that you will receive a hand consisting of all 13 hearts, or the second hand, which lacks any obvious feature? The naive answer is that the nondescript hand is more likely, because it is an "ordinary" hand that no one but a bridge expert could remember for five minutes, let alone the next day, whereas if you picked up all 13 hearts you would be telling your friends about it for the rest of your life. A more sophisticated answer is that as long as the cards were properly shuffled before dealing them, the two hands are exactly equally likely and you should receive each of them once on average in every 635,013,559,600 deals; we perceive one as more improbable than the other only because there are vastly more of the boring hands than there are of the interesting ones.

The second answer is also naive and is also wrong. The correct answer is that you are more likely to receive the 13 hearts, and several hands of this kind are reported to have been dealt in the history of bridge, even though it is unlikely that 635,013,559,600 hands of bridge have yet been dealt. As Zia Mahmood commented in the original bridge column, "There is a small chance that someone will have stacked the deck so that you get all 13 hearts, while nobody is going to arrange for you to be dealt hand b!" The point is that for calculating the odds, I slipped in a qualification ("as long as the cards were properly shuffled before dealing them") that I hoped might sound so reasonable that it would pass unnoticed. This sort of qualification is *not* reasonable, either in games of bridge or in the evolution of life, because there are always some biases that cause apparently equivalent outcomes in a random test to be unequally likely. You should always be suspicious of statistical calculations and search for possible biases that can cause one-in-a-million events to occur much more often than once in every million trials. This is not intended as an argument against making estimates of probabilities: It is far better to base

[16]They do not use special mechanisms to alter the protein after it has been made, in contrast to certain other unusual amino acids, which are not coded in the DNA: An example is hydroxyproline, an essential component of the fibrous protein collagen that will form a major part of the subject of Chapter 11.

your decisions on some sort of calculation of chances than on a hunch or the pattern of tea leaves in a cup, but you need to avoid placing too much trust in the calculations. I wrote these words for the first edition a few days after the rocket Ariane V exploded, destroying some two billion dollars' worth of investment in about a second: Clearly someone put too much confidence in their risk calculations!

When I was calculating above how likely it was that a bacterium could survive a change in code, I assumed that the two codons for aspartate were equally used, so that a change in the code would affect about half of all the aspartates in the proteins of the organism that suffered the change of code, but codons that code for the same amino acid are by no means equally used. There is no doubt that codons are unequally used, and although the reasons for this have not been thoroughly worked out, there is evidence that organisms prefer certain codons for proteins that are being made in large quantities, but different ones for proteins being made in small amounts.

Mitochondrial genetic codes

It is quite possible, then, that variations in codon usage might cause a particular codon not to be used at all in a particular organism. This is not purely hypothetical: For example, the mitochondria of *Candida glabrata* appear not to use four codons at all. Why this should happen in the first place is unclear, but it might, for example, permit some simplification of the recognition machinery. Once it has happened, it has one immediate consequence: A subsequent evolutionary event causing the unused codons to be used again for any amino acid at all will have no effect on the correct coding of any existing proteins, and thus will have no deleterious effect on the viability of the organism. The "universal" genetic code begins to look rather less frozen than it once did, and numerous variations have been found in mitochondria.

There are two reasons why we should not be surprised to find these variations among mitochondria rather than elsewhere. First of all, the number of different proteins coded by mitochondrial DNA is small—14 in the human, for example. This is far fewer than the 4000 assumed above, and, as the calculation showed, the risk incurred by changing the code increases steeply as the number of proteins concerned increases.[17] Second, the small number of mitochondrial genes makes it much more likely than it would be for the nuclear DNA that certain codons should fall into disuse as a simple statistical fluctuation, thereby

[17]To avoid implying something that is not true, I repeat a point made briefly in Chapter 2, that mitochondria get most of their proteins from the host cell, and these are coded in the nucleus with the usual "universal" code. In most organisms, the proteins coded by the mitochondrial DNA represent a small fraction of the proteins needed for mitochondrial function. In humans, they account for about 14 out of more than 600, but some species have more, such as the protozoan *Tetrahymena pyriformis*, which has 44, and others have fewer, such as another protozoan, *Plasmodium falciparum* (the infective agent of malaria), which has only three. In all of these we see a general evolutionary trend to remove protein-coding genes from the mitochondria.

facilitating the sort of mechanism I have just discussed. These sorts of considerations suggest that it is quite possible that the mitochondrial variations arose after the universal code became established, though they do not exclude the alternative supposition that they are survivors of extremely ancient versions of the code.

I have spent most of this chapter discussing examples of accidents rather than of adaptation, and without even mentioning the main theme of this book, to emphasize at the outset that this is not a Panglossian book. I will not be searching for optimality everywhere, and I will not suggest optimality without providing evidence of how it is assessed. In other words, I am going to describe examples where we can show beyond any doubt that, although other metabolic designs might be possible, and might work adequately, the designs that successful organisms use are the best possible ones. Because I have already discussed in this chapter that accidental choices can plausibly be frozen in evolution, I have to show how in many cases organisms have nonetheless sought out the best solution, even though replacing a sub-optimal design that worked quite well may have required complicated adjustments. Finally, by studying where in the living world we may expect to find exceptions to the general rules, organisms that survive while retaining sub-optimal solutions to biochemical problems, I believe that we can shed light on some unsolved questions of evolution.

Nonetheless, I ask you to approach what I have to say with skepticism, to look for the guiding hand of Dr. Pangloss in forming my opinions, and to ask whether the facts I will describe can be explained without invoking optimality principles. To encourage such a healthy skepticism, I will devote Chapter 7 to examining *entropy–enthalpy compensation*, a false example of optimization that beguiled many distinguished biochemists in the 1970s (and still retains a few today). I hope I will convince you that my examples of optimization can survive all your skepticism.

4. Metabolism and Cells

Nothing in biology makes sense except in the light of evolution.

Theodosius Dobzhansky[1]

Living cells carry out a multitude of functions. Some are more specialized than others: Some have specialized in contractility, others in transmitting stimuli, storing fats, regulating the temperature, emitting light or accumulating electricity, but all are in constant activity: They move, control what passes through their membranes, degrade substances to produce energy, use that energy to accumulate reserve materials, for the continuous synthesis of their own materials, for their own internal contractile activity, and for many other functions.

Enrique Meléndez-Hevia, *La Evolución del Metabolismo*[2]

IN THE EARLY 1980s Departments of Biochemistry around the world started to rename themselves Departments of Biochemistry and Molecular Biology, a reflection of the changing status of classical biochemistry, with its roots in physiological chemistry, as compared with the growth of gene-related biochemistry—a change in status hinted at in the quotation at the beginning of the Preface. In the days when biochemists were not ashamed to call themselves biochemists, you could go into any biochemistry laboratory in the world and find a chart of metabolic pathways on the wall. You can still find these old relics, the paper brown and curling, in forgotten corners of today's biotechnology institutes, but they are liable to have "© 1972" printed on them and to have avoided being replaced by maps of the human genome only by luck.[3]

Metabolism

This is a pity, because even the most abbreviated charts of metabolic pathways (and even the largest are to some degree abbreviated) impress you immediately with the sheer number of different chemical reactions that are in progress in any living organism, many of them going on simultaneously in the same cell. In an organism like

[1]T. Dobzhansky (1973) "Nothing in biology makes sense except in the light of evolution" *American Biology Teacher* **35**, 125–129. This sentence is often quoted, and you may have recognized it when it appeared in garbled form in Chapter 1.

[2]E. Meléndez-Hevia (1993) *La Evolución del Metabolismo: Hacia la Simplicidad* (The Evolution of Metabolism: Towards Simplicity), Eudema, Madrid: My translation.

[3]Things have improved in the years since I first wrote these words. The twenty-first century has seen a steady revival of interest in enzymes and metabolism. A famous paper of 1913 [L. Michaelis and M. L. Menten (1913) "Kinetik der Invertinwirkung" *Biochemische Zeitschrift* **49**, 333–369] was cited more times in 2011 than in *any* previous year. The total was exceeded in 2013 and 2014, but many of the citations were in papers commemorating the centennial.

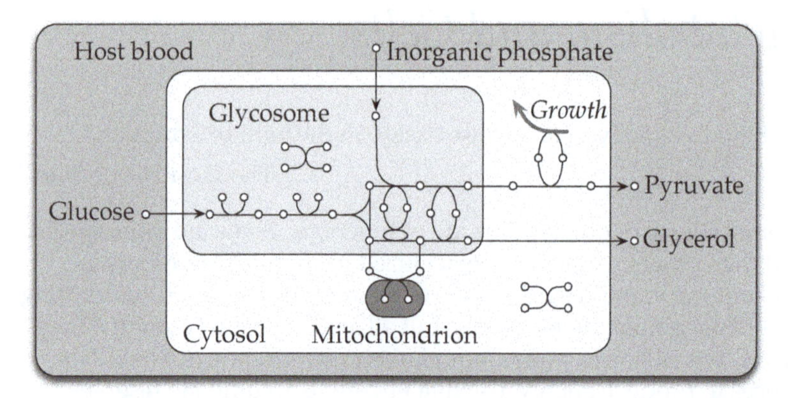

Figure 4.1 Simplified representation of the main metabolic activity of the form of the parasite *Trypanosoma brucei* found in the bloodstream. The different degrees of shading distinguish the four different compartments in the system, including the blood of the host, which supplies the glucose and inorganic phosphate needed by the parasite for its metabolism and growth, and receives the pyruvate and glycerol that it excretes. The white circles represent the different metabolites, connected by lines to indicate enzyme-catalyzed reactions.

a bacterium, in which one cell constitutes the whole organism, one cell has essentially the same capabilities as any other; but in organisms such as ourselves, in which the individual is a collection of about 37,000,000,000,000 (3.7×10^{13}) cells,[4] these are by no means all equivalent, as noted in the quotation at the beginning of this chapter.

A simplified example is shown in Figure 4.1, in which the different substances involved are just represented by white balls, with lines between them representing chemical reactions; in a real chart for teaching metabolism, the names and other details would be shown explicitly. This diagram may look complicated, and you may suspect that I have chosen an unnecessarily complicated example, but the opposite is true: The organism represented, the form of the parasite *Trypanosoma brucei* that exists in the bloodstream of people infected with African sleeping sickness, has possibly the simplest metabolism known, and only its major chemical activity is included. In the bloodstream, it depends on its unfortunate host for most of its needs, and it excretes pyruvate, a substance that most organisms treat as being much too valuable to throw away. Yet even in this grossly oversimplified example, you can see that

[4]Published estimates span a wide range. This one comes from a recent and careful study: E. Bianconi, A. Piovesan, F. Facchin, A. Beraudi, R. Casadei, F. Frabetti, L. Vitale, M. C. Pelleri, S. Tassani, F. Piva, S. Perez-Amodio, and P. Strippoli (2013) "An estimation of the number of cells in the human body" *Annals of Human Biology* **40**, 463–471. An idea of the huge size of this number can be obtained by reflecting that it implies about one cell for every hour that has passed since the formation of the Earth. It does not include the bacterial cells that inhabit the same space, which are about 10 times more numerous, but also much smaller (Figure 4.2).

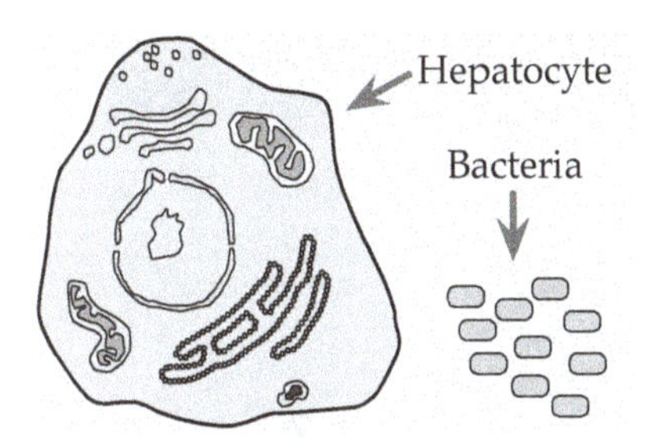

Figure 4.2 Bacterial and human cells. A cell of *Escherichia coli*, the most abundant species of bacteria in the human digestive tract, occupies about 2×10^{-15} liter, whereas a *hepatocyte*, the most abundant kind of cell in the liver (not "typical," because there is no such thing as a typical human cell, but representative) occupies about 3.5×10^{-12} liter.

numerous reactions are in progress, and that they do not even take place in the same place, but in three different compartments of the cell, together with the environment provided by the host.

All of this chemical activity, far more complicated in most cells than what I have shown, constitutes *metabolism*. Its reactions are highly specific, because each enzyme is itself highly selective for the substrate of its reaction. Inside a cell the many chemical reactions proceed at once, at similar rates, without interference or obstruction. The organization of all this chemistry is, in effect, what constitutes life, a topic that is explored more deeply in Chapter 14. Understanding the principles of its organization has been the major focus for some years of the research of Enrique Meléndez-Hevia in Tenerife. He has set out his ideas in the book quoted at the beginning of this chaper, but this is not available in English, so I will deal with them in some detail in this and the following chapter.

Theodosius Dobzhansky's statement that nothing in biology makes sense except in the light of evolution remains the central biological truth that has to be understood by anyone who seeks to rationalize biology. The question of how life has evolved is open to scientific investigation, both with experiments and by mathematical analysis. Experiments may show, for example, whether environmental conditions postulated to have existed at the origin of life could have resulted in the spontaneous appearance of amino acids and the bases of DNA or (more likely) RNA. In some cases mathematical analysis can show whether the actual way in which metabolism is organized is the best possible way.

THEODOSIUS GRYGOROVYCH DOBZHANSKY (1900–1975) was born in Ukraine and emigrated in 1927 to the USA, where he worked at CalTech, Columbia, and the Rockefeller University. A geneticist and evolutionary biologist, he worked especially with the genetics of fruitflies, and was a central figure in shaping the unifying modern evolutionary synthesis.

Given the apparent completeness and perfection[5] of modern metabolism, we may fear that there is no hope of deducing how it evolved to its present state, if all traces of its more primitive states have disappeared. However,

[5]As we will see in Chapter 5, "perfection" is an exaggeration, because it is becoming increasingly clear that many enzymes are less than ideally specific and produce "wrong" metabolites that need to be eliminated by other enzymes.

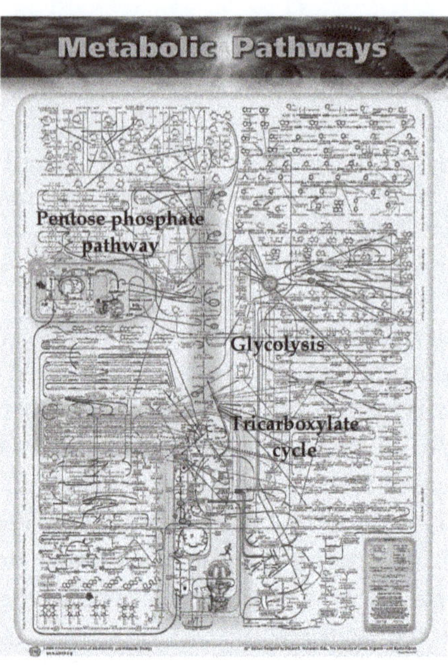

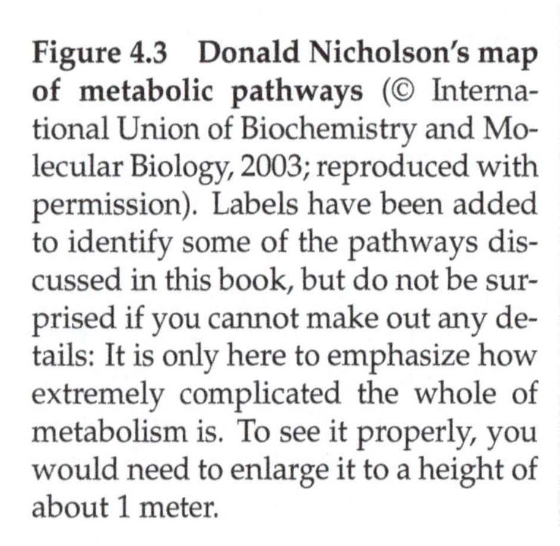

Figure 4.3 Donald Nicholson's map of metabolic pathways (© International Union of Biochemistry and Molecular Biology, 2003; reproduced with permission). Labels have been added to identify some of the pathways discussed in this book, but do not be surprised if you cannot make out any details: It is only here to emphasize how extremely complicated the whole of metabolism is. To see it properly, you would need to enlarge it to a height of about 1 meter.

just as mitochondrial genetic codes yield some clues as to how the "universal" code may have evolved, as discussed in Chapter 3, so we may hope that some organisms preserve some traces of more primitive types of metabolism. For example, an enzyme that was needed to catalyze certain reactions in earlier stages of evolution may have survived in some organisms even though its reaction no longer fulfills any function and has been eliminated from most organisms.

Metabolic pathways

As I have emphasized, the metabolic chart illustrated in Figure 4.1 is unusually simplified. A more realistic one would not only include names of the molecules transformed in the reactions, but it would probably include their chemical structures as well and would identify the enzymes that catalyze the different reactions. More important, unless it was clearly intended to represent just a small part of the metabolism of a cell, it would include many more reactions than the 20 or so in Figure 4.1. It is unfortunately difficult to put much more on a small page without making the type too small to read, so you will have to imagine how Figure 4.3 would look if it occupied an area of the order of a square meter on the wall of a laboratory. The designer does not show all of the chemical reactions as having equal status, any more than a modern road atlas shows high-speed roads and farm tracks with equal prominence and does not scatter them arbitrarily over the paper. This is especially true if you look at a chart designed by someone like Donald Nicholson, who spent many years thinking about how to convey the central ideas in the clearest possible way. (His charts were published for years by the Koch–Light Chemical

Company, but the copyright is now owned by the International Union of Biochemistry and Molecular Biology, which makes them available on the web, at `http://www.iubmb-nicholson.org/`.) Instead, certain groups of reactions are given higher status by being placed more centrally or printed in larger or heavier type, whereas others are made to appear more peripheral.

To some extent this may be subjective, imposing a human interpretation on the observations, but it also reflects an objective reality: Some reactions carry a greater flux of metabolites than others, some are active in a wider range of cell types than others, and so on. To understand the entire chart, therefore, it is useful to collect the reactions into groups of transformation sequences, or *metabolic pathways*. The number of steps considered to be one pathway can be small if few steps are needed to convert one important metabolite into another. For example, serine biosynthesis is a three-step pathway, in which the amino acid serine is synthesized from 3-phosphoglycerate. At the other extreme, β-oxidation, the process that converts fatty acids from the form in which they are stored in fat cells into the form in which they are metabolically active, involves seven repetitions of the same four types of step, making an unbranched pathway of nearly 30 reactions.

DONALD ELLIOT NICHOLSON (1916–2012) was a biochemist at the University of Leeds who devoted much of his professional life to the production of charts of metabolic pathways. He drew his first combined chart by hand and had it printed in an architect's office. When he was 90 years old, however, he acquired his first personal computer and joined the digital revolution: The last years of his life were highly active and resulted in numerous new charts.

The pentose phosphate cycle

The *pentose phosphate cycle* is the pathway that Meléndez-Hevia and his colleagues have analyzed most thoroughly in their efforts to understand the design of metabolism. It consists of 11 successive reactions, in two distinct phases: In the first (oxidative) phase, glucose 6-phosphate (a *hexose,* or sugar with six carbon atoms in each molecule, with a phosphate group attached to it) is converted into ribulose 5-phosphate (a *pentose*, or sugar with five carbon atoms in each molecule, again with a phosphate group attached to it); in the second (non-oxidative) phase, the carbon atoms of six pentose phosphate molecules are rearranged to produce five hexose phosphate molecules, which allow the cycle to begin again. Figure 4.4, which represents this pathway, doubtless appears complicated and unmemorable, and that is no illusion: It *is* complicated and unmemorable. For the student trying to learn metabolism, it is the archetype of an arbitrary and meaningless collection of reactions, but it will appear much less arbitrary and meaningless (though still perhaps unmemorable) if we take the trouble to analyze it.

Functions of the cycle

The pentose phosphate cycle has various functions, which are different in different types of cells. One of these is its coupling with the synthesis of fatty acids,

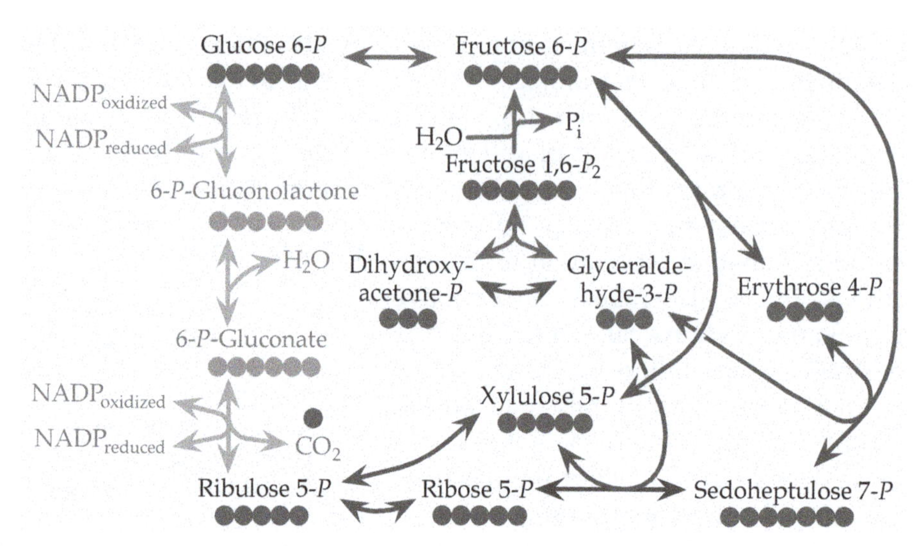

Figure 4.4 The pentose phosphate pathway. The balls under the chemical names, in which italicized *P* represents a phosphate group, indicate the number of carbon atoms that each molecule contains. The names shown in gray at the left of the diagram belong to molecules in the oxidative part of the pathway, which is not considered in the analysis in this chapter. (It is important nonetheless, because healthy human cells require it as a major source of reducing power in the form of reduced NADP. As we will see in Chapter 13, cancer cells are different.) The remainder, the non-oxidative part, can be interpreted as a scheme to allow the hexoses and pentoses to be converted into one another.

which constitutes a good example of the harmony in the coupling of metabolic pathways and of the specialization of cells for it. The oxidative phase of the cycle, converting glucose 6-phosphate into ribulose 5-phosphate, produces "reducing equivalents" in the form of a molecule commonly known by its initials as reduced NADP, which is needed in synthetic processes, in particular for the synthesis of fatty acids; it is a necessary coupling, given that the pentose phosphate cycle is one of the main sources of reduced NADP in the cytoplasm of cells where fatty acids are synthesized. The mammary gland, adipose tissue, and the liver are the organs in which the synthesis of fatty acids is most highly developed, and they are also the organs with the highest activity of the pentose phosphate cycle.

The red blood cells are the simplest cells in the mammalian body; they have lost the nucleus, and because they cannot renew their material, their lifetime is short, only four months. They do not synthesize fatty acids and have no respiratory activity, because they have no mitochondria. The energy that they need—primarily to maintain their membranes in the correct electrical state and for transporting substances across them—they obtain by converting glucose into lactate by *glycolysis*, a process that does not require oxygen. The energetic yield from glycolysis is feeble, but the red cells are in the bloodstream itself, so they have no problem with fuel supply. It may seem paradoxical that the cells that have the most oxygen available to them make no use of it in their own metabolism, but this can be explained in terms of the total specialization of red cells for transporting hemoglobin: The fewer other proteins and other components of all kinds they contain, the more space they have for packing in the maximum quantity of hemoglobin.

Certain genetic deficiencies that affect the pentose phosphate cycle produce severe anemias, however, thereby demonstrating that the pentose phosphate cycle is essential in red cells, but what is it for, if there is no synthesis of fatty acids? It produces reduced NADP, which is needed for other things apart from this: In the red cell it is needed for repairing oxidation damage to hemoglobin. Although hemoglobin is a protein, it contains a non-protein part, *heme*, and it is to the heme that oxygen becomes attached. In the center of each heme is an iron atom, which is liable to destruction by oxygen in a process akin to rusting, and reduced NADP is necessary to repair this damage. It is no coincidence that when blood is dried in the air it turns a rusty color, which it owes to the same type of oxidized iron that gives rust its color.

The pentose phosphate cycle has other functions in other cells, such as the production of precursors for the synthesis of various compounds in plants and bacteria, or to feed the photosynthetic cycle in green plants, but to understand how it came into existence in evolution, we need to know what its original function was. The answer comes from recognizing that in nature there are just two kinds of sugars that exist in significant quantities, the pentoses, with five carbon atoms in each molecule, and the hexoses, with six. Other sugars exist, with three, four, or seven carbon atoms per molecule, but these are just transient metabolic intermediates, present in tiny amounts compared with the hexoses and pentoses. The two major kinds of sugar have clearly different functions. The major pentoses, ribose and deoxyribose, are components of nucleic acids and thus constitute part of the information system of life; the hexoses are used as fuels, both free as glucose for immediate use and accumulated into polymers as glycogen or starch for long-term energy storage. In addition, they are constituents of quite different polymers whose role is structural: Cellulose, the principal building material in plants, is a polymer of glucose; chitin, used for making the shells of crustaceans, molluscs and insects, is a somewhat more elaborate polymer of molecules closely similar to glucose.

Two worlds of sugars

Hexoses and pentoses would be totally separate in nature, forming two pools of organic carbon, if there were no metabolic pathway capable of interconverting them. Although we can conceive of a living system organized in this way, it would represent substantial duplication of effort, and would make nutrition a more complicated science than it already is. Instead of just labeling processed

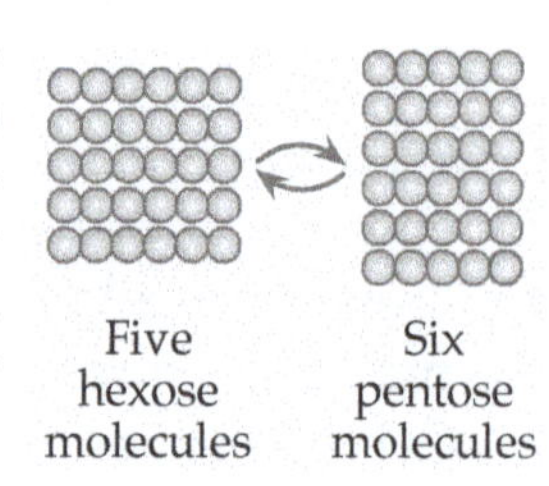

Figure 4.5 Two worlds of sugars. Glucose and the other hexoses are needed in metabolism primarily for energy management, whereas ribose and the other pentoses form essential parts of the nucleic acids, RNA and DNA. Their functions are quite separate and they would constitute two unconnected groups of metabolites if the pentose phosphate pathway did not allow them to be interconverted. In real metabolism, food that is rich in one class of sugars can be used to supply the other when necessary.

food with their carbohydrate content, as at present, we should need separate labels for hexoses and pentoses, to ensure that the two kinds of carbohydrate were properly balanced. The two classes are similar enough for each to represent a potential reservoir of the other if needed, however, and it is precisely the function of the pentose phosphate cycle to provide a route between them. This interface between the two great classes of sugars was without doubt its original and primary function.

Evolution and the development of life require continuous adjustment of the quantity of carbon dedicated to information and the quantity of carbon dedicated to the energy and structure. *Photosynthesis*, the process whereby green plants use light energy to transform carbon dioxide in the air into organic carbon derivatives, involves sugars of various sizes—trioses, pentoses and hexoses—but the final result, starch, the storage carbohydrate in plants, is a form of hexose, and when sugars are broken down, it is always in the form of hexoses. The pentose phosphate cycle, functioning in all organisms, continuously regulates the quantity of glucose 6-phosphate that has to be converted into ribose 5-phosphate for synthesizing nucleic acids, or the reverse pathway, when nucleic acids are being degraded. This need must have existed near the beginning of the development of life, because the management of energy and information is fundamental in any kind of life that we can imagine.

Some primitive organisms, sea urchins for example, consist of a reproductive system, a protective shell, and not much else; others, such as ourselves, lead much more complex lives, with so many other components that the reproductive organs represent a small proportion of the total mass. Nonetheless, we can eat sea urchins, and other pentose-rich foods like fish roe and yeast extract, without burdening our systems with excess pentoses, because we can convert the unwanted pentoses to hexoses.

Put at its most simple, the reactions of the pentose phosphate pathway can be considered as a series of shuffling steps in which units are exchanged between five-unit and six-unit structures. In this form the pathway lends itself readily to mathematical analysis, because we can ask whether the particular set of exchanges is the most efficient that can be conceived. This will be the subject of the next chapter.

5. *The Games Cells Play*

> The road to simplicity has not been easy. In reality, it has been a twisted path, full of obstacles and mistakes, including wrong turnings, bad solutions that were only revealed to be bad when better ones were found.
>
> Enrique Meléndez-Hevia, *Evolución del Metabolismo*[1]

ANALYSIS OF THE PENTOSE PHOSPHATE PATHWAY needs techniques similar to those used for analyzing mathematical games, of which one class is highly pertinent to metabolism. This involves minimizing the number of steps needed to transform one arrangement of objects into another, which is well illustrated by the classic problem of the farmer crossing a river.

Fox, goose, and oats

A farmer must cross a river in a small boat with a fox, a goose, and a sack of oats. The fox cannot be left alone with the goose, and the goose cannot be left alone with the oats, but the boat is too small to carry more than the farmer and one of his three burdens at once. How can he get everything across the river without mishap? The solution is simple, as illustrated in Figure 5.1, but it is worth analyzing in detail, to illustrate points useful for more difficult problems. For the first river crossing, the farmer must take the goose. Anything else will allow the fox to eat the goose (if the farmer takes the oats), or the goose to eat the oats (if the farmer takes the fox), or the possibility of either (if the farmer goes alone). The second step is almost equally clear: Although the farmer could bring the goose back with him, this would just recreate the original state and thus cannot be a step toward solving the problem. So, the second step must be for the farmer to return alone. At this stage, for the one and only time in the game there are two different but equally good choices, to cross with the fox or to cross with the oats. Whichever he does, he is faced with only one possibility for the next step, which is to take the goose back with him, because otherwise the fox will eat the goose or the goose will eat the oats. In the next step, returning with the goose restores the previous state, so it is not useful, and so he must take whichever out of the fox or oats is left. The two pathways inevitably come together at this point to yield the same state in which the goose is the only item left on the original side of the river. The farmer then returns to collect the goose.

There are many variants of this problem. We can make it more symmetrical (as well as making it a more realistic problem for a real farmer to want to solve) by replacing the fox by a second sack of oats or by a second goose. All of them have essentially the same solution, all of them involving more river crossings

[1]E. Meléndez-Hevia (1993) *La Evolución del Metabolismo: Hacia la Simplicidad* (The Evolution of Metabolism: Towards Simplicity), Eudema, Madrid: My translation.

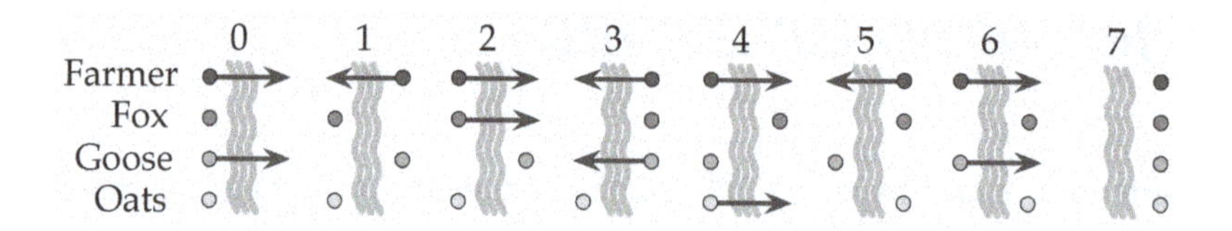

Figure 5.1 A simple combinatorial game. At stage 0 all four items are on the left, and the object is to arrive at stage 7 with all of them on the right, subject to the following conditions: The farmer can take only one item across the river at each stage, the fox cannot be left alone with the goose, and the goose cannot be left alone with the oats. The only transition that is not obvious is between stages 3 and 4, where bringing the goose back to the left appears to be undoing an earlier move toward the solution.

that appear at first sight to be necessary. The last variant, with two geese and one sack of oats, and thus a single resource that needs to be protected from multiple dangers, makes a convenient introduction to the problem to be considered next.

Sending goods or information securely

Suppose that I want to send you a chest containing some valuable jewels. I am unwilling to entrust it to the post, for fear of theft. I have a padlock of high quality that I am confident is impossible to open without the use of a key that is in my possession. That seems to be of little use, though, because if I lock the chest you will not be able to open it unless I send the key as well, but then I run the same risk as before, that the key may be stolen. So, the existence of a secure padlock does not seem to solve the problem, and if you have a similarly secure padlock that only you can open, that will not solve it either, because I will not be able to put the jewels into a chest that I cannot open. How can I send you the chest without risking that it can be opened in transit? This does not have the same logical structure as the problem of the farmer and the river, but it is similar enough that you should not have too much difficulty in deducing the answer—which I will come to shortly.

The problem of the farmer crossing the river was regarded in the 1950s as simple enough for children's magazines, but a problem logically equivalent to that of the jewels was believed to be insoluble for 2000 years, and was not solved at all until the middle 1970s. In the form usually presented, it concerns methods for sending and receiving secure messages—that is, messages that cannot be read by anyone apart from the intended recipient, without requiring either sender or recipient to possess the key needed for deciphering messages enciphered by the other. Exchanging keys

GEORGES JEAN PAINVIN (1886–1980) was a French cryptanalyst during World War I. His most notable achievement was the breaking of the ADFGVX cipher in June 1918, and he is regarded as the founder of modern cryptanalysis. Before the war, he taught paleontology and geology. His cryptanalytic work for the French army arose from a chance encounter with a member of the French *Bureau du Chiffre*.

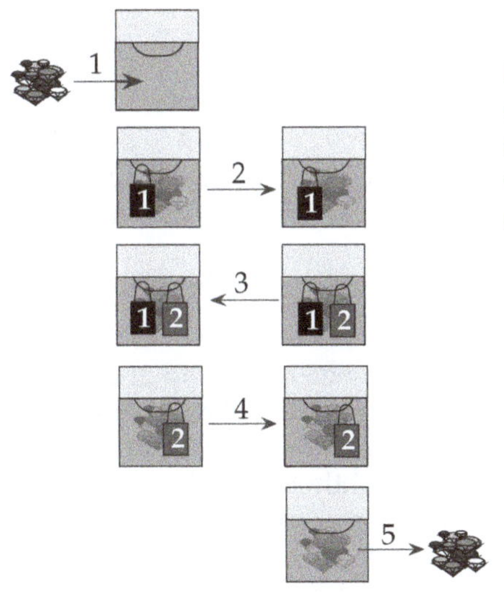

Figure 5.2 Sending locked goods without sending the key. Use of two different locks allows goods to be sent in a locked container without any key leaving the possession of its owner. The steps are as follows:

1. The sender places the jewels in the chest, locks it with padlock 1, and
2. Sends the locked chest to the recipient.
3. The recipient adds padlock 2 and sends it back to the sender, who
4. Removes padlock 1 and sends it again.
5. The recipient removes padlock 2 and takes out the jewels.

is expensive for governments, armies, financial institutions, and so on, that send large numbers of secret messages, but, in the words of Simon Singh,[2] "It seems that the distribution of keys is unavoidable. For 20 centuries this was considered to be an axiom of cryptography—an indisputable truth." Even if we discount the first 19 of these centuries on the grounds that before the work of Georges Painvin in France during World War I the modern scientific study of cryptography had hardly begun, it remains true that for at least 60 years the greatest minds in cryptography believed it impossible.

Yet with the example of the farmer and the river in mind, the solution to the problem of the jewels seems quite easy, as illustrated in Figure 5.2: I send you the jewels in a chest locked with my padlock, and you send it back to me unopened but now locked with your padlock as well as mine. I remove mine and send the chest back to you, which you open without difficulty after removing your padlock. This was the problem that was solved by Whitfield Diffie, Martin Hellman, and Ralph Merkle in the 1970s as the first (and in retrospect easiest,[3] though it did not seem like that at the time) step in the development of public key ciphers—now the standard method of sending secret information without risk of having it stolen.

Problems of this kind illustrate several points that are worth stating explicitly. First, the solutions involve steps that appeared to be in the "wrong" direction, such as bringing the goose back to where it started: Realizing this is the whole point of the original problem, because everything else is obvious. Second, although the solution can be found simply by thinking logically, it

[2]Simon Singh (1999) *The Code Book: The Science of Secrecy from Ancient Egypt to Quantum Cryptography*, Doubleday, New York.

[3]Much more difficult was the need for ciphers that could be applied in any order and reversed in any order (Figure 5.3).

				i =	12345	i =	12345
	START		START		START		START
Caesar 3	VWDUW	Caesar 3	VWDUW	Caesar i	TVDVY	Caesar i	TVDVY
Caesar 5	ABIZB	Caesar 5	ABIZB	Cycle 2	DVYTV	Cycle 2	DVYTV
Caesar –5	VWDUW	Caesar –3	XYFWY	Cycle –2	TVDVY	Caesar –i	CTVPQ
Caesar –3	START	Caesar –5	START	Caesar –i	START	Cycle –2	VPQCT
1a. Reverse order		1b. Original order		2a. Reverse order		2b. Original order	

Figure 5.3 Reversible ciphers. If a trivially simple cipher is used to encipher a message once, and then a similar (but different) one is applied, it makes no difference whether the deciphering is done in reverse order (1a) or in the original order (1b). This is the case of the locks in Figure 5.2, but for only slightly less trivial ciphers the deciphering must be done in reverse order (2a), not in the original order (2b). Devising a nontrivial method of enciphering that did not require deciphering in reverse order proved to be very difficult. A Caesar 3 cipher (so called because Julius Caesar used ciphers of this kind in his private correspondence) is one in which each letter is replaced by the letter three places later in the alphabet—that is, replacing A with D, and so on.[4] In this figure, Cycle 2 means a cycling of the text by two positions.

can also be found by brute force—that is, by just listing all the possibilities at every stage and not thinking at all. Inelegant though it is, this is often the best method for dealing with complicated problems, and it is the only one that is easy to implement in a computer. Third, problems of this kind always have an infinite number of solutions, because you can always return to earlier states by reversing the previous steps or by going around a cycle in a branched part of the solution. Consequently, it is often insufficient to ask for a solution: We ought to find the best solution, which implies that "best" must be defined. In general, I will take "best" to mean the solution with the smallest number of steps, because this implies the least cost in terms of energy and materials, but I may need to introduce additional criteria in cases where this definition produces more than one solution.

It may also be helpful to amplify a remark in the preceding paragraph, about the ease of implementing brute-force methods in the computer. We are often so impressed by how effective computers are at certain kinds of task that we may forget how bad they are at ones that involve any real thinking, even thinking at a level well within the capacity of a human baby. Understanding why computers are good at some tasks and bad at others is useful for its own sake, but it has an additional relevance for this book because it helps us to

[4]The ROT13 cipher often used for obfuscating text is a Caesar 13 cipher. The choice of 13 has two convenient features: Because it is half the number of letters in the alphabet, decoding is the same as encoding, and it is a big enough shift that most people cannot decode the text instantaneously in their heads.

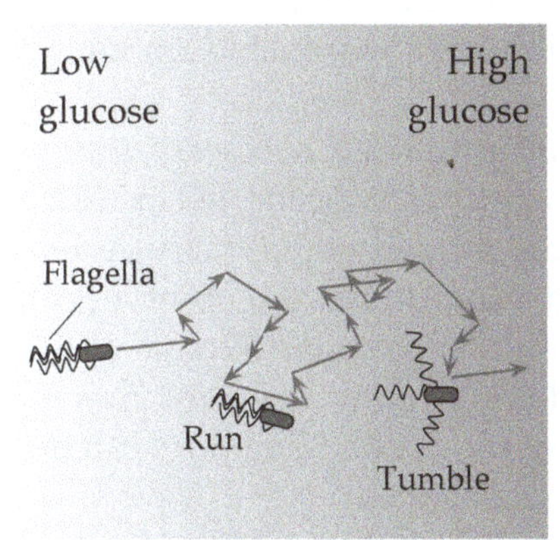

Figure 5.4 Bacterial chemotaxis. Cells of bacteria such as *Salmonella typhimurium* use whip-like appendages called *flagella* to swim in a chemical gradient, either toward a food source, such as glucose, or away from a harmful one, such as phenol. They cannot detect which direction they should take, as explained in Figure 5.5, but they can tell when things are getting better or worse, so they continue for a longer time in a "good" direction than they are moving in a "bad" direction. Between straight runs, they "tumble" and then continue in a completely random direction. This type of progress is called a *biased random walk.*

understand how evolution finds the solutions to problems of optimization.

Machine intelligence

Many dogs have learned to do such tasks as make their way along a busy street to collect a newspaper from a shop and return with it to their homes. If necessary, they can cross the road, avoiding any cars that may be approaching and without interfering with any other users of the road that they may encounter. No computer-controlled robot is yet able to do this—at least, none for which the results have been made publicly available. I am not privy to any secrets the armed forces may have, but even there some speculation may be possible: Despite all the boasting that one may hear during military actions, there is little to suggest that anything resembling a weapon with the intelligence of a trained dog yet exists. As long ago as 1957, Herbert Simon, an expert on artificial intelligence at Carnegie-Mellon University, predicted that a computer would be able to defeat the world chess champion within 10 years. It actually took about four times as long as that, because the early developers of chess programs greatly underestimated the difficulties.[5] Moreover, a program that can play chess cannot normally do anything else, whereas a human chess player can do a great many other things, such as going to a shop to buy a newspaper.

Despite great efforts to move away from a brute-force approach to computing, most programs still rely heavily on it. What does this mean? It means that analyzing a problem involves examining a huge number of possible solutions, testing them according to some preset criteria, and finally selecting the solution that gives the highest score according to these criteria. However,

[5]Machine translation is another example of a field that has advanced far more slowly than its advocates predicted. For as long as I can remember, virtually perfect translation has been expected to be about five years in the future. However, as the years go by, it continues to be about five years in the future. Today the results are remarkably good for some surprising language combinations, such as from Hausa to English, and remarkably bad for others that one might expect to be easier, such as from French or Spanish to English.

even a "huge number" is not infinite, and in complex problems there may be far more solutions to be studied than can be done in a reasonable amount of time, so ultimate success is not guaranteed.

Unlike a human, a computer program can do all the necessary testing accurately, does not forget which possibilities have already been tried or which solutions have yielded the best scores, and does not get bored or frustrated. Moreover, it can do all this at enormous speed—so fast that the observer can be fooled into thinking that some thought or intuition has gone into the result. But none of this requires any thought at all, and until we can take it for granted that robots can do the sort of tasks that we take it for granted that dogs can do, there will be no reason to attribute any intelligence to them.

Natural selection works in a similar sort of way, but it differs from a modern computer in two respects. It is extremely slow, but enormous amounts of time have been available—some 4.5 billion years since the formation of the Earth, and perhaps half that time since the first living organisms appeared. Natural selection also has no memory, and thus no capacity for knowing that possible solutions have already been tried and found to fail. Instead, it tinkers, changing things[6] haphazardly here and there. If, as is usual, the modified organism is less capable of surviving than the original, it is discarded, but when, rarely, it is just as good or better, it has some chance of leaving more descendants than the original and eventually replacing it.

Bacterial chemotaxis

This type of behavior is seen in the response of bacteria such as the pathogen *Salmonella typhimurium* (responsible for some kinds of food poisoning) to chemicals in the mixture of water and nutrients in which they grow. When a colony of cells grows in a mixture containing a gradient of glucose, which they consume as food, it moves, apparently steadily, in the direction of the glucose, a property known as *chemotaxis*. This picture is misleading, however, as you can see by watching a single cell under a microscope. It can swim in a straight line, using its *flagella* (whip-like tails), but it has no idea which is the right direction to find more glucose, and even if it has once found a "good" direction (increasing glucose concentration), it does not remember it afterwards. After each period of swimming, it "tumbles" or rotates at a single point, and then sets off again in a completely random direction—the new direction is not related either to the previous direction or to the direction where the highest glucose concentration can be found; it is just arbitrary.

How, then, is it possible for the population as a whole to move in the right direction? Each individual bacterium does move on average in the right direction, albeit in a drunken way: It does so because it is able to recognize

[6]"Changing things" perhaps gives the wrong impression, implying some conscious intention to test different possibilities. In reality there is no intention, just mistakes, but these have the same effect.

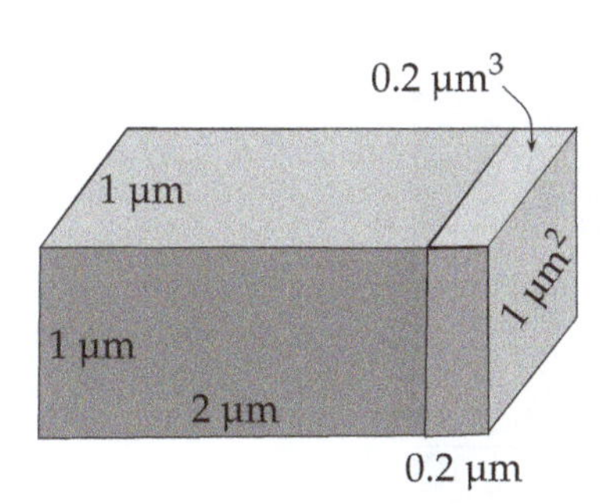

Figure 5.5 Detecting a gradient. A cell of *Salmonella typhimurium* has a cross section of about 1 µm^2 and a length of about 2 µm. If it can count the glucose molecules in a volume extending 10% of the length of the cell, the sampling volume is 0.2 µm^3, and it can detect a gradient when this volume contains as few as 60 molecules of glucose. However, statistical fluctuations mean that this has to be written as 60 ± 8 molecules, and it is impossible to determine the gradient by comparing the concentrations at the two ends of the cell.

whether its present state is better (more glucose) or worse (less glucose) than it was a few seconds ago. If things seem to be getting better, it swims for a longish period, whereas if they are getting worse, it stops quickly and tumbles. This kind of movement, a *biased random walk*, is illustrated in Figure 5.4: For an intelligent organism, it may seem a clumsy way of reaching a goal, but it works, and that is all that matters. It is a little like the procedure used by people who have lost their way in a wood, trying different directions at random in the hope of judging whether they are getting closer to or further away from where they want to go. It works better for a simple organism than for an intelligent one, however, because intelligent organisms are liable to go around in circles, whereas an organism without memory does not fall into that trap. Likewise it is not necessary for natural selection to have any foresight or capacity to analyze problems for it to be able to advance steadily forwards.

Moving prisoners

Returning now to combinatorial games, the problems mentioned are sufficient for introducing the essential idea of metabolic optimization and looking at the pentose phosphate pathway in the light of it. First, I will treat this pathway as the game defined in Figure 5.6, but afterwards we will see how the apparently arbitrary rules apply to the actual biochemistry.

We start with a prison camp for dangerous prisoners that consists of six widely separated blocks that house five prisoners each—30 prisoners in total. Each block is large enough to accommodate more than five prisoners and can hold six on a permanent basis, or more than that for short periods, during the process of reorganization, for example. The authorities have therefore decided that it is uneconomic to use six blocks for 30 prisoners: One block is to be closed, and each of the remaining five is to accommodate six prisoners.

You have to organize the transfers. Because the blocks are widely separated, you need a vehicle for the transfers, but this presents no problem because you have a suitable one that can accommodate a driver and up to three prisoners. In doing so, you have to take account of the violent behavior of the prisoners, which means that you cannot leave fewer than three together in a

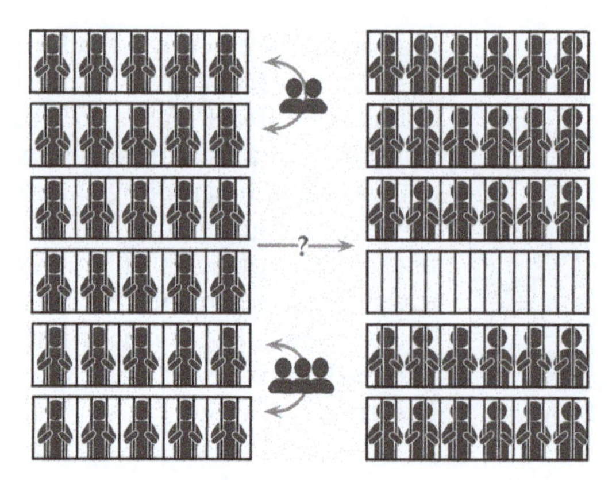

Figure 5.6 Redistribution of prisoners. The aim of the game is to find the smallest number of steps to redistribute the 30 prisoners into one empty block and five with six prisoners in each. Two or three prisoners must be moved in each step. No block can contain fewer than three prisoners at any moment unless it is completely empty, but a block is allowed to contain more than six prisoners during the redistribution process.

block at any time, because two of them left alone will certainly fight, and any one left alone is liable to commit suicide. At this point, the problem appears quite easy: You move one prisoner from the first to the second block, then another prisoner from the first to the third block. This leaves three, six, and six prisoners in the first three blocks. Then you do the same thing with the other three blocks, leaving three, six, and six prisoners in these three blocks. If you move the three remaining prisoners from the first block to the other one that contains just three, the first block will be empty and each of the other five blocks will have six prisoners each. The problem is thus solved in just five transfers, and no solution involving fewer than five transfers can exist, because each of the five blocks occupied at the end of the operation must have appeared at least once as the receiving block of a transfer.

Unfortunately, you not only have to contend with the genuine constraints in the prison, but you also have to follow rules imposed by the bureaucracy responsible for the prison service. The financial director has decided that it is inefficient to use the vehicle to move fewer than two prisoners at a time: You can move two or three at a time, but not one. It is no use arguing that following this rule will make the whole reorganization more expensive. Bureaucrats are impervious to such arguments and insist that they know best and that you just have to obey the rules.

The problem is now appreciably more difficult, and it may not be obvious at the outset that it can be solved at all, even inefficiently. I will give the solution shortly, but before reading on you may like to reflect on how you might set about solving it. In this way you will arrive at an opinion about how difficult the problem is, and how likely it is that the best solution could be found by trying out different possibilities haphazardly.

We can represent the initial state as 555555 and the desired final state as 666660. You cannot begin by moving three prisoners, because this would produce a block with only two prisoners left in it, so the first step can only be to move two prisoners from one block to another, giving a state that can be represented as 375555 (or something equivalent like 575553). After this, there are several possibilities, in which I have written 10 as X to make it a single digit:

Move two prisoners:	55 → 73	375555 → 373755
	75 → 93	375555 → 393555
	75 → 57	375555 → 357555 (synonymous)
	53 → 35	375555 → 393555 (synonymous)
	73 → 55	375555 → 555555 (regressive)
Move three prisoners:	75 → 48	375555 → 348555 (unphysiological)
	53 → 80	375555 → 078555
	73 → X0	375555 → 0X5555

Two of these possibilities are labeled "synonymous," meaning that the new state is equivalent to the first, and another is labeled "regressive," meaning that it recreates a state that existed earlier. None of these three kinds of moves is helpful, and we can ignore them, but it is not obvious which of the other five is best. Another is labeled "unphysiological," because although not forbidden by the rules as I have stated them, it turns out that in the biochemical equivalent of the problem, moving three prisoners at a time always involves a block with three prisoners in it. For solving the nonbiochemical version of the problem we will allow moves of this kind.

We can continue with each of the five new states in the same way, examining what new states can be created from them, and in the brute-force approach to the problem, the computer programmer's approach if you like, that is the easiest thing to do. We just write down all the possibilities, following each pathway through until we eventually reach 666660, and select the shortest. To simplify matters, you can make a preliminary guess that it is not necessary to pass through a state with more than seven prisoners in the same block, so you can ignore any move that leads to a code with an 8 or a 9 in it. If you do this, you will find that five seven-step pathways exist, as well as many longer solutions. The number of solutions is infinite if we allow solutions that involve aimless cycling around the parts of the network where cycling is possible.

All of the seven-step solutions start with 555555 → 755553, and all finish with 666543 → 666633 → 666660, but they differ in how they get from 755553 to 666543:

(a) 755553 → 455556 → 635556 → 633756 → 663456
(b) 755553 → 773553 → 746553 → 766353 → 466653

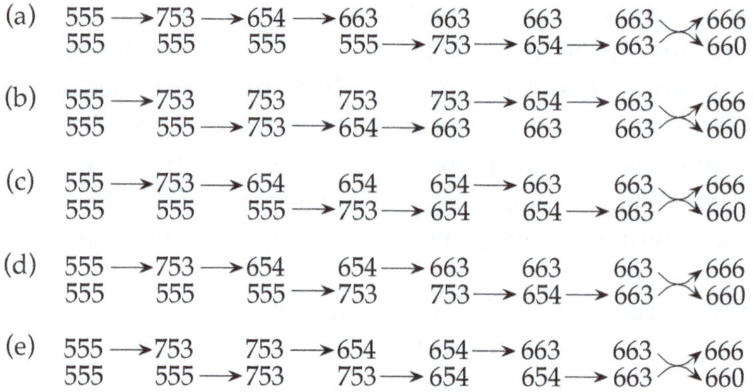

(a) 555 → 753 → 654 → 663   663   663   663 ↘↗ 666
    555   555   555   555 → 753 → 654 → 663 ↗↘ 660

(b) 555 → 753   753   753   753 → 654 → 663 ↘↗ 666
    555   555 → 753 → 654 → 663   663   663 ↗↘ 660

(c) 555 → 753 → 654   654   654 → 663   663 ↘↗ 666
    555   555   555 → 753 → 654   654 → 663 ↗↘ 660

(d) 555 → 753 → 654   654 → 663   663   663 ↘↗ 666
    555   555   555 → 753   753 → 654 → 663 ↗↘ 660

(e) 555 → 753   753 → 654   654 → 663   663 ↘↗ 666
    555   555 → 753   753 → 654   654 → 663 ↗↘ 660

Figure 5.7 Equivalence of the five seven-step solutions. Although the five solutions listed in the text appear at first sight to be different, they are all equivalent. Each set of six can be grouped into halves, initially 555, and each separately undergoes the transition 555 → 753 → 654 → 663. In line (a), the first set of transformations is completed before the other set begins; in line (e), the two halves are changed in alternation; in the other lines, the changes occur in arbitrary orders. When both halves have undergone the three changes, they arrive at 666633, which is transformed in one final step to the desired 666660. This analysis is concerned with combinations, not permutations—that is, the order of the numbers is not important (and, in the illustration, all the numbers are listed in decreasing order), so $\genfrac{}{}{0pt}{}{465}{636}$, for example, is not different from $\genfrac{}{}{0pt}{}{654}{663}$.

(c) 755553 → 455556 → 473556 → 446556 → 646356
(d) 755553 → 455556 → 473556 → 673356 → 646356
(e) 755553 → 775533 → 745563 → 445566 → 645366

According to the rules defined at the outset, these five solutions are equally good, because they have equal numbers of steps. Is there any sense in which we can say that one of them is "simpler" than the other two? We can answer this by examining the penultimate state 666633, which is the same in all five solutions. This can be regarded as two sets of 663, so that we may ask whether the problem to this point can be regarded as the conversion of 555 to 663 twice over. Solution (a) is plainly of this form, because it consists of 555 → 753 → 654 → 663 carried out first on one 555 group and then on the other. At first sight the other four solutions appear not to fit this pattern, but in reality they do, because they all contain two 555 → 753 → 654 → 663 sequences, but unlike solution (a), they do not complete the first of them before starting the second. This analysis can be understood more readily by studying the version in Figure 5.7. It follows from it that all five solutions are equivalent and we cannot say that any is better than any other.

As I have mentioned, I have preferred to solve this problem by brute force, to better illustrate the sort of approach a computer program or natural selection would use, but you can arrive at the right solution much less laboriously by using a little intelligence. First, you may deduce that a 33 → 60 step will be needed at some stage to produce the empty block and a block with six prisoners, and you may then recognize that it makes sense for this to be the last step so that the rest of the problem is symmetrical, requiring 555 to be converted twice into 663. All that then remains is to find the simplest way of converting 555 into 663.

The pentose phosphate cycle

Later we will see which solution has been adopted by living organisms for the equivalent biochemical problem, but before we can do this we need to see how the problem as I have presented it relates to the pentose phosphate pathway (Figure 4.4). This consists of both oxidative and non-oxidative phases, but for the moment (until Chapter 13 and Figure 13.3), we will consider only the non-oxidative phase, which involves exchanging the carbon atoms of sugars so as to transform six pentose molecules into five hexose molecules; symbolically we can express this as converting 6C5 into 5C6. The exchanges are brought about by enzymes that transfer a certain number of carbon atoms from one sugar to another. The mechanisms available to the cell are the transfer of two carbon atoms, catalyzed by the enzyme *transketolase*, and the uniting of two sugars (one of them always with three carbon atoms) to make a single one, catalyzed by *transaldolase*.

The problem that presents itself at this point is how to organize these reactions so as to bring about the complete conversion (6C5 → 5C6) in the least number of steps. A little reflection will show that this is exactly the same problem as the one we have already solved, expressed in different words, apart from the additional point that transaldolase always uses at least one sugar of three carbon atoms.[7] In terms of the original problem, transketolase is the enzyme that catalyzes moves of two carbon atoms, and transaldolase is the enzyme that catalyzes moves of three. The rule in the original game that you could not have a block with one or two prisoners in it derives from the empirical observation that there are no sugars with just one or two carbon atoms. This is partly a matter of definition: The poison formaldehyde (the main component of the disinfectant formalin) has only one carbon atom and can be considered a carbohydrate, but it is not considered to be a sugar because it is not used as a sugar by any organism (probably it is too reactive as a chemical to be kept under the degree of control needed for a metabolite). Likewise glycolaldehyde (Figure 5.8), a carbohydrate with two carbon atoms, exists and participates in some metabolic reactions, but is not regarded as a sugar by biochemists (though it does appear to be regarded as a sugar by astronomers

[7]This additional point is the reason for labeling one of the steps listed above as "unphysiological."

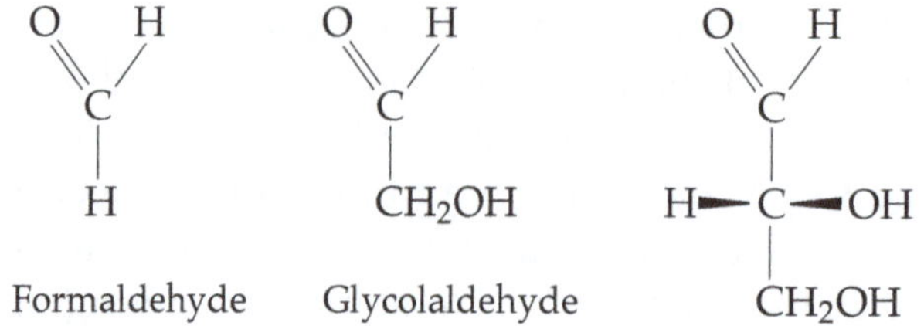

O H, C, H — Formaldehyde

O H, C, CH_2OH — Glycolaldehyde

O H, C, H—C—OH, CH_2OH — Glyceraldehyde

Figure 5.8 Simple carbohydrates. All of the structures illustrated satisfy the generic formula $C(H_2O)_n$, but only glyceraldehyde, with $n = 3$, has the chemical properties that qualify it to be considered a simple sugar. The other two, formaldehyde, with $n = 1$, and glycolaldehyde, with $n = 2$, do not have these properties, do not participate in metabolism as metabolites, and are not regarded by biochemists as sugars.

anxious to find evidence of extraterrestrial life[8]).

Let us now express the solution to the combinatorial game in terms of the real biochemical problem: In effect, we found that we could accomplish the conversion of three pentoses into two hexoses and one triose by the following steps:

$$555 \xrightarrow{\mathrm{TK_2}} 753 \xrightarrow{\mathrm{TA_3}} 654 \xrightarrow{\mathrm{TK_2}} 636$$

where $\xrightarrow{\mathrm{TK_2}}$ is the transketolase reaction, and $\xrightarrow{\mathrm{TA_3}}$ is the transaldolase reaction. In the earlier discussion we called the transketolase reaction a move of two prisoners, and the subscript 2 is a reminder that it transfers two-carbon fragments. In the same way for the transaldolase reaction, the subscript 3 is a reminder that it transfers three-carbon fragments and corresponds to moving three prisoners. Because it does not matter which position is which in the sequence 636, we can write it as 663 to bring it into closer relationship with the discussion above, so the three-step conversion be written as follows:

$$555 \xrightarrow{\mathrm{TK_2}} 753 \xrightarrow{\mathrm{TA_3}} 654 \xrightarrow{\mathrm{TK_2}} 663$$

If each of two groups of three pentoses is transformed in this way, the two trioses left over can be made into one hexose with a final transaldolase step:

$$33 \xrightarrow{\mathrm{TA_3}} 6$$

This is not the only seven-step solution that exists, because there is an alternative way of getting from 753 to 663 in two steps that we did not consider because we restricted the discussion to stages with no more than seven carbon

[8]J. K. Jørgensen, C. Favre, S. E. Bisschop, T. L. Bourke, E. F. van Dishoeck, and M. Schmalzl (2012) "Detection of the simplest sugar, glycolaldehyde, in a solar-type protostar with ALMA" *Astrophysical Journal Letters* **757**, L4.

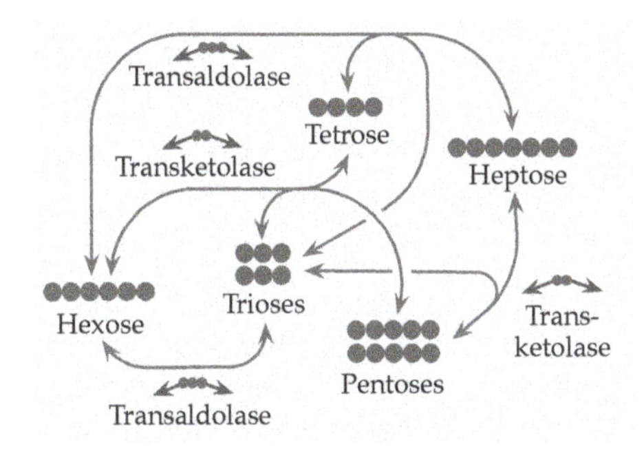

Figure 5.9 Schematic representation of the pentose phosphate pathway. The pathway was shown in a more chemical way as Figure 4.4. This is more schematic, to emphasize the similarity between the pathway as it exists and the solution to the problem as set out in this chapter.

atoms in one molecule:

$$753 \xrightarrow{TK_2} 933 \xrightarrow{TA_3} 663$$

This involves exactly the same kinds of step but puts the transketolase step before the transaldolase step. It results in a solution just as short as the first one, but is it any simpler? Apparently not, because it involves all of the same molecules as the first solution (3, triose; 5, pentose; 6, hexose; 7, heptose) with one additional one, a 9-carbon sugar. So, it is a more complicated solution, because it not only requires a new molecule not needed in the first solution, but this new molecule must be of a larger and more complicated kind than any of the others. We are left with the conclusion that the original solution is indeed the simplest.

It remains to ask how living cells have solved this problem: Which solution is found in real metabolism? If you study the scheme of the pentose phosphate pathway that appeared as Figure 4.4, you will see that the answer is that the solution that mathematical analysis proves to be the simplest is indeed the way that pentoses are transformed into hexoses in cells. The relationship is easier to follow in the schematic version of the pathway shown as Figure 5.9.

Earlier in the chapter we saw that the problem of how to send a tamper-proof message was once thought impossibly difficult, but became almost trivially easy when approached in the right way, and we have a similar case here. Before Enrique Meléndez-Hevia decided to investigate whether metabolism is optimally organized, most biochemists would have doubted whether the question made sense and would have seen little hope of answering it. Certainly, no one had previously thought of treating it as a mathematical game.

As I discussed in Chapter 1, biochemistry is much the same whether we study it in *Escherichia coli* or an elephant. How did it happen, then, that the right questions were first asked on an island off the coast of Africa and not in one of the main centers of the subject? Probably the crucial step was to realize that the problem could be investigated at all, and a relative isolation from the centers of power allowed calmer reflection, independent of current fashions, than would have been possible in Madrid or Barcelona. Moreover, Meléndez-

Hevia's father was a distinguished paleontologist (and his brother is another), and he grew up in a household where problems of evolution were a matter of everyday discussion. As an accomplished amateur composer of music, he also doubtless has less difficulty than most of us in keeping mental track of all the interconnections of the pentoses and hexoses.

ENRIQUE MELÉNDEZ-HEVIA (1946–) is a Spanish biochemist working in Tenerife, in the Canary Islands. He invented the analysis that forms the theme of this chapter, and his encyclopedic knowledge of metabolism allowed him to originate many of the ideas discussed in this book, not only the optimization of metabolic pathways discussed in this chapter, but others such as the structure of the storage carbohydrate glycogen (Chapter 6), the causes of osteoarthritis and obesity (Chapter 11), and the evolution of metabolism (Chapter 15).

Before accepting too readily the conclusion that natural selection has arrived at the best possible solution to an evolutionary problem, we need to answer an accusation that the rules as I have given them are arbitrary; rather than examining how nature solved a genuine problem, I have apparently just constructed an artificial problem whose solution happens to be the one found in nature.[9] This objection is not altogether unfounded, and there is one rule in particular that cries out for justification. It was one thing in the problem of moving prisoners to ascribe the rule that the car could not be used to move one prisoner at a time to bureaucratic silliness, but such an argument will hardly do for the organization of metabolism, where there are no bureaucrats insisting on adherence to pointless rules (notwithstanding the title of an influential paper[10]).

So, we need to explain the lack of an enzyme to move one-carbon fragments in a different way. This has to do with the fact that although we have treated all pentoses as a single entity, and similarly with trioses, tetroses, hexoses, and heptoses, the reality is that each class of sugar contains several types. In the scheme to illustrate the pathway, for example, you can find two different trioses, glyceraldehyde 3-phosphate and dihydroxyacetone phosphate, and the three different pentose phosphates shown in Figure 5.10. In the whole pathway, the number of distinct sugars that appear is much smaller than the total numbers of sugars in the various classes. If all possible trioses, tetroses, pentoses, hexoses, and heptoses appeared in the pathway, it would be vastly

[9]That is exactly what happened two centuries ago in the theory of statistics, with results that continue to confuse textbook authors to this day. Karl Friedrich Gauss is commonly credited with proving that the ordinary mean is the best kind of average because it follows from the normal (or "Gaussian") distribution of errors. He quite explicitly did the opposite, deciding at the outset what conclusion he wanted to reach and then working out what properties the world would need to have for it to be valid. See K. F. Gauss (1809) *Theoria motus corporus coelestium in sectionibus conicis solem ambientium* (English translation by C. H. Davis, 1857), Section 177, pages 257–259. Dover, New York, 1963.

[10]E. Meléndez-Hevia, J. Sicilia, E. I. Canela, and M. Cascante (1996) "Molecular bureaucracy: Who controls the delays?" *Journal of Theoretical Biology* **182**, 333–339.

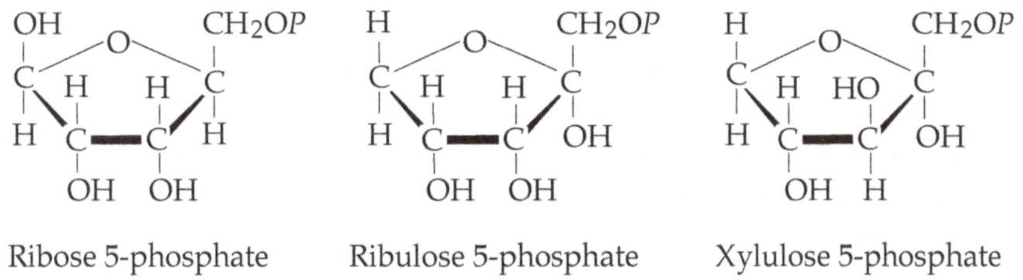

Ribose 5-phosphate Ribulose 5-phosphate Xylulose 5-phosphate

Figure 5.10 Pentose phosphates. The pentose phosphate pathway involves three different pentose phosphates, illustrated here, as well four different hexose phosphates, two different triose phosphates, a tetrose phosphate, and a heptose phosphate. The three structures shown here, in which italicized *P* represents a phosphate group, are quite similar to one another, and also to the structures of other pentose phosphates that do not participate in the pathway, and thus demand a high degree of specificity on the part of the enzymes that act on them.

more complicated than what is shown, and that would not just be a problem for students; it would also be a problem for the cell, which needs to keep track of all the different components. Keeping the total number of participants within manageable limits requires some degree of specificity in the enzymes that catalyze the reactions, accepting some sugars as possible substrates, but not others. When this is taken into account, together with the fact that chemical principles do not allow an enzyme to have a completely arbitrary specificity,[11] it turns out that an enzyme capable of moving one-carbon fragments between sugars could not be specific enough to generate a complete pathway as simple as the one that exists.

It is, in fact, an almost universal problem in metabolism that enzymes are less specific than they would be in an ideal world. We have already seen in Figure 2.1 that base pairing in DNA is less precise and error-free than popular accounts and elementary textbooks suggest, and additional errors arise even when an exactly correct DNA sequence is used to synthesize a protein, because the different amino acids are not all different enough from one another to ensure that the "wrong" one is never incorporated. For example, valine and threonine are about the same size and shape, and the enzyme that is supposed to recognize valine accepts threonine about once in 300 times. That is too much to be tolerated, and in the "double-sieve" mechanism proposed by Alan Fersht, a second enzyme removes the wrong amino acid. This also makes an error about once in 300 times, removing valine when it should be removing threonine. Overall, errors occur about once in $300^2 \approx 100,000$ times,

[11]Why not? The limits on what specificity is possible arise from the fact that the enzyme must have a structure that matches that of its substrates. Even though it may easily reject substrates that are completely different, it may have difficulty in rejecting ones that are approximately right.

a frequency low enough to be acceptable. Error correction and "proofreading" in DNA replication and translation have been known for several decades, but it has become increasingly clear that enzymes throughout metabolism make specificity errors, sometimes producing metabolites that are not only unable to fulfill their tasks, but are positively harmful. These need to be recognized and removed by additional enzymes, in mechanisms reminiscent of the double-sieve mechanism.[12]

I have taken some trouble to present the solution to the problem of the pentose phosphate pathway as simply as possible, first showing that brute force will inevitably lead to the right answer eventually, albeit with a lot of work, and then showing that with a little logical analysis you can arrive at the same result much more quickly. All this may give the impression that the fact that cells have adopted the simplest solution has no importance, because the best solution may appear obvious once it has been pointed out. It is not obvious at all, as you may readily check by setting the combinatorial game as a problem for someone who has not encountered it before and does not know what the solution is. In practice, such a person will normally flounder around for a considerable time before finding the best solution. Remember, too, that by presenting the public-key cipher problem in terms similar to those used in the problem of the farmer and the river, just after discussing that problem we made its solution seem simple and obvious, although historically the solution escaped some of the greatest experts in cryptography. Almost any great and original discovery can be labeled "obvious" once its solution has been found, explained, and understood.

Once we accept that living organisms have found the best solution to a nontrivial problem, we need to ask what steps were followed during evolution to arrive at it, because we cannot easily suppose that the primitive organism that first adopted the pentose phosphate pathway just happened to organize it in the best possible way. Human players use their intelligence to improve their strategy, because they reflect and learn lessons for the next time, but this approach is not open to a mechanism that has no intelligence and no foresight. Answering that natural selection works by tinkering, rejecting the modifications that fail, and selecting the ones that work better, is inadequate, because even if it is true, it leaves some important questions unanswered: How does an organism decide that one way of organizing the pentose phosphate pathway is better than another, and in what sense is a simpler solution better than a complicated one? I will try to answer these later in this chapter, but first we need to consider whether the successful analysis of the pentose phosphate pathway is just a one-off chance result, or whether there is any evidence that

[12]E. Van Schaftingen, R. Rzem, A. Marbaix, F. Collard, M. Veiga-da-Cunha, and C. L. Linster (2013) "Metabolite proofreading, a neglected aspect of intermediary metabolism" *Journal of Inherited Metabolic Diseases* **36**, 427–434.

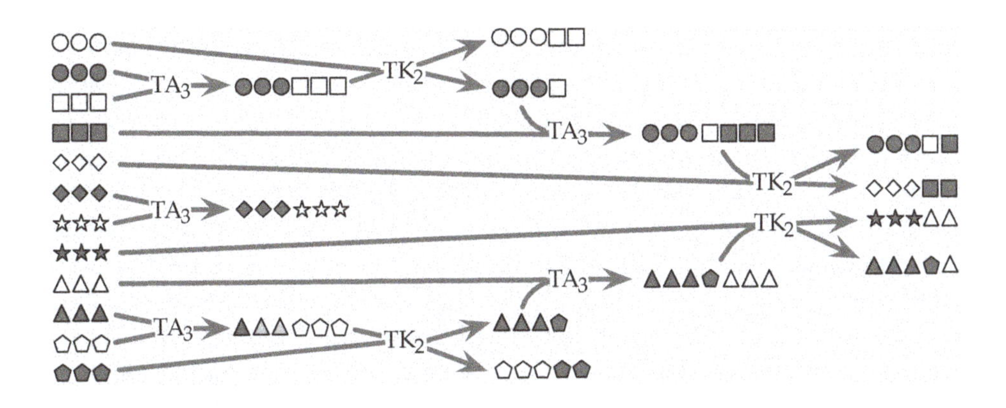

Figure 5.11 Solution to the Calvin cycle game. The aim of the game, similar to the one shown in Figure 5.6, is to redistribute elements two or three at a time such that 12 groups of three elements are converted into one group of six elements and six groups of five.

metabolism in general is optimized, rather than just one pathway that happens to have been studied.

The Calvin cycle

Photosynthesis, the process that green plants use for harnessing the Sun's energy, provides another example. It involves a metabolic pathway called the *Calvin cycle*, which, like the pentose phosphate pathway, contains a portion that provides a mechanism for converting one set of sugars into a different set, again using the transketolase and transaldolase reactions. The requirements are sufficiently different that it is not just the same problem in a different guise, and the fact that the problem of the pentose phosphate pathway has been solved by living organisms gives us no reason to expect the corresponding problem for the Calvin cycle to have been solved as well, unless we accept the existence of a general tendency in evolution to find the simplest solutions to metabolic problems. If this problem is expressed in the same terms as those we used for the pentose phosphate pathway, the objective is to convert 12 trioses (three-carbon sugars) into one hexose (six carbons) and six pentoses (five carbons each).

Applying the same rules as before (transfers restricted to fragments containing two or three carbon atoms; no sugars with fewer than three carbon atoms), the pathway as it exists in plants has the organization illustrated in Figure 5.11. You may like to search for a simpler way of achieving the result, and if you do you will find that you cannot, so the Calvin cycle, like the pentose phosphate pathway, is optimized in living cells.

After noticing that evolution has arrived at the simplest ways of organizing these pathways, we may feel that it is obvious that the simplest solution is the best solution and that no discussion of the advantages of simplicity is necessary. This is naive, however, because if we look at metabolism as a whole we can find many examples where it is not obvious at first sight that the simplest approach has been adopted.

One of the best known pathways is glycolysis, the anaerobic (independent of oxygen) part of the mobilization of glucose: This is often the first pathway that students are taught, and often the one that they know in most detail; it usually occupies the central part of a metabolic pathways chart, the part that is printed most prominently. (In Figure 4.3 it is the part just left of center, shown with a gray background.) Examination of the reactions of glycolysis reveals a peculiarity, however. Glycolysis begins with the conversion of glucose to glucose 6-phosphate—that is, by the transfer to glucose of a phosphate group from ATP, the energy currency of the cell. A little later in the pathway another phosphorylated sugar, fructose 6-phosphate, receives a second phosphate group in the same way, making it into fructose 1,6-bisphosphate. Yet, in the later steps of glycolysis, these phosphate groups are removed, regenerating ATP, and additional ATP is made with inorganic phosphate ion as the source of the phosphate group.

What is the point of transferring two phosphate groups from ATP in the early stages of glycolysis, only to transfer them back again later on? Surely it would be simpler and more efficient to dispense with these apparently unnecessary steps? This is a little like asking the point of requiring passengers in a moving vehicle to attach their seatbelts even though we know they will have to detach them again at the end of the journey! The phosphorylated sugars cannot cross biological membranes without specific mechanisms to take them across, and so they cannot leak into parts of the cell where they are not wanted. The phosphorylation also allows the process to be forced to go in the required direction, and makes it easier for the regulatory mechanisms to determine the rate at which it proceeds.

It follows, then, that in considering whether a particular pathway is optimized for simplicity, we must avoid asking the question in excessively simple-minded chemical terms: Apparently pointless steps may fulfill necessary functions. Incidentally, all of the sugars we considered earlier in the chapter are also phosphorylated, but because they do not become unphosphorylated in any of the reactions considered, we could discuss the organization without taking account of the phosphorylation.

Simpler is better?

We cannot assume that simplicity will have been selected because of the intellectually pleasing results that it produces. To understand why simplicity should be selected, we must understand why it is advantageous to an organism to have a simpler pathway, why such an organism will live longer and have more descendants than a rival with a more complicated arrangement. In short, the answer is that a shorter pathway is more efficient for converting metabolites into other

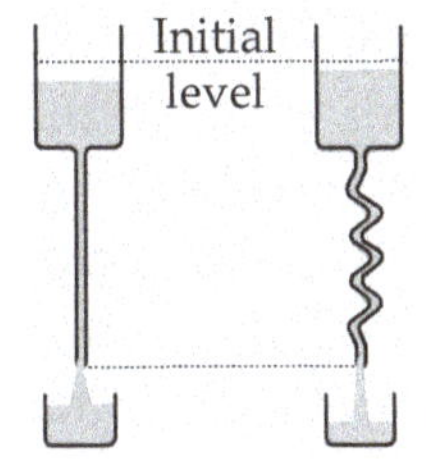

Figure 5.12 Emptying two tanks. The exit tubes have identical internal diameters and they end at the same level, but one is straight and the other is curly and thus much longer. The flow through the shorter tube is greater, even though the hydrostatic pressure from top to bottom was the same for both when the experiment was started.

metabolites than a longer pathway: It uses less materials, whether these are chemicals stored as intermediate metabolites in the pathway, or protein needed to make the enzymes that catalyze the reactions; more important, it is faster. Other things being equal (and this may be an important qualification, because there is no strong reason to be sure that other things will actually be equal), a pathway catalyzed by two enzymes will transform a given concentration of starting material into product faster than a pathway of three enzymes with the same kinetic properties as those of the first. Let us examine this in terms of a simple experiment illustrated in Figure 5.12, which is easy to set up with equipment available at home or from a hardware shop.

Take two identical tanks with outlets at the bottom, to each of which a section of plastic tubing can be attached in a water-tight way. For one outlet, use a tube of a meter or so in length, draining into a sink; for the other, do the same thing but use a tube about twice as long, with its outlet at the same level as for the short tube (so that that the initial water pressure at the outlet of each tube is the same). Then fill the tanks with water to the same intial level, and collect the water in two sinks. If all this is done correctly, the only difference between the two at the beginning is the length of tube that needs to be traversed by the water on its way from the tanks to the sink. Which tank will empty faster?

Two reasonably plausible answers may be given: Either there will be no difference or the tank with the shorter tube will empty faster. (Not many people suggest that the tank with the longer tube will empty faster, and that is certainly wrong.) Whichever you think, you are quite likely to think that it is obvious, and that the other possibility is absurd, but if you cannot make up your mind, then do the experiment. It is easy to do, and if you do it, you will find that the second answer is correct: The tank with the shorter tube empties more quickly.

If you do not believe this, please do the experiment. Do not do what a colleague of mine did after reading about it. He told me that the experiment did not work as described, so I was worried, because at that stage I had not tested it, but just believed Meléndez-Hevia's account of it in his book. I thought my colleague meant that he had tried it for himself and found it to fail. But no, that was not what he had done. Despite being an experimental biochemist (whereas I am sometimes accused of being an armchair biochemist, too much interested in theory and not enough in the experimental realities), he consulted a physicist who knew about hydrodynamics. The physicist did not do the experiment either, but made some calculations on the back of an envelope and reported back that the flow would be exactly the same in both tubes! Since then, I have done the experiment (it only takes a few minutes) and found that the flow through with the shorter tube is greater, as I said above. So much for theory when inappropriately applied.

The tank experiment also illustrates, in a way that is so obvious that simply describing the arrangement is sufficient, the two other advantages of the shorter tube apart from flow rate: It uses less material for achieving the connection between tank and sink, less plastic for the analogy, and less protein for the metabolic pathway; less material is tied up in the pathway itself during the process, less water in the tube at any moment, and less chemical resources in the pathway intermediates. This type of consideration played a part in the decline of the canals and the replacement of their functions by railways in the second half of the nineteenth century. Canals were efficient for transporting nonperishable goods, and much cheaper in principle than railways. Moreover, for goods like coal that are not subject to sudden wild fluctuations in supply and demand, they could ensure that the output rate always matched the input rate, so their inherent slowness might seem to have little importance. However, the slowness meant that at any particular moment a large amount of material was in transit, so the supplier would suffer if it had not been paid for, and the buyer would suffer if it had been paid for but had not yet been delivered.

Returning to the pentose phosphate pathway, it follows that an organism with a shorter sequence of reactions to accomplish the conversion will have some economic advantages over a rival that uses a longer sequence to achieve the same result. This much is clear, but what is perhaps less clear is the evolutionary route by which a less efficient system can be transformed into something shorter and better. Once a given set of reactions is in place, it may seem to require major reorganization to replace the chosen reactions by a different set: New catalytic activities need to evolve, the now redundant older enzymes need to be eliminated, and so on. All this requires some improbable events to occur, whereas somewhat less improbable events may improve the performance of the organism that retains the longer pathway. For example, mutations that alter the structures of proteins occur from time to time in all lineages, and although many of these are harmful or at best neutral, occasionally these can result in an enzyme that catalyzes its reaction more effectively than the unmutated enzyme.

Because these events are purely random, and have no special tendency to improve performance, it is perfectly possible that an organism that is on its way toward improving its metabolic design may compete with another that retains the less efficient design but happens to have mutated some of its enzymes to be more effective catalysts. In such a competition, the organism with a less efficient design may still be able to live a more economic life and may overrun the other. However, just as there is no reason for the improving organism to be specially favored with valuable new mutations, there is equally no reason for the opposite to be true. On average, therefore, and given enough time, none of these chance effects will favor one organism over another, and ultimately we may expect the more efficient design to win. Thus, today we find that in all organisms that have been studied, the pentose phosphate pathway and the Calvin cycle are arranged to use the smallest number of reactions possible for the conversion.

The organism that retains the less efficient pathway may seem to have an easier way of increasing fluxes than selecting rare mutants that improve the catalytic activities of the component enzymes: Instead of improving the enzymes, it can simply make more of them, because, in general, more catalyst means a higher rate. As a way of competing with an organism that has found a more efficient pathway, this is ultimately ruinous, however: More enzyme means more protein, and protein is expensive; an organism cannot just make unlimited amounts of protein without considering the cost in terms of dietary sources of amino acids and the energy resources needed for assembling them into proteins. Some enzymes, indeed, are already present in such large concentrations that it is hard to imagine that their amounts could be increased. Ribulose bisphosphate carboxylase, an enzyme necessary for photosynthesis, is the most abundant protein on Earth, and accounts for about half of all the protein in the leaves of green plants. Even a 5% increase in its concentration would imply severe strains on the supplies of many other proteins needed for the health of the plant. This parallels the example in Chapter 3, where we saw that the blood of a mammal is packed so full with hemoglobin that there is no room for any more. More generally, just increasing the quantities of enzymes in an inefficient pathway is not an option for competing with an organism that has found a better one.

The pentose phosphate and Calvin cycles are just two rather similar metabolic pathways out of many known in biochemistry. For the moment, therefore, it would be dangerous to generalize to claim that all of metabolism can be analyzed in the same sort of way and is correspondingly optimal. More work will have to be done, but the idea that metabolism has evolved to an optimal state needs to be regarded as a serious possibility. What of extraterrestrial life, if it exists? I will discuss this is a general way in Chapter 12, but meanwhile there is no reason to doubt that if optimality has been achieved in terrestrial metabolism, it should also be a feature of metabolism elsewhere in the Universe, though the detailed chemistry may well be different.

6. *The Perfect Molecule*

> The pursuit of perfection, then, is the pursuit of sweetness and light....
> He who works for sweetness and light united, works to make reason and the will of God prevail.
>
> Matthew Arnold, *Culture and Anarchy* (1867–1868)

> Perfection is the child of Time.
>
> Bishop Joseph Hall, *Works* (1625)

> One has to rely on the principle of Occam's razor; that is to choose the simplest mechanism that is consistent with the results whilst being prepared to increase the complexity if further results require this and remembering that "God does not always shave with Occam's razor" (G. D. Greville, unpublished observation).
>
> Keith Tipton (1974)[1]

WHEN I CHOSE A TITLE for the original edition of this book, I did not realize that I was quoting Matthew Arnold, still less that placing his words in context would make them rather inappropriate for the idea that blind chance operating over several billion years has produced the appearance of perfection in many aspects of biochemistry that we can recognize today. The idea is conveyed rather better by the second quotation above, though because it comes from the hand of a bishop, he probably had something different in mind from what concerns us here. But no matter, in the preceding two chapters I have discussed Enrique Meléndez-Hevia's demonstration that the metabolic pathways he examined are organized in living organisms as well as they possibly could be. He has also been interested in other questions of optimality in biochemical systems, and here I discuss one of these, the structure of the storage carbohydrate *glycogen*. Before doing so, we should note that although nature may often choose the simplest solution to a problem, Keith Tipton in the quotation above advises biochemists to apply Occam's razor, but he points out that this does not always work, and so I begin with an example of a structure that is surely not optimal.

[1]K. F. Tipton (1974) in *Companion to Biochemistry* (edited by A. T. Bull, J. R. Lagnado, J. O. Thomas, and K. F. Tipton), page 249, Longman, London.

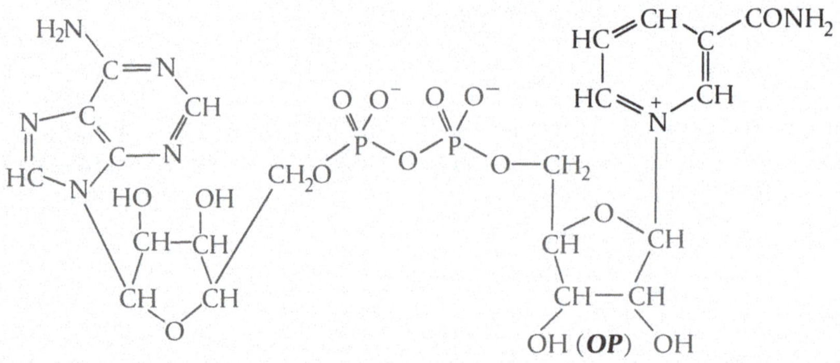

Figure 6.1 The structure of NAD and NADP. Only the atoms shown in black, fewer than one-quarter of those in the molecule, play any role in the chemical function of NAD and NADP. Some of the others are necessary for specific binding to enzyme molecules, but it is hardly possible to believe that this function could not be fulfilled by a simpler structure. NADP differs from NAD at the position shown as OH (*OP*), with a hydrogen atom in NAD and a phosphate group in NADP.

NAD: A degraded fragment of RNA?

In principle we could ask of almost any molecule selected for a prominent biochemical role whether its structure is well suited to its function, or could be designed better. Steven Benner[2] has asked this question about various molecules, and has come to the conclusion that several are far from optimal, looking more like frozen accidents of evolution: For example, the coenzyme NAD looks more like a degraded fragment of RNA than the result of selection of the ideal structure (see Figure 6.1). As I will be taking a different view in this chapter about another molecule, glycogen, it is fair to ask why one example should be different from another. The point is that the structure of glycogen could be optimized by *quantitative* changes in the properties of the enzymes concerned, whereas replacing NAD by something that would work better would require much more radical changes.

For many structures it is not obvious how to put the question in a concrete testable form. Until we know a great deal more than we do now about how the details of protein structure are related to the fine details of enzyme activity, it will be difficult to know whether the enzymes we have are the best we could possibly have. Certainly, there have been reports that artificially modified enzymes can have higher catalytic activity than their natural counterparts (though not usually by large factors), but this does not show that they are "better," because a real enzyme has more to do than just catalyze a reaction. It has to have the required degree of stability: not so stable that it fails to disappear when it is unwanted; not so unstable that it needs to be continuously

[2]S. A. Benner (2009) *Life, the Universe and the Scientific Method*, Ffame Press, Gainesville, Florida.

synthesized to replace the molecules that are being lost. It must be capable of being recognized by other large molecules that interact with it, either to modify its catalytic activity (or their own) or to move it according to metabolic signals from one location in the cell to another. Proteins are just too complicated in structure and function for us to have any confidence at the moment in estimates of whether they have optimal structures or not.

Glycogen

There is, however, a large molecule with a function that can be expressed precisely, that is built from a vast number of copies of a single kind of building block, and that needs to satisfy functional constraints that can be defined precisely. It is possible, therefore, to assess the extent to which the structure that exists in nature is optimal. This is glycogen, and Meléndez-Hevia has carefully studied its structure from the point of view of optimization.

It is sometimes called *animal starch,* and although it has a tighter, more compact structure than starch, the corresponding polymer of glucose in plants, its function in animals is similar to the function of starch in plants. It allows considerable quantities of glucose to be stored in liver and muscle in such a way that it makes no detectable contribution to the osmotic pressure of the blood or the cell water, but still allows rapid access to the glucose when it is needed. The need for rapid access is clear enough, and I will come back to it shortly, but what is *osmotic pressure,* and why is it important that glycogen does not contribute to it?

Hydrostatic pressure

Before trying to answer these questions, I will look at a more familiar kind of pressure, such as the pressure of gas that keeps a car tire inflated or prevents a balloon from collapsing. This pressure depends on the *number* of molecules in the gas (and not, for example, on the weight of the gas), and forcing more molecules into a confined space causes the pressure to increase. This can be done mechanically, as for example when a tire is inflated, causing the rubber to be stretched in response to the pressure. It can also be done by boiling a confined liquid, as when a sealed container bursts after being overheated, and it can also be done chemically, as in the explosion of a substance such as nitrocellulose, when a small number of nitrocellulose molecules in the solid state are converted almost instantaneously into a large number of gas molecules—mostly carbon dioxide, water vapor, and nitrogen.

Osmotic pressure

Osmotic pressure is similar in origin and can have similar catastrophic effects if not properly regulated. Although it is less obvious in everyday life than other sorts of pressure, it does have some easily observable effects. For example, it causes flowers to take up water from a vase. A plant fills with water from its roots because of capillary action, an effect due to surface tension, which has the even more familiar consequence that water can wet objects immersed in it:

When water touches your hand, for example, it does not fall right off, leaving your hand dry, but tends to stick to it. However, although capillary action can fill a plant with water, even a tall tree, it does not generate a continuous upward flow: If capillary action were the only process, it would stop once the plant was saturated with water. What prevents it from stopping is osmotic pressure. When water evaporates from the leaves or petals, the amount of water left in the leaf cells decreases, and all of the material dissolved in it becomes more concentrated. Now, just as the pressure of the air in a tire depends on the number of gas molecules inside it, the osmotic pressure in a water solution depends on the number of molecules dissolved. Even the numerical details are almost the same: A given number of molecules dissolved in a liter of liquid make almost the same contribution to the osmotic pressure as the same number of gas molecules would to the gas pressure if pumped into a closed container of one liter. So, when water evaporates from a leaf, the osmotic pressure in the leaf cells increases, and the only way the plant can balance the pressure is by bringing more water from elsewhere.

A less familiar example, but important for the treatment of patients in hospitals, is provided by the red cells in the blood, which burst when blood is mixed with water. This bursting is directly due to osmotic pressure, and people have occasionally died after being injected or perfused with water. This effect is obvious if you examine the blood and the blood–water mixture under a microscope, but not with the naked eye. There is, however, a similar phenomenon that is easy to see in your kitchen. If you boil potatoes in *very* salty water (dissolve as much salt as you can in water that is just enough to cover the potatoes in the saucepan), then they will not become waterlogged even if you boil them for an hour or more, but instead will acquire the character of potatoes baked in an oven. In other words they will lose water during cooking, not absorb it, because the osmotic pressure of the water is much higher than it is in the cells of the potatoes. Considerations of this kind explain why the fish of the sea need to protect themselves with waterproof skin: Despite the apparent abundance of water all around them, it is so salty that they would rapidly become dehydrated if water could flow freely through their scales. Incidentally, potatoes cooked in the way I have described are called *papas arrugadas*,[3] and are popular, appropriately enough, in Tenerife, the island where Meléndez-Hevia works. In the light of this discussion you should have no difficulty in understanding the reason for adding salt when boiling vegetables (Figure 6.2).

Osmotic pressure is crucially important in systems containing semipermeable membranes, and although these do not feature much in everyday life, they are widespread and important in the bodies of living organisms. A semi-

[3]Similarly cooked potatoes in Syracuse, New York, are called salt potatoes.

	Why add salt?	***Comments of "a scientist colleague"***
1.	It keeps the beans green.	*Only the acidity and calcium content of the water affect the color of the beans.*
2.	It raises the boiling point of the water so the beans cook faster.	*Adding salt does increase the boiling point of water, but by such a small amount that it will make no difference to cooking times.*
3.	It prevents the beans from becoming soggy.	*Vegetables will go soggy if cooked for too long regardless of whether salt is added or not.*
4.	It improves the flavor.	*Very little salt is actually absorbed onto the surface of a bean during cooking—typically* 0.0001 *gram of salt per bean, which is too little to be tasted by most people.*

Figure 6.2 Why add salt when boiling vegetables? A web page of the Royal Society of Chemistry (`http://tinyurl.com/keqlzz7`) suggests several answers to the question of why salt is added when boiling vegetables, and adds comments of "a scientist colleague." Three of the comments are certainly correct, but the third one completely ignores osmotic pressure and is at best misleading.

permeable membrane is a divider between two liquids that small molecules like water can pass through and large ones like proteins cannot. What happens with medium-sized molecules depends on the specific characteristics of the particular membrane considered, but for the purposes of this discussion I will assume we are dealing with a membrane that is permeable only to solvent molecules like water and not to any of the molecules that are dissolved in it. Connecting two children's balloons inflated to different extents, or two soap bubbles of different sizes, causes air to flow from one to the other until the pressures are balanced,[4] and putting two solutions in contact via a semipermeable membrane causes water to flow across the membrane from the solution with lower osmotic pressure into the solution with higher osmotic pressure.

In principle this flow continues until the osmotic pressures are balanced, but if one or both of the compartments in contact is sealed, and if the sealed compartment has the higher osmotic pressure (as with a red blood cell suddenly diluted into pure water) it will burst when its membrane can no longer withstand the increase in volume. A healthy red blood cell has rather a flat structure, and its volume can increase quite substantially before it bursts;

[4]The statement as I have made it is true for both balloons and soap bubbles, but it may appear not to be, because the larger balloon will lose air to the smaller, whereas the smaller soap bubble will lose air to the larger. How can we explain this?

nonetheless, it inevitably bursts in pure water because the osmotic pressure of pure water is zero and its own osmotic pressure can never fall to zero, no matter how much water flows in. In principle the same sort of thing could happen to any cell immersed in pure water, but many, most plant cells for instance, are much more protected than red cells, because they are encased by a strong protective walls that make up for the structural weakness of the membranes.

The point of this digression about osmotic pressure is to emphasize that although most people can pass their entire lives without ever becoming conscious of its existence or noticing its effects, it does have a crucial importance inside the living body and it is essential to guard against its potentially catastrophic effects. Membranes are everywhere in the body, and the osmotic pressure needs to be balanced across every single one of them. Consider what happens to someone who rapidly eats three 5 g glucose tablets (prescribed to diabetic patients, but also often recommended to healthy people as a source of "energy" by unscrupulous advertisers). Fifteen grams of glucose is about 0.08 mole, a mole being, roughly speaking, the amount of a substance that contains the same number of molecules as there are atoms in a gram of hydrogen.

Measuring quantities in moles may seem bizarre if you are more used to weights or volumes, but it makes a lot of sense in chemistry and biochemistry, where we are often more interested in how many molecules there are than in how much they weigh or how much space they occupy (though these can be important as well).[5] As I have mentioned, osmotic pressure is determined primarily by the number of dissolved molecules, so it is especially important in this context to measure quantities in moles.

Returning to the three glucose tablets, if the 15 grams (about half an ounce) became instantaneously dispersed over the whole volume (about four liters) of the blood, it would cause an increase of about 0.02 mole per liter. This would be a tremendous increase, because the normal concentration of glucose in the blood is about one-quarter of this, around 0.005 mole per liter. It would bring about a correspondingly large increase in the osmotic pressure of the blood—not a fourfold increase, certainly, because there are many other things dissolved in the blood apart from glucose, but about a 15% increase, enough to cause serious problems for the person.

Fortunately, the effects of eating too much glucose in a short time (or adding

[5]The small molecule 2,3-bisphosphoglycerate is important for regulating the properties of hemoglobin. Its concentration in blood was measured with reasonable accuracy in the 1920s, but its importance went completely unnoticed because, when expressed in grams per 100 milliliters, the ordinary units of that time, it seemed negligible compared with that of hemoglobin. Only 40 years later, when Reinhold and Ruth Benesch converted the numbers to moles per liter did anyone notice that its concentration was similar to that of hemoglobin. Susan Benesch, their daughter, told me that the recalculation was prompted by a student who asked what the function of the 2,3-bisphosphoglycerate was.

too much sugar to your tea or coffee) are not as dramatic as this calculation may suggest, in part because glucose taken by mouth does not pass instantaneously to the bloodstream. Nonetheless, it does pass quite fast,[6] and to avoid excessively perturbing the sugar concentration and osmotic pressure of the blood, it is essential to have an efficient and rapid way of converting glucose into something that does not contribute to the osmotic pressure. This something is glycogen, which can contain more than 50,000 glucose units linked together in a *β-particle*. In muscle, the β-particles are stored as such, but in liver, with which we are mainly concerned here, they are organized into larger assemblies, *α-particles*, each of which consists of several β-particles.

Because the osmotic pressure depends on the number of particles, not on their size, this system allows the liver to store the equivalent of a huge concentration of glucose, up to 400 millimoles per liter, or 80 times the concentration in the blood. (This would be enough to produce an osmotic pressure of around 10 times the pressure of the atmosphere, the sort of water pressure a diver feels at a depth of about 100 meters.) Storing a large amount of glucose is not by itself enough, and the structure also needs to be capable of being made and broken down fast, so that the glucose is available when needed. This is especially important for the glycogen stored in muscles, which may need to convert much of it into mechanical work in a short time. In the liver, speed of release is much less important, but it does no harm to use essentially the same structure in the liver as is needed in muscles.

In mammals, the glycogen-rich white fibers occur in the same muscles as the myoglobin-rich red fibers, but in birds and fish (Figure 6.3), they occur in different muscles and are thus easier to distinguish. Red muscle is relatively slow to respond but energy-efficient, and thus suitable for routine swimming for long periods. White muscle responds fast, but can only be used for a short time—as long as the glycogen lasts—and is used in emergencies, such as fleeing from a predator.

Mobilization of glycogen

To allow rapid mobilization, the whole surface of the glycogen molecule needs to have many points for attack by the enzymes that remove glucose from it. We can think of these enzymes like animals that graze by eating the leaves from a tree and can only reach the exposed ends. (I am thinking here from the point of view of the grazers, not from that of the tree, which has certainly not adopted a particular form in order to be eaten most efficiently.) To expose the maximum number of leaves, the tree needs to have a highly branched structure, and that is also how glycogen is arranged, with more than 2000 *non-reducing ends* available on the surface of a β-particle.

[6]From this point of view, carbohydrate eaten in the form of sugar stresses the organism much more than an equivalent amount of starch, for example in the form of bread or potatoes, because starch is converted to glucose much more slowly.

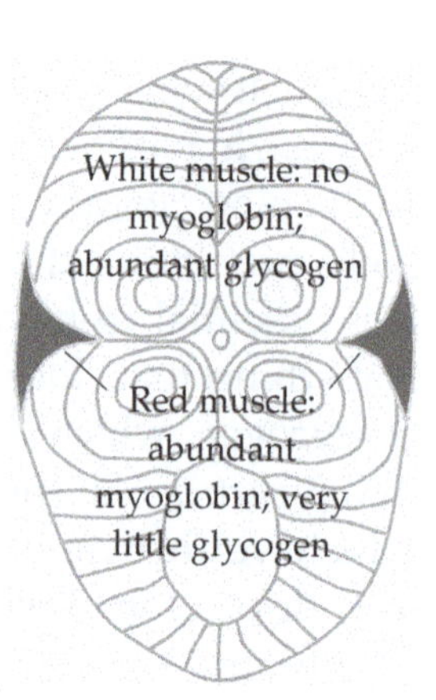

Figure 6.3 Cross section of a white fish. In mammals, the aerobic red and anaerobic white fibers occur together and are difficult to distinguish with the naked eye, but in birds and fish, they occur in different muscles. The white muscle, abundant in glycogen and without myoglobin, is fast and is used by the fish in emergencies. The red muscle, low in glycogen and rich in myoglobin, is much slower to respond, and is used for routine swimming.

Structure From this description it is clear that glycogen has a good structure for fulfilling its function, but that is not the same as saying that it is optimal. How can we be certain that tinkering with the structure—making the branch points closer together or further apart than they are in the real structure, for example—might not result in something even better? To establish this, we need to make some quantitative comparisons, and we need to base them on the real structure of glycogen, not the fantasy structures illustrated in most textbooks of biochemistry.

CLAUDE BERNARD (1813–1878), a French physiologist, made many major contributions to the subject, including elucidation of the role of the pancreas in digestion and that of the liver in mobilizing glycogen. He is particularly remembered for the concept of *homeostasis*, the capacity of organisms to regulate their cellular processes to maintain approximately constant internal conditions.

From the time when glycogen was discovered by Claude Bernard in the nineteenth century, it was clear that it was a device for storing glucose and that it must therefore consist mainly of glucose molecules linked together. However, the structure of glucose is such that there are numerous different ways of linking the molecules together to make a polymer. Cellulose, for example, the principal structural material for making plants, is also a polymer of glucose, but it has a different structure from glycogen and starch and has different properties: For example, starch is easily digestible in the mammalian gut, whereas cellulose is completely indigestible. There are two different kinds of links between glucose units in glycogen, called $\alpha(1 \rightarrow 4)$ and $\alpha(1 \rightarrow 6)$ links. The numbers here refer to the six different carbon atoms that each glucose molecule contains, so $\alpha(1 \rightarrow 4)$ means that the atom labeled 1 in one molecule is attached to the atom labeled 4 in another glucose molecule. Another kind of link, called a β link, occurs in other biologically important molecules such as cellulose, the main structural material in plants. Although α and β links are superficially rather similar, they lead to dramatically different properties: Starch and glycogen, on the one hand, and cellulose, on the other, are all made primarily of $(1 \rightarrow 4)$-linked glucose units, but you can nourish yourself by eating potatoes (mainly

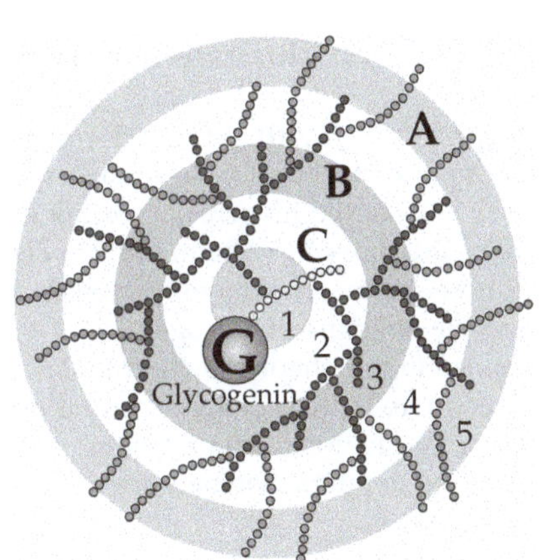

Figure 6.4 Structure of glycogen. The diagram illustrates the structure of the storage carbohydrate glycogen established by William Whelan and colleagues. It is not the same as the (wrong) structure illustrated in most biochemistry textbooks and in Figure 6.5. There is a small protein called *glycogenin* at the center connected to three kinds of chain of glucose units, labeled A, B, and C. A-chains are linear, but B- and C-chains are branched. There is only one C-chain, which differs from the B-chains by being connected to the central glycogenin. To avoid making the figure too complicated to understand, only five layers are shown, but the real structure has about 12. All the A-chains reach the surface.

starch) but not by eating paper (mainly cellulose).

Because glycogen has both $\alpha(1 \rightarrow 4)$ and $\alpha(1 \rightarrow 6)$ links, we see immediately that carbon atom number 1 is used in every link. Any glucose unit can therefore be used once only at the left-hand end of an arrow, but it can appear once or twice at the right-hand end. This allows a branching structure to be built, but not a more complicated sort of network with lines joining as well as separating. Within the glycogen structure, there are many linear arrays of several glucose units joined by $\alpha(1 \rightarrow 4)$ links, which come in two varieties, known as A-chains and B-chains: The B-chains are branched and the A-chains are not, and the branches in the B-chains are produced by $\alpha(1 \rightarrow 6)$ links, with each such link making the start of an A-chain. In addition to these two main kinds of chain, there is a single C-chain, which is similar in structure to a B-chain except that it is attached to a small "primer" protein, *glycogenin* (small compared with the glycogen molecule but, like any protein molecule, large compared with a glucose molecule). This glycogenin is located at the center of the entire structure.

In Figure 6.4, glycogenin appears as a circle labeled G and the three kinds of chains are labeled A, B, or C as appropriate. Notice that all the chains are the same length (have the same numbers of glucose units) and each B-chain is the starting point for two other chains. Notice also that this branching system produces a series of layers, as illustrated by the gray circles in the background. In the complete β-particle, there can be as many as 12 layers, but this would be almost impossible to draw in an intelligible way (50,000 glucose units!), so only the five innermost layers are shown.

Understanding of this structure did not come instantaneously, and unfortunately a wrong structure became thoroughly entrenched in the textbooks before the correct one was known. In the early years there were numerous speculative structures that were consistent with the sparse knowledge of the facts. Only one of these, proposed by Kurt Meyer and Peter Bernfeld in 1940, survived the discovery in the 1950s that there were approximately equal numbers of A- and B-chains. It is the only one you will find illustrated in many standard biochemistry textbooks.

While writing the first edition of this book, I checked in 10 current and widely used university textbooks of biochemistry for science students. Not a single one of them showed the correct structure. Seven of them plainly showed the obsolete Meyer–Bernfeld structure illustrated in Figure 6.5; of these, five mentioned glycogenin, albeit in another context, but the other two ignored it. Of the others, two did not show sufficient information for me to know whether their text referred to the real structure or not, and the last drew a complete structure in such a vague way that I could not be certain what structure it was supposed to be, but most likely the Meyer–Bernfeld structure was intended there as well. The situation is slowly improving, however.

WILLIAM J. WHELAN (1925–) grew up in Stockport (near Manchester), and started his career at Birmingham, but moved to Miami in 1967. He is noteworthy not only for establishing the structure of glycogen discussed in this chapter, but perhaps even more for his energy in setting up three international organizations devoted to interaction between biochemists.

This correct structure (initially without glycogenin, which was discovered later) was established by William Whelan and his group in 1970,[7] long enough ago for the information to have filtered through to the authors of textbooks. There is, incidentally, no controversy about this: The structure determined by Whelan and colleagues is accepted by everyone who works on glycogen. The two essential points for analyzing whether the structure is optimal or not are that all of the chains are of the same length and all of the B-chains, apart from those at the outermost layer, have two branch points, neither more nor fewer. Instead of the rather regular arrangement shown in Figure 6.4, the incorrect structure has a more haphazard appearance, with considerable variation in chain lengths and with more or less than two branch points in each B-chain. The wrong structure is, incidentally, more difficult to draw than the correct one and more difficult for students to remember, so it is doubly hard to understand why textbooks persist in showing it in preference to the real structure.

Optimizing the structure

We can now analyze whether the function of glycogen could be improved by modifying its structure, as described by Meléndez and collaborators.[8] There are three characteristics that we would like to maximize: The number of chains on the surface available for attack by the degrading enzymes, the num-

[7]Z. Gunja-Smith, J. J. Marshall, C. Mercier, E. E. Smith, and W. J. Whelan (1970) "Revision of the Meyer–Bernfeld model of glycogen and amylopectin" *FEBS Letters* **12**, 101–104.

[8]R. Meléndez, E. Meléndez-Hevia, F. Mas, J. Mach, and M. Cascante (1998) "Physical constraints in the synthesis of glycogen that influence its structural homogeneity: A two-dimensional approach" *Biophysical Journal* **75**, 106–114.

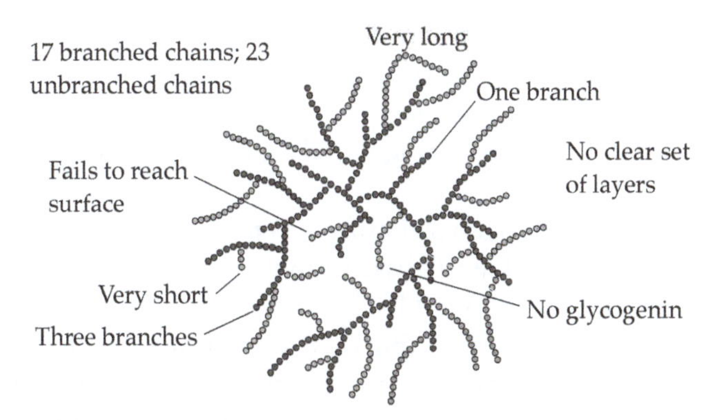

Figure 6.5 ***Incorrect*** **structure of glycogen.** Most textbooks illustrate the structure of glycogen with some approximation to the obsolete Meyer–Bernfeld structure, though not all textbooks include all the faults illustrated here. Several points should be noted. The number of branch points in each chain should be two (B-chains) or zero (A-chains), but some chains here have one or three, so there is no clear distinction between A- and B-chains. The unbranched chains should reach the surface. There should be a reasonably clear set of layers. The chain lengths should be almost uniform. The protein glycogenin should be at the center of the molecule.

ber of sites for attack, and the total number of glucose molecules stored in the complete structure when it reaches its maximum size. In addition, we would like to minimize the total volume of the assembly. This last is, like the need for rapid mobilization, especially important in muscle. It may not matter much if the liver is somewhat larger than it would be in a perfect world, but every unnecessary milliliter devoted to muscle represents a significant cost in terms of less efficient production of work. Taking account of all these characteristics, we can define a mathematical function that we would like to optimize, so that we can say that the best structure is achieved when the function is maximized.

Glycogen is not a protein, and so there is no gene as such for it. Instead, there are genes for the enzymes that make it, and the normal evolutionary mechanisms can act on these enzymes so as to vary the final structure of the glycogen that they make. They can vary, for example, the numbers of glucose units in each chain and the frequency with which new branch points appear.

So far as the degree of branching is concerned, the conclusions are clear. With fewer than two branch points in each B-chain, the molecule would not fill the space available as it became larger; with more than two, it would fill it up so quickly that it would run out of space for adding further layers at quite a small total size. A structure with two branch points per B-chain is the only one that allows efficient use of the space available. If this is fixed at two, there is then just one variable that can be adjusted by natural selection, and that is the length of each chain. This cannot be less than four glucose units per chain, because any shorter chain would not have room for two branch points, and at exactly four the structure is far from optimal, with a poor capacity for storing glucose and making glucose available at the surface. As the chain is made longer, the quality function initially increases steeply, reaching a rather flat maximum at a chain length of 13 and then declining slowly as the chain length becomes still longer.

It follows that if the glycogen structure has truly been optimized by natural selection, we should find it with two branch points in each B-chain and a chain length in a range of about 11–15. This is exactly what we do find in many organisms—mammals, fish, birds, invertebrates, bacteria, protozoans, and fungi. The only important exception is an interesting one: Oysters have a substantially sub-optimal glycogen structure with a chain length that varies wildly from two to about 30, averaging about seven. The oyster is an animal that has not changed morphologically since the Triassic era—that is, in 200 million years. It is often called a "living fossil," and although we cannot assume that a conservative morphology necessarily implies a conservative biochemistry, it does appear that the oyster has not kept up with progress in optimizing its glycogen. Given that it does not lead an exciting life, remaining fixed to a rock and rarely having occasion to make rapid or sudden movements, it is plausible to suppose that the selective pressure to improve its glycogen has been feeble compared with that felt by other species. Still, fungi do not lead exciting lives either, so we should not get carried away by this argument, and it may be that oysters have some as yet undiscovered reasons for possessing a deviant glycogen structure.

Enrique Meléndez-Hevia does not accept the view of some paleontologists that it was mainly luck that decided which species escaped the major extinctions like the one that caused the dinosaurs to disappear and to be replaced by mammals as the principal large animals on the Earth. He believes that when the full facts are known, we will find that the lineages that became extinct did so because of major inefficiencies in their biochemistry that made them unable to compete when it became necessary. He thinks that by studying the few species that escaped the extinction—crocodiles, for example, in the case of the dinosaurs—we may discover what defect the extinguished group had, for example, that they had a sub-optimal pentose phosphate pathway. In the case of glycogen, he argues that the oyster may be just such a remnant of an earlier group of animals that did not manage to evolve a glycogen molecule that was good enough for successful competition.

One feature of the glycogen structure (in animals like mammals, not in oysters) that I mentioned but did not develop is that the maximal structure has 12 layers. Why does it stop there? Is the number of layers optimal? Is there a mechanism to ensure that the molecule grows to that size and no more? In their celebrated and in many respects excellent textbook *Biochemistry*,[9] Jeremy Berg, John Tymoczko, and Lubert Stryer said that elongation stops when the enzymes responsible for elongation are no longer in contact with the glycogenin, and they described this as "a simple and elegant molecular device for setting the size of a biological structure." They did not explain how a small protein molecule buried deep inside a molecule more than 25 times as large could remain in contact with enzymes working on the surface, and without this we can only regard this explanation as a fantasy without any basis in reality.

The maximum size is determined by the branching number and the chain length. With a branching number of two, the space is filled more and more tightly as the molecule expands, so that the inner layers are rather loosely packed and the outer ones tightly packed. This inevitably means that the molecule cannot grow forever, and that the number of layers compatible with function is limited to about 12: Modeling in the computer shows that although a molecule with 13 layers could exist, it would have a smooth surface with around two-thirds of it occupied by glucose.[10] This might be good if the aim were simply to build the biggest possible molecule, but in the living organism, enzyme molecules, which are much larger than glucose units, need to have space to work on the surface, either adding more glucose while it is growing, or taking them off when it is being degraded. A hypothetical 13-layer glycogen would not leave enough space for enzyme molecules to work and so the molecule cannot grow as large as that.

[9]J. M. Berg, J. L. Tymoczko, and L. Stryer (1981) *Biochemistry*, 5th edition, W. H. Freeman, New York. This explanation of the function of glycogenin has been dropped from more recent editions of the book, and the structure of glycogen corrected.

[10]The 12 layers should not be confused with the chain lengths of 11–15 mentioned earlier, because these refer to two different quantities.

7. *Fear of Phantoms*

> Jim was a little more brash, stating that no good model ever accounted for all the facts, since some data was bound to be misleading if not plain wrong. A theory that did fit all the facts would have been "carpentered" to do this and would thus be open to suspicion.
>
> Francis Crick, *What Mad Pursuit*[1]

CHAPTER 3 ENDED WITH A PROMISE to devote a later chapter to an example of a Panglossian absurdity that illustrates how biochemists have sometimes believed things that were too good to be true, finding adaptations where no adaptation exists. The time has come to make good this promise, and I will begin by examining why biologists cannot apply the same criteria as physicists for deciding how much trust to place in their theories.

Physics and biochemistry

Quantum electrodynamics is the part of physics that deals with electromagnetism, which is due to interactions of electrons and protons with light, and underlies the whole of chemistry. The theory predicts that the magnetic moment of an electron should differ from the value given by a simpler theory by a factor known as *Dirac's number*. Richard Feynman told us in his book *QED*[2] that the experimental value for Dirac's number was 1.001,159,652,21, whereas theory gave a value of 1.001,159,652,46. So, the measured value of the magnetic moment agreed with the theory to better than one part in a billion. In the years since the book was written, the agreement has improved by several decimal places, but no matter: Feynman's value is good enough for what I want to say.

It is easy to find examples in the biochemical literature that present numbers more precisely than common sense would justify, such as giving a value as 12.38 when the observations would justify saying "about 12." To find truly gross examples, we need to look to advertisements, statements by politicians or journalists, or popular science books. For example, Bjørn Lomborg asserted in *The Skeptical Environmentalist*[3] that "overall forest cover in 1961 was 4.375,086e9 ha or 32.66 percent of the world's land area." Or, we could consider Abraham Lincoln's prediction in 1860 that the population of the United States would be

[1]F. H. C. Crick (1988) *What Mad Pursuit: a Personal View of Scientific Discovery*, Basic Books, New York.

[2]R. P. Feynman (1990) *QED: the Strange Theory of Light and Matter*, Penguin Books, London.

[3]B. Lomborg (2001) *The Skeptical Environmentalist*, Cambridge University Press, Cambridge, note 767.

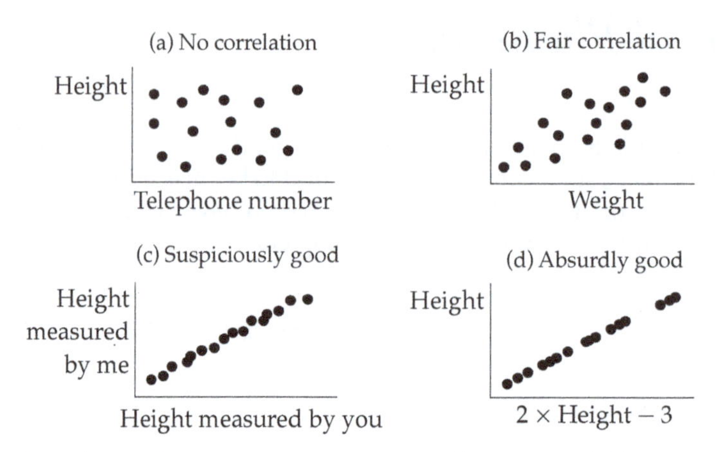

Figure 7.1 Correlation between measurements. When two different properties are measured for a set of entities, one can plot one of them against the other. (a) If there is no relation between them at all, the points should be scattered with no evident pattern. (b) If the correlation is typical of quantities measured in biology, a definite trend should be evident. (c) An excellent correlation is, in biology, typical of the agreement between two sets of measurements of the same thing. (d) A perfect correlation is most likely to arise from one set of measurements transformed mathematically to give the impression that two different things have been measured. This last is the sort of correlation we will try to understand in this chapter.

251,669,914 in 1930.[4] Probably more typical than either of these is an estimate from the Department of Agricultural Economics of Oklahoma State University that 80.44% of Americans would support a mandatory label on food containing DNA.[5] Unless the survey polled 10 million or more people (which seems unlikely, and the technical note just says "at least 1000"), this degree of precision is absurd. The question asked is also absurd, but I will not pursue that point.

Too good to be true?

When physicists obtain an impressive agreement between experiment and theory, they are naturally delighted and take it to mean that their theory is on the right lines. Biologists, by contrast, should be deeply suspicious: All our experience tells us that few biological numbers can be measured precisely, and, even if they can, values for different organisms never agree exactly because sampling variations and all sorts of complications lead to differences between reality and expectation that are never less than a few percent. For example, if theory teaches us that the sizes of the testicles of the males of different animals ought to be inversely correlated with the degree of monogamy of the species, we might test this by

[4]Gary Smith (2012) *Essential Regression, and Econometrics*, Academic Press, Waltham, Massachusetts. As it turned out, Lincoln's estimate exceeded the true value by more than two-fold (or perhaps I should say by 104.61%), so even an estimate of "about 250 million" would have been seriously in error.

[5]`http://agecon.okstate.edu/faculty/publications/4975.pdf`.

measuring the testicles of numerous animals and plotting them against some measure of monogamy.[6] If we find that the points on the resulting graph are scattered without any hint of any relationship, as in panel (a) of Figure 7.1, we should reject the hypothesis, whereas if they are scattered, but still lie within 20 or 30% of a trend line, as in panel (b), we should think the degree of agreement quite good.

If the agreement between numerous species is within 1 or 2%, as in panel (c), we should start to worry: This is looking too good to be true; surely we cannot have measured the values as accurately as this? Even if we have, can this be the whole story? Surely there must be something more than whatever we have chosen as our index of monogamy that contributes to testicle size? If the agreement is even better, so good that the eye can detect no difference between the plotted points and the line, as in panel (d), we should *know* that something is wrong; nothing in biology is as reproducible or predictable as that.

When the agreement between two biological measurements is better than it ought to be, one possible explanation is that although we think we have measured two different things, closer study will reveal that we have measured the same thing twice in different ways. Even this will not do if the agreement is perfect, because it does not allow for our inability to measure any biological quantity without error. In this case, all we are left with is the possibility that we have measured one thing in one way and then done some mathematical transformations to convert it into two different numbers that we fool ourselves into thinking of as two different biological variables.

Human nature being what it is, we are not always as skeptical as we ought to be, and even quite reputable scientists are sometimes seduced by results that are too good to be true. In this chapter I will use as an object lesson a particular example of results that were too good to be true.[7] This may seem to contradict the general thesis of the book, but it does not: We should not be interested in finding optimality wherever we look; we should be interested in finding genuine and demonstrable cases of optimality. So, although I believe that some examples of biochemical optimality are established beyond any doubt, we must always approach any new example with an appropriately critical eye. To show that any given system is optimal, we must define appropriate criteria and we must admit the possibility that we could be wrong. A system is not optimal if its properties are inevitable: If any conceivable system would have the same properties and there is no possibility of changing them during evolution, then it cannot be an adaptation. The number of toes that you have

[6]This is not just hypothetical: Chimpanzees, for example, are highly promiscuous and have large testicles; gorillas are not and have small ones.

[7]In Chapter 9 we will see that even as great a scientist as Gregor Mendel has sometimes been accused of publishing data too good to be true.

is exactly equal to the number of feet that you have multiplied by the average number of toes on a foot. You cannot ascribe this remarkable concordance to natural selection, because it follows inevitably from the way the average is defined.

Entropy and enthalpy

An example is provided by the phantom phenomenon of *entropy–enthalpy compensation*, which attracted the attention and support of some of the most able enzymologists during the 1970s and undermined the scientific value of at least one important book on biochemical adaptation. It is not dead even today.

Entropy and enthalpy are not familiar concepts to most readers, and their meanings are by no means as obvious, even in general terms, as those of testicle size and monogamy. So, I need to begin by explaining them, both obscure in sound, and one obscure in meaning as well, but both of them necessary for the discussion. Enthalpy is the less difficult of the two. We will not go far wrong if we take it as synonymous with heat or energy. If we worked at constant volume (as chemists interested in reactions of gases often do), we could just call it the internal energy and leave it at that, but biochemical reactions are nearly always studied at constant pressure, not at constant volume. The small volume changes that accompany chemical reactions in liquids at constant pressure mean that when a reaction mixture expands, it compresses the rest of the Universe slightly, and when it contracts, it allows the rest of the Universe to expand slightly. These small changes imply that such reactions have to do work on the rest of the Universe or have work done on them by it. Consequently, the total energy involved in the process is not quite the same as the change in internal energy in the reaction mixture itself. To allow for the difference, it is convenient to define a separate quantity to measure the energy involved in a process at constant pressure, and this is the enthalpy. As I have said, it will not matter much if you just take it as an obscure word for energy.

Entropy is another matter. Not only is the word obscure, but its meaning is difficult as well. The nineteenth-century physicist Rudolf Clausius invented the word, deliberately making an obscure choice. He thought that if he chose a word sounding vaguely like "energy," based on Greek roots meaning "transformation," he would have a term that meant the same to everyone whatever their language. As the physicist Leon Cooper drily remarked,[8] he succeeded in this way in finding a word that meant the same to everyone: nothing. This is unfortunate, because it is an essential concept, fundamental for understanding much of engineering and chemistry. For the purposes of this chapter, we can think of the entropy of a chemical process as a measure of the probability that it will occur when all energetic requirements have been

[8]L. N. Cooper (1968) *An Introduction to the Meaning and Structure of Physics*, Harper & Row, New York, page 331. He suggested that "lost heat" would have been a better term, but it is too late for that.

taken care of. In other words, the probability that any process will occur can be partitioned into energetic and non-energetic considerations. If we ask what is the probability of throwing a six with a die, the answer depends both on whether the die is biased, an energetic question, and on how many possible results there are, a non-energetic question, which has the answer six in this case.

The calculation of how far a chemical reaction will proceed, if left sufficient time, can thus be partitioned into an energetic component, measured by the enthalpy change, and a non-energetic component, measured by the entropy change. This is a problem of equilibrium, and the science that deals with it is *thermodynamics*. This is important for much of life chemistry, because much of life chemistry is just ordinary chemistry; nonetheless, it is of limited value for understanding what is special about life, because this is mainly a question of rates, not equilibria. All of the equilibrium positions are exactly the same in living systems as they are in nonliving systems, because a catalyst, whether an enzyme or anything else, has no effect whatever on the position of an equilibrium, it only affects how quickly it is reached. All living systems are far from equilibrium, and without catalysts they would reach equilibrium so slowly that for practical purposes they would not proceed at all. As a result, their states at all moments are determined almost entirely by rates, and hence by enzyme activity, and not by equilibria.

Despite this, thermodynamic ideas are useful for understanding rates because, with some additional assumptions that I will gloss over, effects on rates can still be expressed in terms of enthalpy and entropy. When a reaction occurs, the substances that react are supposed to pass through a high-energy state, the *transition state*. Because this is a high-energy state, the reactants can only reach it if they have sufficient energy, which means that they must be moving fast enough when they collide, and this in turn depends on the temperature: The hotter a system, the faster all of the molecules that compose it are moving and so the more energetically they collide. This can be measured as an *enthalpy of activation* and is the reason why nearly all chemical reactions proceed faster as the temperature is raised.[9] Even if two molecules have enough energy to react when they collide, they may nonetheless fail to react, because there are questions of orientation that need to be correct as well: These are measured by the *entropy of activation*.

The general idea may be illustrated by reference to a game like tennis in which a ball has to be hit over a net in such a direction that it will land (if not

[9]It is clear enough why this should be the general rule, but perhaps less clear why there should be exceptions. This becomes an important consideration for enzyme-catalyzed reactions, which nearly always become slower if the temperature is raised too much. This is because proteins are subject to *thermal denaturation*, or inactivation by heat. At high temperatures, the molecules that are not yet inactivated do indeed react faster, but there are fewer of them.

hit by an opponent first) within limits set by the rules of the game. One can imagine a small child with the potential to become a great player but lacking the physical strength needed to hit the ball over the net. Such a player cannot impart enough energy to the ball to satisfy the enthalpy of activation. For illustrating the entropy contribution, there is no need for imagination because one can find suitable players to illustrate the idea on any public tennis court: players like me, for example, who can easily impart enough energy to the ball, but who are inefficient for sending it in the right direction, so it has a low probability of landing in the intended place. Such players impart too much entropy to the ball, and here we say "too much" rather than "too little" because entropy increases with the degree of randomness: Entropy is a measure of disorder, not of order.

Estimation of the enthalpy of activation of a chemical reaction is not difficult, because we know that we are making more energy available as we heat a reaction mixture, and if we determine how much faster it gets as more energy becomes available, it is not particularly hard to convert this into a measure of the enthalpy of activation. Experiment shows that many reactions, whether uncatalyzed or catalyzed by enzymes or other catalysts, increase their rates by about a factor of two for each 10°C increase in temperature, for temperatures around the everyday temperatures at which we live. This could mean that many reactions of all kinds have similar enthalpies of activation, which would have important implications about the sort of events that are taking place at the level of the atoms involved. It could also just mean that the reactions that behave in this way are the ones that are easiest to study at convenient temperatures, and if so we may be talking about observer bias rather than anything more fundamental about the world.

It is less obvious how we might measure the entropy of activation. As this concerns how fast the reaction could go if all energetic requirements were taken care of, we can expect to get closer and closer to the pure entropy term by studying the rate at higher and higher temperatures, ultimately eliminating enthalpy considerations entirely at infinite temperature. In reality, we cannot produce temperatures of more than a few hundred degrees in an ordinary laboratory, let alone an infinite temperature, and anyway the reactants will boil or burn long before we reach even the modest temperatures we can manage. But if we treat it as a mathematical exercise, we can try to estimate what the rate at an infinite temperature will be if we continue the trend observed at ordinary temperatures. Given that ordinary temperatures fall infinitely far short of being infinitely hot, this may seem a calculation of dubious validity, and so it is for enzyme reactions.

It is not quite as bad as it sounds, because the detailed theory predicts not a straight-line dependence of rate on temperature, but a straight-line dependence of the logarithm of the rate on the reciprocal of the absolute temperature—

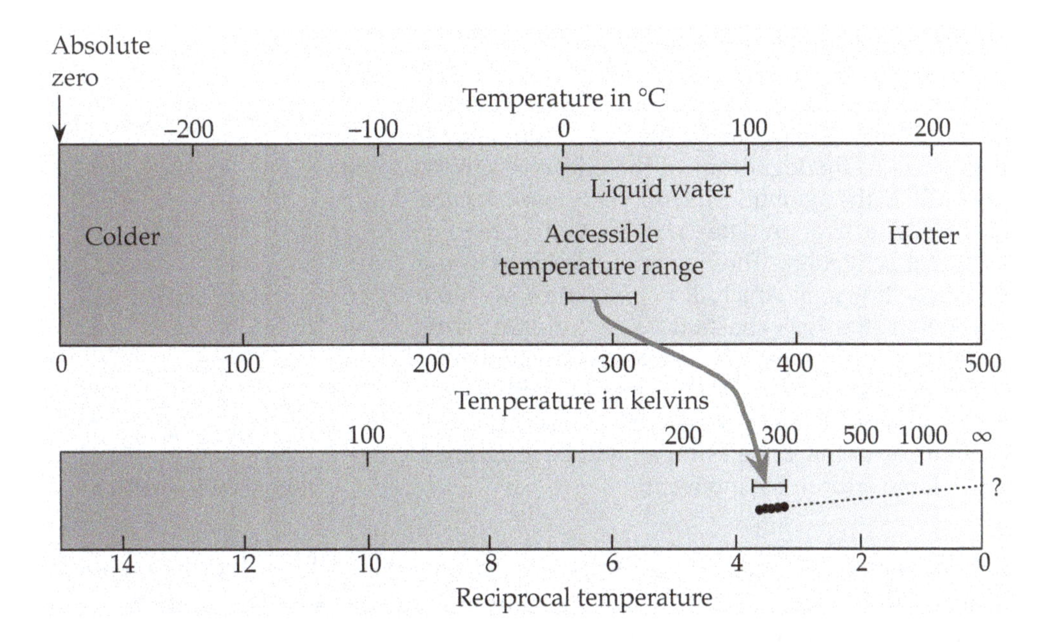

Figure 7.2 Extrapolating measurements at different temperatures to infinite temperature. The range of temperature over which water is liquid (at ordinary pressures) is 0–100°C, but the scale really starts at absolute zero, –273°C, because nothing can be colder than that. Physicists often prefer to use a scale measured in kelvins, each step of 1 kelvin being the same as a step of 1°C, with absolute zero equal to 0 kelvin. The range of liquid water is then 273–373 kelvins. Even within this limited range, the range in which most enzymes can easily be studied is quite small, typically about 5–40°C, or 278–313 kelvins. Estimating the entropy of a process means estimating what value the property would have at an infinite temperature, and to make it easier to do this, it is convenient to obtain a reciprocal temperature by dividing 1000 by the temperature in kelvins. Even on this scale, estimating any enzyme property at infinite temperature typically involves extrapolating an observed trend for more than 10 times its observed range.

that is, on the reciprocal of the temperature above absolute zero, which is at −273°C. Now, although ordinary temperatures are infinitely far from infinity, their reciprocals are not infinitely far from zero. In the units commonly used, everyday temperatures correspond to reciprocal temperatures of about 3.3,[10] so we might hope that measurements of a rate over a range of, say, 2–4 in the reciprocal temperature might allow us to estimate what the rate would be at a reciprocal temperature of zero. This is rather a wide range of temperature, from about 25°C below freezing to about 230°C: not impossible for some chemical reactions, but far outside the range in which we could hope to measure an

[10]The reciprocal of the temperature in kelvins (K) is typically about 0.0033, but it is usual to multiply by 1000 (Figure 7.2).

Figure 7.3 Estimation of an entropy of activation. Note the extremely long extrapolation and the fact that left and right in the horizontal scale are interchanged with respect to Figure 7.2. The vertical axis refers to the logarithm of the enzyme activity, $\ln k$, $\Delta S^{\ddagger}$ is the symbol conventionally used for the entropy of activation, and $\Delta H^{\ddagger}$ for that of the enthalpy of activation. The line is supposed to fit the equation shown, in which, $R = 8.31$ joules per mole per kelvin (the "gas constant") is a constant. This sort of experiment may give a reasonable estimate of the slope, and hence of $\Delta H^{\ddagger}$, combined with a very poor estimate of the intercept, or $\Delta S^{\ddagger}$, because any small error in the slope estimate will be magnified into a large error in the intercept.

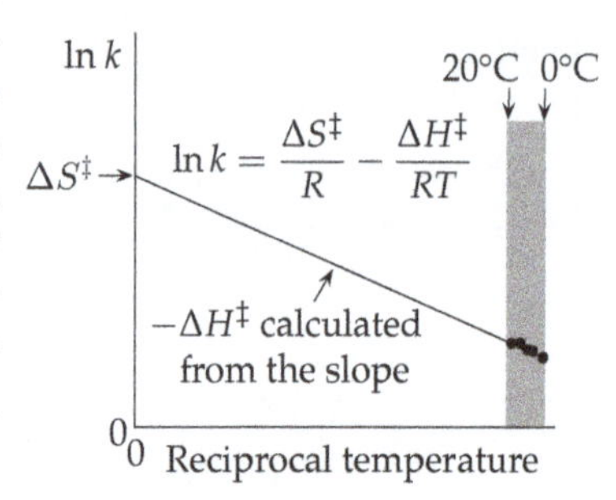

enzyme-catalyzed reaction; for such a reaction we should be lucky to manage 0–40°C, and a range of 0–20°C is sometimes necessary, because attempts to use wider ranges often results in complications (failure of the basic assumptions) that I will touch on briefly at the end of the chapter. A range of 0–20°C is a mere 3.41–3.66 in reciprocal temperature, implying an extrapolation of about 13 times the range of measurements.

These considerations are illustrated in Figure 7.2, which concentrates on the ranges of temperature involved in ordinary measurements and the relationship between temperature and reciprocal temperature. Figure 7.3 illustrates the same ideas, but with emphasis on the specific case of estimating the entropy of activation from observations in the range 0–20°C.

The Arrhenius equation

For this discussion we do not need the equation that is shown in Figure 7.3, which is known as the *Arrhenius equation*, and you will not miss anything important if you skip this section, but I mention it anyway to illustrate a point that is usually overlooked when students are taught kinetics and thermodynamics. When it is written as in the figure—that is, as follows:

$$\ln k = \frac{\Delta S^{\ddagger}}{R} - \frac{\Delta H^{\ddagger}}{RT}$$

it is clear that the enthalpy is in the temperature-dependent term, whereas the entropy is in the temperature-independent term, as we should expect from the way we introduced these properties. However, it is more usual to write the equation as

$$RT \ln k = T\Delta S^{\ddagger} - \Delta H^{\ddagger}$$

which is equivalent. It has the advantage of avoiding fractions, but it puts the entropy in the temperature-dependent term, something I found confusing

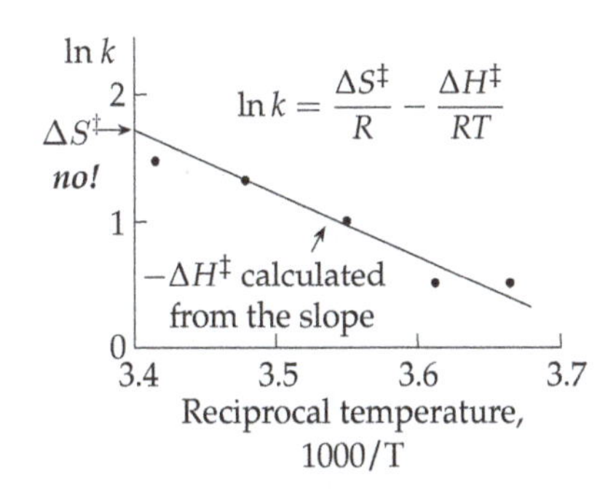

Figure 7.4 Estimating an entropy of activation in practice. Published plots usually resemble this rather than Figure 7.3: The horizontal scale does not extend to zero, and the long extrapolation is not made obvious. Worse, the intercept mislabeled $\Delta S^\ddagger$ is not the entropy of activation.

when I was a student. It should not confuse anyone who understands the topic, but I suspect that others will have had the same difficulties that I had.

Extrapolation

It follows, then, that trying to estimate the rate at infinite temperature from observations in the range 0–20°C is a task for an optimist, and a realistic person will not expect to get more than a rough idea, at best, of the rate at infinite temperature, and hence of the entropy of activation, from such observations. Even a long extrapolation is not always a bad thing to do, because sometimes there is no choice. If you shoot at a target 150 meters away with a rifle of barrel length of around three-quarters of a meter, you are in effect making an extrapolation of 200 times the range of "measurements": The first three-quarters of a meter is fixed by the position of the barrel, the rest is an extrapolation based on a hope that the rifle was pointing in the intended direction when fired, that it is well made enough for the bullet to continue in exactly the same direction when it emerges, and for any effects of wind or gravity to be properly taken into account.

If you are an excellent shot, you may well be able to hit a target in these conditions, but most people are not. Why are they likely to fail in such an experiment? First of all, the rifle is likely to be pointing in a slightly different direction from what is intended, and at a distance of 150 meters, any small error is magnified 200-fold. Second, a poor shot is likely to move the rifle when pressing the trigger. Third, if the rifle is of poor quality or is not intended for use at such a distance, the direction in which the bullet travels may not be precisely determined by the angle at which it is held when it is fired. With most scientific experiments, we must think of a low-quality rifle fired by a nervous person with poor vision out of a vehicle moving along a bumpy road on a day with a strong gusty wind. This is not because most experiments are badly or carelessly carried out, but simply because it is usually impossible to control all the possible sources of error to the point where you can expect to hit a target more than three or four barrel lengths away.

Fish ATPases

Nonetheless, let us look at what happens if you brush aside all the difficulties and estimate the enzyme activity at infinite temperature anyway. Suppose you extract an enzyme from

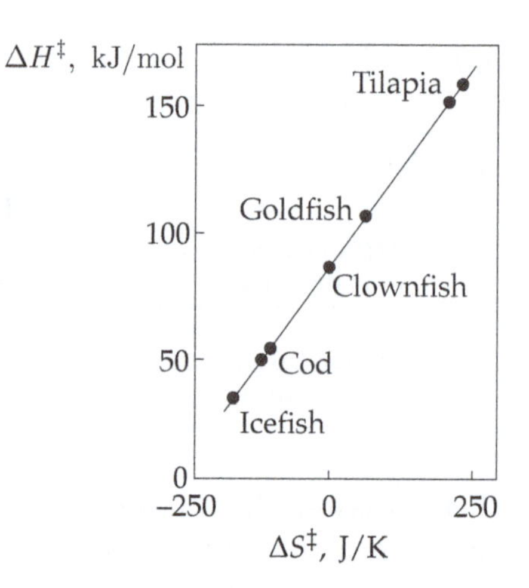

Figure 7.5 Fish ATPases. The figure shows a plot of enthalpies ($\Delta H^{\ddagger}$) against entropies ($\Delta S^{\ddagger}$) of activation of the ATPases from seven fishes. Two different species of Tilapia (*Tilapia grahami* and *Tilapia negra*) and two different species of cod (*Gadus morhua* and *Gadus virens*) are included. Note that the habitats of the fish span a wide range, from the Antarctic to tropical lakes in Africa. The plot is drawn from data of Ian Johnson and Geoffrey Goldspink.[11]

several fishes that live at different temperatures, from polar seas at around 0°C to tropical lakes at around 40°C, measure the temperature dependence of the activity of each enzyme over a range 0–20°C (completely outside the living range of some of the fishes), and estimate the enthalpy and entropy of activation of each. If you then draw a graph of the enthalpy of activation as a function of the entropy of activation for all the different fishes, you might expect to be lucky to see any relation at all. You need to remember three things: First, the estimation of the entropy of activation cannot possibly be better than extremely crude; second, the range of temperatures used for the measurements is far too cold for the tropical fishes to live and its upper range far too hot for the Antarctic fishes; third, fishes living in African lakes and Antarctic oceans are not likely to be close relatives, and so there are all sorts of reasons why their enzymes might differ beyond the fact that they live at different temperatures.

Despite all this, when the experiment is done, the correlation is not merely good, but unbelievably good, as illustrated in Figure 7.5. Experiments of the general kind I have described have been done by various people, but the particular example that I have in mind comes from a study of the enzyme that powers the muscles of fish (and ourselves, for that matter). Now, it is perfectly reasonable to suppose that fishes that live at different temperatures may have evolved enzymes that are adapted for the appropriate temperatures, and perfectly reasonable to try to identify the properties of the different enzymes that constitute the adaptations, but to do this we have to keep within the bounds of what can realistically be measured.

In this and other similar experiments, all of the points lie on a perfect straight line, so close to it that the eye cannot detect any deviations! Such a perfect result in biology has to be an artefact: There is no way in which it could result from any biological relationship, so it has to be generated by the

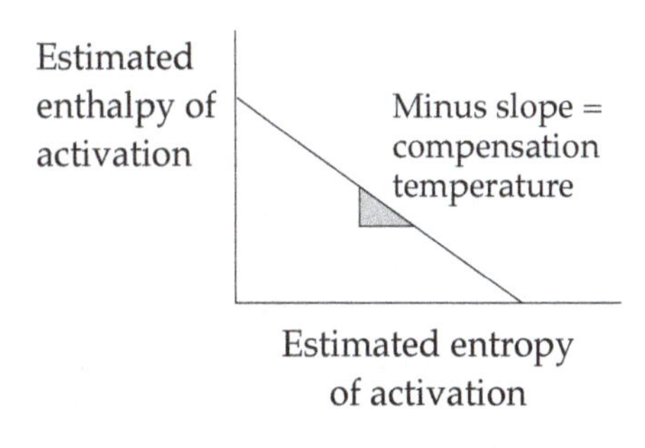

Figure 7.6 The compensation temperature. If values of the enthalpy and entropy of activation are estimated as indicated in Figure 7.3 and plotted against one another, the slope has the dimensions of a temperature, and minus this slope is the *compensation temperature* or *isokinetic temperature.*

mathematics. Figure 7.3 shows how this comes about. The experimental points in the range 0–20°C are too tightly clustered together to give much idea about the best line to be drawn through them. They give a good idea of a central point corresponding to about 10°C that the line has to pass through, but that is about it. We may get a rough idea of the slope to calculate the enthalpy of activation, but any small error in the slope will be translated into a large error in the intercept on the vertical axis; in other words, we cannot determine anything useful about the entropy of activation. In effect, we are trying to estimate two supposedly independent numbers from measurements of one, the activity at 10°C. If the plot is drawn as in Figure 7.3, with the horizontal scale extending to zero, the problems are obvious; at least, they ought to be. In practice, plots showing how entropies of activation are determined are almost always drawn, even in papers in highly regarded journals, with scales that stop far above zero, and when this is done, the extent of the extrapolation is far from obvious. Figure 7.4 is far more representative of what happens in practice, and in the worst cases, people think that the intercept (wrongly) labeled $\Delta S^{\ddagger}$ is the real entropy of activation.

Entropy–enthalpy compensation

Unfortunately, many of the people who have done such experiments have not only failed to be amazed but have also invented the fantastical idea of *entropy–enthalpy compensation,* whereby evolution in its infinite wisdom has compensated for the changes in the enthalpy of activation that result from working at different temperatures by making changes in the entropy of activation that are just right to produce the same activity at a mystical temperature called the *compensation temperature*. This can be estimated as indicated in Figure 7.6, and, wonder of wonders, it always turns out to be within the range over which the measurements were made. Whole branches of the study of biochemical adaptation have been built on this sort of nonsense.

Before accepting that it is all an artefact of the mathematics, you will need some justification for the claim. Perhaps I can explain it in terms of an analogy involving more concrete things than entropies and infinite temperatures. Liv-

[11] I. A. Johnson and G. Goldspink (1975) "Thermodynamic activation parameters of fish myofibrillar ATPase enzyme and evolutionary adaptations to temperature" *Nature* **257**, 620–622.

ing as I do in a region susceptible to forest fires, I have sometimes in summer time had the depressing (but striking) experience of watching the fire-fighting airplanes, or "Canadairs," flying overhead on their way to deal with fires in the hills a kilometer or two from where I live (Figure 7.7). Let us suppose that I cannot see where the fire is, for example because it is hidden by a tree in the foreground, but I know that the ridge of the hills is two kilometers away, and that two Canadairs overhead are 150 meters apart. I am interested in knowing both the direction of flight (the "enthalpy") and the location of the fire (the "entropy"). I have no sophisticated measuring instruments and can estimate the positions of the Canadairs only roughly, but I have maps, graph paper, and mathematical knowledge. So, assuming that they are flying one behind another, I can estimate their direction; continuing the line connecting their two positions at the moment of observation, I can estimate where the fire is. Doing this on several occasions during different fires, I notice a curious fact: If I make a graph of the direction of flight as a function of the point along the range of hills where I estimate the fire to be, I find a perfect relationship between them, far better than I could reasonably expect in view of the crudeness of the original measurements and the fact that the Canadairs have to fly more than 13 times further than the distance over which I have observed them. Forgetting that both measurements come from exactly the same observations, transformed in two different ways, and that I would not have seen the Canadairs at all unless they flew over a point close to me, I postulate a compensation effect whereby the controller of the fire-fighting team organizes matters in such a way that the direction of flight and location of the fire are correlated to make all of the Canadairs fly over a special point, the "compensation point," which turns out to be close to the point at which I made my observations. This is a surprisingly accurate analogy to the evidence for entropy–enthalpy compensation.

What is being "optimized"?

You might suppose that this was all that needed to be said about entropy–enthalpy compensation, but there is more, because to make a usable theory about biochemical optimization we have to be able to see some biological value in the property that is supposedly being "optimized." What if the phenomenon in question were not just a mathematical artefact? What if there really were a compensation temperature at which the enzymes from fishes living at different temperatures from one another (and different from the compensation temperature) all had the same activity? What would be the advantage to a fish that spent its entire life in tropical waters at 40°C if its muscle enzymes, once removed from the killed fish, proved to have the same catalytic activity when cooled to 18°C as the corresponding warmed enzymes from a quite different fish that spent its life in the near-frozen waters of the Antarctic? The absurdity of this is illustrated in Figure 7.8. Why should such a correspondence have been selected by evolution? I can see no reason, and so even if the phenomenon

Figure 7.7 Analogy of entropy estimation. How accurately can one locate a fire by observing airplanes on their way to it? The airplane pictures are superimposed on a photograph taken from my balcony on July 22, 2009, and the picture as a whole is not to scale. Although one center of fire is clearly visible, the airplanes appear to be heading toward a different one hidden behind the tree.

were based on a much firmer basis of experimental fact than it is, I should be deeply suspicious of it until some plausible advantage to the fishes had been identified.

The story does not end even there. As I have mentioned, studying the temperature dependence of enzymes is often restricted to even narrower ranges of temperatures than the 0–40°C that we might suppose accessible for almost any group of enzymes. This is because the estimation of the enthalpies and entropies of activation requires that the changing temperature affects only the rate, but in practice this is often true only in a narrow range; for example, a phase change (melting or freezing) in the associated biological membranes may cause a large and abrupt change in behavior at some temperature. In the study of fish muscle enzymes that we considered earlier, such an abrupt change occurred at 18.5°C, and that is why all of the measurements were done below that temperature. This means that if entropy–enthalpy compensation really does represent an evolutionary optimization, then we must suppose that certain tropical fishes have "optimized" the behavior of their enzymes at a temperature that is not only well below the temperature at which they spend their entire lives, but on the other side of a discontinuity. It is as if the old-world monkeys had optimized an anatomical feature valuable for defence against piranhas—no use to them in Africa, but it might become useful if some of their descendants found themselves in Brazil.

I can hardly emphasize too many times that this is not how evolution

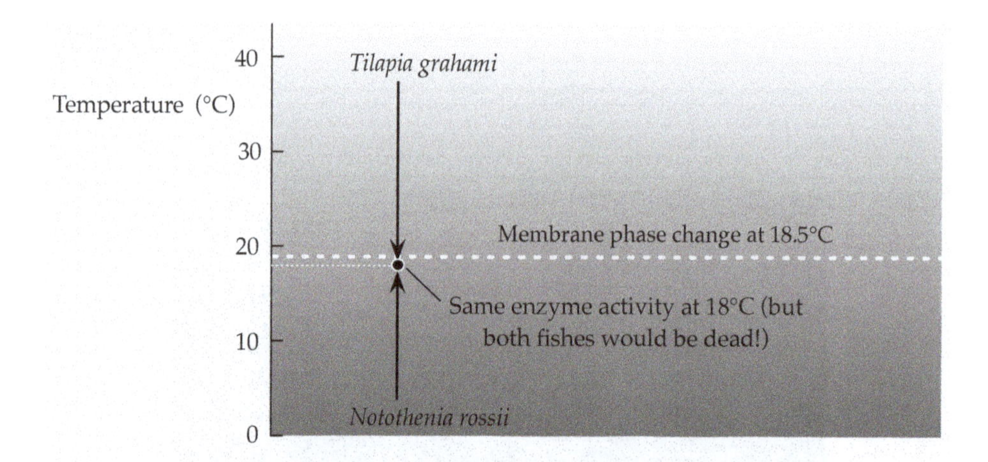

Figure 7.8 Fish ATPases. *Tilapia grahami* is a tropical fish that lives in water at about 40°C, whereas *Notothenia rossii* is an Antarctic fish that lives at nearly freezing temperatures. Both have an enzyme ATPase with the same activity at 18°C, a temperature at which neither fish can live, and on the wrong side of a phase change for one of them.

works: Evolution has no foresight; evolution prepares nothing for a hypothetical rainy day; evolution works only in the here and now. Remembering the slowness with which evolution proceeds, we should not take "here and now" as too short a period of time. Anything an individual encounters between birth and parenthood is within its present, and because the fixation of a gene takes far more than one generation (see Chapter 2), the "here and now" for a species lasts for many generations. That is why individuals retain the detoxification pathways (Chapter 10) they can use to cope with poisons that they may never meet. As long as a chemical threat occurs often enough during the long march toward gene fixation, the means to combat it will not get eliminated from the gene pool by natural selection. One characteristic of humans, not widely shared by other animals, is that they take care of their elderly individuals when they become unable to look after themselves. This is a characteristic that goes back at least 1.8 million years, as an almost toothless skull found in Dmanisi, in Georgia,[12] showed on examination to have belong to a man who had lost all but one of his teeth several years before he died, and could not have survived without care. What could possibly be the adaptive value of such behavior? One suggestion, which should be treated with as much caution as any other explanation of human behavior in terms of adaptive advantages, is that very old individuals in a preliterate society can advise their companions on how to survive a catastrophe, such as widespread famine, that occurs too rarely for a

[12]D. Lordkipanidze, A. Vekua, R. Ferring, G. P. Rightmire, J. Agusti, G. Kiladze, A. Mouskhelishvili, M. Nioradze, M. S. Ponce de León, M. Tappen, and C. P. E. Zollikofer (2005) "Anthropology: the earliest toothless hominin skull" *Nature* **434,** 717–718.

younger person to remember the previous occasion.

If an "adaptation" is only useful in some hypothetical future environment, it is not an adaptation and will not be selected. Every single change along the chain from primitive to modern must be at least as good as the currently available alternatives, or it will not be selected. This is the famous eye problem—what is the use of half an eye?—that delights creationists so much, because they think that evolutionary theory has no answer to it. (I will discuss the question of design without a designer, and creationism in general, in Chapter 15.)

To guard against similar errors, I have tried in the earlier chapters of this book to give examples of biochemical properties that are not only demonstrably optimized according to some defined criteria, but that also bring some definite biological benefit to their possessor. In discussing the pentose phosphate pathway (Chapter 5), for example, it was insufficient just to show that the results appeared to satisfy the rules of an arbitrary game; I also had to show that this allows the organism to grow faster and to waste less of its resources, and thereby compete more effectively with other organisms that had played the game less well.

To end this chapter on a more positive note, I should repeat that searching for adaptations that make particular enzymes best suited to their owners' environments is not by any means a pointless exercise. Many years ago, Ernest Baldwin wrote a book[13] that attracted the attention of many biochemists to the interest and importance of comparing the ways different organisms cope with their environments. Since that time, Peter Hochachka and George Somero have devoted much of their careers to this subject and have discovered numerous interesting examples. In their book[14] they claimed entropy–enthalpy compensation as one of them, but it is not.

[13]E. Baldwin (1964) *An Introduction to Comparative Biochemistry*, 4th edition, Cambridge University Press, Cambridge.

[14]P. W. Hochachka and G. N. Somero (1984) *Biochemical Adaptation*, Princeton University Press, Princeton.

8. *Living on a Knife Edge*

It has often been said that "pools must be maintained at their proper levels" or that there is a "normal" level, "excess" of which would either be uneconomic or would upset "the delicate equilibrium" so necessary to "integrate the different metabolic functions." Natural selection has been invoked as being responsible for this amazing feat of juggling. Those who are aware of the forces responsible for coming to a steady state realise, of course, that this is a fanciful delusion. Almost any set of enzymes will generate a steady state with all fluxes in operation. The existence of the vast array of genetic variation shows that there are very many different "delicate equilibria" which are just right. As Mark Twain observed, while marvelling at our amazing adaptation: "Our legs are just long enough to reach the ground."

Henrik Kacser and Jim Burns (1973)[1]

Again, hear what an excellent judge of pigs says: "The legs should be no longer than just to prevent the animal's belly from trailing on the ground. The leg is the least profitable portion of the hog, and we therefore require no more of it than is absolutely necessary for the support of the rest. Let any one compare the wild-boar with any improved breed, and he will see how effectually the legs have been shortened."

Charles Darwin, *The Variation of Animals and Plants*[2]

One of the distressing aspects of modern genomics and molecular biological studies is that they are almost entirely kinetic-free zones.... Only when we effectively study functional interactions in a kinetic way will we begin to understand what we are doing.

David Horrobin (2003)[3]

THIS CHAPTER DEALS with another false trail in biochemical optimization, but a different one from that of the previous chapter. Here we will discuss *metabolic regulation*, the general term given to all the effects on enzymes that prevent them from acting when they are not needed and ensure that they

[1]H. Kacser and J. A. Burns (1973) "The control of flux" *Symposia of the Society for Experimental Biology* **27**, 65–104. I have quoted the last sentence accurately, but I have not succeeded in identifying where Mark Twain wrote it. It seems more likely that it is based on Abraham Lincoln's answer to a question of how long a man's legs ought to be: "I should think they ought to be long enough to reach from your body to the ground." The source for this attribution is a book written many years after Lincoln died: F. T. Hill (1906) *Lincoln the Lawyer*, The Century Co., New York, page 106.

[2]C. Darwin (1868) *The Variation of Animals and Plants under Domestication*, John Murray, London.

[3]D. F. Horrobin (2003) "Modern biomedical research: an internally self-consistent universe with little contact with medical reality?" *Nature Reviews Drug Discovery* **2**, 151–154.

act at the rates that are appropriate for the circumstances. We will be concerned now with real observations, not fantasy, and the biological advantages will be clear enough; after all, it would be quite inconvenient if our legs were not long enough to reach the ground. However, these characteristics are not enough to save an example of optimization from being trivial. If a property, no matter how advantageous, is inevitable, so that we cannot easily imagine matters being otherwise, it is not an example of adaptation or optimization and requires no evolutionary explanation. This has an unfortunate echo of Dr. Pangloss, who thought, at the beginning of Chapter 3, that because the world was perfect, it could not be otherwise than that spectacles existed to fit over noses, and so on. I am not asking you to imagine a perfect world here, but an imperfect one in which not everything is necessarily for the best, but in which some consequences follow inevitably from the starting conditions. Because we live in a world where gravity is the most obvious everyday force that affects our lives, we cannot conceive of evolving legs that are too short to reach the ground; the fact that they do reach the ground is thus not an example of an adaptation.

The study of metabolic regulation has many useful things to tell us as long as we are careful to separate the fantasy from the reality. Apart from anything else, a better understanding of the way in which metabolic rates are regulated could have saved the vast expenditure wasted on hopeless biotechnological wild goose chases in recent decades. We will see an example of this in Chapter 10, with great efforts being made to increase the activity of the enzyme phosphofructokinase in organisms like yeast, in the belief that this would increase the rates of industrially useful processes. Such beliefs are not based on a real understanding of how metabolic systems behave and do not produce the benefits expected.

Enzyme regulation

Starting with the discovery of feedback inhibition by Umbarger[4] and by Yates and Pardee[5] in 1956, and during the 1960s in particular, biochemists learned a great deal about how enzymes are "regulated"; they found that many enzymes did not obey the simplest rate laws possible, but instead responded to various kinds of signals that told them whether particular products were appearing in sufficient amounts, or too much or too little. All of this reminded the researchers of the time of feedback control systems designed by engineers, and they adopted much of the same terminology for what came to be seen as the perfect design of metabolic systems at the hands of natural selection.

[4]H. E. Umbarger (1956) "Evidence for a negative-feedback mechanism in the biosynthesis of isoleucine" *Science* **123**, 848. Notice that an important discovery can be reported, together with its supporting evidence, in a paper of less than one page.

[5]R. A. Yates and A. B. Pardee (1956) "Control of pyrimidine biosynthesis in *Escherichia coli* by a feed-back mechanism" *Journal of Biological Chemistry* **221**, 757–770.

There was much that was good in this, and few would seriously doubt that natural selection is indeed at the heart of the properties discovered. There was also a tendency to be overimpressed and to see living organisms as being far more precariously balanced on a knife edge than they are. As Henrik Kacser and Jim Burns pointed out in the passage quoted at the beginning of the chapter, almost any set of enzymes, with almost any reasonable kinetic properties (including an inhibitory effect of each metabolite on the reaction that produces it), will lead to a steady state with all processes operating.

This is a state similar to what you can see in a river in calm conditions: Water is constantly being supplied from above, but it is flowing away at exactly the same rate, with the result that not only does the flow rate remain unchanged as long as you continue watching, but also the level of water at the point of observation remains unchanged as well. This example of a river underlines that a steady state does not imply complete inactivity and is perfectly consistent with continuous flow. In a place where drivers are highly disciplined, an aerial film of a freeway in good conditions will show a similar picture, with plenty of motion but no accumulation, cars moving away from any point just as fast as they arrive, with all moving at constant speeds. In a natural system, such as a river in calm conditions, the steady state is stable, in the sense that if a system is perturbed it will tend to return to the same state as before. Thus, we do not have to be especially impressed when we see that mixtures of real enzymes, whether in the cell or in artificial conditions, reach stable steady states without difficulty.

Normal organisms can support considerable variation in the activities of their enzymes away from "normal" levels. Individuals are far from uniform in their genetic composition, yet nearly all are "normal," in the sense that they can live in a healthy state long enough to leave offspring. The proportion of individuals that reach this level of maturity varies enormously between species: Among humans living in prosperous conditions, for example, most individuals reach childbearing age, and this is also true of some animals living in the wild. But among most fish, by contrast, the overwhelming majority are eaten by predators long before they reach maturity. In the latter kind of species, the few that escape predators may do so in a few cases because they are better at surviving, but it is far more likely that they are just lucky. In species where comparatively few individuals die before maturity, it is more likely that the ones that die are less healthy, or less good at running away, but even then luck plays a part. But despite the complicating factors, it is fair to say that the substantial genetic variations between individuals, reflecting substantial variations in the activities of particular enzymes, result in little or no variations in the capacity to survive and leave offspring.

50% just as good as 100%?

As a specific example, the human disease *phenylketonuria* is caused by the lack of an essential enzyme, phenylalanine 4-monooxygenase. I will return to this in the next chapter, and for the moment it is sufficient to say that people who lack the enzyme completely have the disease, but people who have only half of the normal amount of enzyme have no related health problems at all:[6] Half the normal amount of enzyme appears to be just as good as the full amount. How can this be? If they have half the normal amount of enzyme (as they do, in this and other similar cases), then the reaction the enzyme catalyzes should proceed half as fast, and this surely ought to have at least some effect? If not, does this not mean that normal individuals have at least twice as much of the enzyme as they need, and that the human species could evolve to become more efficient by decreasing how much they make of enzymes "in excess," thereby releasing precious resources for other purposes?

Phenylketonuria has been thoroughly studied in humans, because it is a serious disease that is easily diagnosed in babies at an age early enough to prevent its symptoms from appearing. It is far from being a unique example, and the level of almost any enzyme can be decreased by half without any observable effect, and this is the biochemical basis of the genetic phenomenon known as *dominance* that I will discuss in Chapter 9. Here I will not be concerned with the specific factor of two, but with the more general observation that the concentration of almost any enzyme can be increased by any factor with no observable effect, and can be substantially decreased, sometimes to as little as one-fifth of "normal" activity, with little or no observable effect. How can we explain this without concluding that organisms are hopelessly wasteful, making much more of most enzymes than they need?

Enzymes in isolation

The answer turns out to be one of mathematical necessity, and explaining it will require some care. There are two biochemical points to consider at the outset. In the first place, most metabolic systems spend a large amount of their time in steady states, and we can go some way in understanding how biochemical systems behave by restricting the discussion to steady states. In doing this, we should keep in mind that it is only a beginning, because many of the most interesting moments in the life of a cell involve transitions from one steady state to another. The second point is that most metabolic reactions proceed at rates that "try to" vary in proportion to the concentrations of the enzymes that catalyze them. By this I mean that if the enzyme concentration were the only thing that changed, all other concentrations remaining the same, then for most reactions the rate would be proportional to the enzyme concentration. In practice, various

[6]If they have children with a similar person, these children do have a one-in-four likelihood of having the disease, as explained in Figure 9.1.

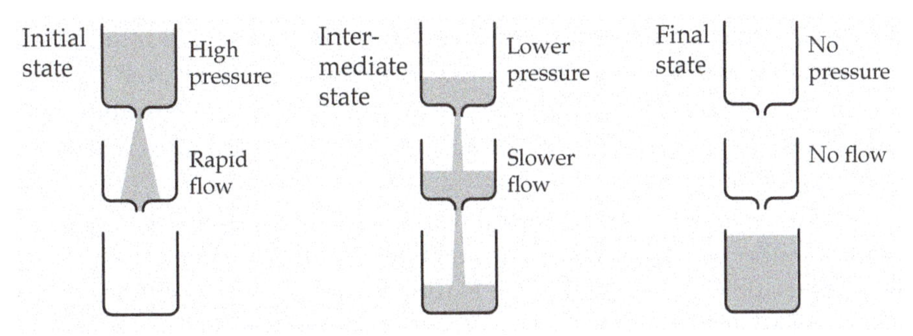

Figure 8.1 The steady state an illusion? When a tank empties and the water in it is not renewed, the level falls steadily, and hence the pressure driving the water out decreases and the flow rate likewise, eventually stopping when the tank is empty. If the water passes first to an intermediate tank and then to the final reservoir, the level in the intermediate tank first rises and then falls back to zero. There is thus no real steady state in these conditions. For most enzymes in metabolism, however, the substrates are replaced as fast as they are consumed, and so a steady state does arise, as illustrated later in Figure 8.3.

complications may prevent a strict proportionality from applying even then, but it is accurate enough for most enzymes to be useful as an approximation. This is easily tested in the laboratory, by studying each enzyme in conditions where no other enzymes are present, and where the concentrations of all the molecules that interact with it are kept constant by the experimenter, and it works fairly well most of the time. The complications that arise in the cell arise because the necessary conditions are not fulfilled: Other enzymes are normally always present, and changing the activity of any of them causes changes in the concentrations of its substrates and products, which affect the rates of other reactions, and so on.

Is the steady state an illusion?

The flow of water in a river corresponds to the chemical flow of metabolites through a series of reactions in a metabolic system; the level of water at a particular point corresponds to the concentration of a particular metabolite. In a metabolic steady state, therefore, the chemical flow through each metabolite is constant and the metabolite concentrations are also constant. Taking a river as an example has the complication that the source of the water, involving a large number of ill-defined tributaries, is rather vague. If we consider water flowing out of a tank instead (Figure 8.1), we can see that in the strictest sense a steady state is an illusion: Although observing it for a short time may give the impression that the flow rate is constant, it is obvious that as time progresses the amount of water left in the tank must decrease, and with it the pressure driving the water out; the flow rate cannot remain constant, therefore, but must decline.

If we study a chemical reaction on its own, without a mass of other reactions occurring at the same time, we find that it behaves much like water flowing out of a small tank, or out of a child's dam on a beach: The concentrations of any intermediates in the process (corresponding to water levels at different points along the route) initially rise fast, pass through a maximum and then fall back to zero again; during the same period, the flow at any point behaves in much the same way. So, although the idea of a chemical steady state arose originally in the context of ordinary reactions, it is unhelpful for analyzing them, because there is no steady state (other than the final state of equilibrium, when nothing is changing), except for an instant when any given intermediate passes through its maximum concentration.

Does this mean, then, that the steady state is just a meaningless irrelevance with no useful application to real systems? No, because in living systems we do find that the flow is virtually unchanging over time and all the way along a pathway. Even if the pathway is branched, the partitioning of the flow at every branch point can remain essentially constant and so the flow within any individual unbranched sequence is likewise constant.

If we think for a moment of a single enzyme-catalyzed reaction—not an entire metabolic sequence—we can obtain a steady state if the number of molecules of enzyme is small compared with the number of molecules of the substrate (the substance to be transformed in the reaction), as we saw in Chapter 1. Even if all of the individual steps within the process are fast, as they usually are in enzyme-catalyzed reactions, the need for every substrate molecule to encounter an enzyme molecule before it can react will ensure that the concentration of substrate decreases sufficiently slowly to be regarded as constant during the period the process is being observed.

This describes quite well the conditions in a typical steady-state study of an enzyme in the laboratory, because the experimenter can ensure that the enzyme concentration is small enough for a steady state to be produced. We have a problem if we want to generalize it to the cell, because it turns out that most enzyme concentrations in the cell are much higher than those used in steady-state laboratory experiments and are often comparable with or sometimes higher than the concentrations of their substrates. Even then we can have a valid steady state if the metabolic system is being supplied from a reservoir of starting materials sufficiently large to be essentially unaffected by the properties of the system we are studying. This is like studying the flow of water through a small stretch of a stream. As long as the water keeps on arriving at the head of the stretch at a constant rate, we do not have to worry about or even know anything about where it comes from. As long as the stream below the lowest point we are studying is capable of removing all the water that arrives, with no danger of becoming blocked and backing up, we do not need to know anything else about it. As long as water arrives at a constant

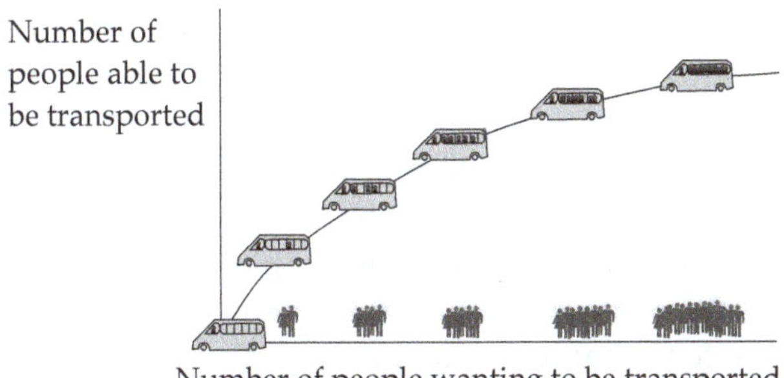

Figure 8.2 Saturation in a public transport system. When the number of passengers wanting transport is much smaller than the number of places available in the bus, the number transported is the same as the number wanting to be transported. As demand increases and the bus becomes more crowded, the number of passengers transported falls below the number wanting to be transported.

rate at the head and disappears at the same rate at the bottom, we can observe that the flow settles down to the same rate at all points in between, and the levels of water at all points in between settle down to levels exactly sufficient to sustain this rate. It is a matter of everyday observation that that is how streams behave: If there are no changes in the external conditions, such as the sudden arrival of flood water from above, the flows and levels remain unchanged for indefinite periods; in these conditions, we do not observe sudden emptying of intermediate pools with overflowing of others; we do not see water rushing rapidly through a particular stretch at one moment and becoming stagnant a moment later.

Metabolic systems can be analyzed in the same way, as long as we accept the existence of certain "external" properties that are fixed independently of the system under study. If we analyze the flow of metabolites between constant reservoirs of starting materials and constant sinks into which the final products flow, and we do not try to explain the constancy of these reservoirs and sinks in terms of the properties of the metabolic processes that connect them, then any metabolic system will settle into a steady state with constant concentrations and flow rates at all points.

Saturation

In writing the last sentence of the previous paragraph, I first wrote "almost any metabolic system," but then removed the "almost" as unnecessary. Why the moment of doubt? As anyone who has studied the elementary properties of enzymes will know, all enzyme-catalyzed reactions are subject to saturation, which means that they have limits that their rates cannot exceed. The same thing happens with transport systems, as illustrated in Figure 8.2. If the number of people wanting to catch the bus is much less than the number the bus can carry, then the number of people transported will be equal to the number wanting to be transported. But the more potential passengers there are, the more crowded the bus becomes, and the more difficult it is for additional people to get into the bus. To visualize this, you need to think of a place where the company regulations are not strictly enforced, so the number of passengers is determined by the degree of squeezing they can manage, not by the number of seats in the bus. I once watched in amazement as 12 additional people climbed into a seven-seater minibus in India that was already, in my eyes, full when I started watching.

It follows, therefore, that if an enzyme is being supplied with substrate at a rate that exceeds its saturation limit, then it is impossible for it to remove the substrate as fast as it is being supplied, so instead of settling into a steady state, the substrate concentration must rise indefinitely. This analysis is valid as far as it goes, but it overlooks two points: First, all enzyme-catalyzed reactions, like all other chemical reactions, are reversible, even if in some cases the forward direction is so strongly favored that we can never normally detect the reverse process; second, the rate at which any substrate is supplied is never fixed by some external decision, but is a consequence of the concentrations of its own starting materials and the properties of the enzymes or other processes that supply it. If these factors are taken into account, then absolutely any metabolic pathway will adjust into a steady state if it is supplied by an reservoir with constant levels of substrates and flows into a sink with constant levels of products.

As noted by Kacser and Burns in the quotation with which this chapter opened, therefore, the appearance of metabolic steady states is a mathematical necessity that does not require natural selection or any other special mechanism to explain it, any more than we need natural selection or the guiding hand of God to explain why storms do not spontaneously develop in small streams in constant conditions. Once we accept this, we can start to look at some of the properties of metabolic steady states that allow us to understand which characters of metabolism can be acted on by natural selection and which cannot. These properties form the core of what is now known as *metabolic control analysis* and were mostly derived by Kacser and Burns in Edinburgh in the early 1970s. Reinhart Heinrich and Tom Rapoport were doing parallel work at the same time in Berlin, and they arrived independently at some of the same conclusions.[7]

HENRIK KACSER (1918–1995) was born in Roumania of Austro-Hungarian parents, educated in Northern Ireland, and spent most of his career in Edinburgh. Trained as a chemist, he regarded himself as a geneticist, but his greatest influence was in biochemistry. He was one of the principal founders of modern systems biology and was the first to argue that the only way to understand whole systems is to study whole systems.

[7]R. Heinrich and T. A. Rapoport (1974) "A linear steady-state treatment of enzymatic chains" *European Journal of Biochemistry* **42**, 89–95.

Rate-limiting enzymes Ever since metabolic regulation started to be seriously studied in the 1950s, accounts in general textbooks of biochemistry have been dominated by a myth. This is the idea that within any metabolic pathway there lurks a *rate-limiting enzyme*,[8] in which all the regulatory properties are concentrated. Change the activity of that enzyme and you will change the properties of the pathway proportionally. For the student of metabolic regulation, this idea is appealing, because it suggests that to understand the regulation of a given pathway you can concentrate all the efforts on a single regulatory enzyme, ignoring all the others.

However, even the most extreme form of this idea is not impossible: We can envisage a pathway for which the rate is exactly proportional to the activity of one enzyme (at least over a limited range) and completely independent of the properties of all the others. The problem is not that it is impossible but that it is not at all likely, as we will see, and also that in practice biochemists have often suggested the existence of more than a single rate-limiting enzyme in each pathway. Depending on whom you talk to, you will find any of several enzymes described as "the" rate-limiting enzyme for glycolysis. Some textbooks offer different candidates on different pages without any apparent awareness of a contradiction. In reality, the flux control in most pathways is shared (unevenly) among all the enzymes, and we need now to examine why this should be so.

Increasing the activity of one enzyme by 5% What will happen to the flux through the pathway if the activity of some enzyme E_1 is increased? In real biochemical engineering projects, we are likely to be interested in large increases, say twofold or tenfold, but the analysis is much simpler if we consider very small changes, so here we will ask what will happen if the enzyme activity is increased by just 5%. If the idea of a rate-limiting enzyme is right, then the answer may be an increase of 5%, if E_1 is the rate-limiting enzyme, or nothing, if it is not. Let us not assume the answer at the outset, but say that a 5% increase in the activity of E_1 brings about a $5C_1^J$% increase in flux:[9] If there is really a rate-limiting enzyme, then C_1^J is either one or zero; if it is not right, then it could be anything.

If we now do the same thing with a second enzyme E_2, producing an increase of C_2^J% in flux, the combined effect will be approximately $5(C_1^J + C_2^J)$%.

[8]This is the term nearly always used, so I use it here, but a more accurate one would be *rate-determining enzyme*. Any enzyme has a *limiting rate* (often but inaccurately called its *maximum rate*), and in any unbranched sequence of metabolic reactions, one of the individual limiting rates will be the smallest: This puts a limit on the rate of the whole sequence, because it cannot exceed this. This would be a correct and reasonable meaning for *rate-limiting enzyme*, but this is not the meaning usually implied.

[9]The symbol C_1^J represents the *flux control coefficient* that expresses the sensitivity of a particular flux J to the activity of enzyme E_1.

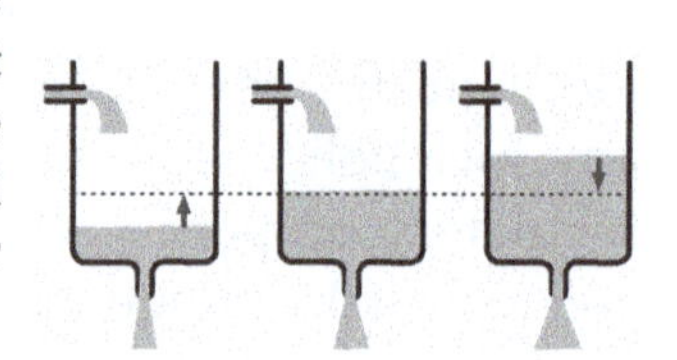

Figure 8.3 Steady state in a water tank. If water enters faster than it exits, the water level rises, increasing the pressure on the outlet, and hence increasing the rate at which it flows out. This continues until the two rates are the same, and the tank is in a *steady state.* If the initial water level is higher than the steady-state level, the water level falls until the same steady state is reached.

We need the qualification "approximately" because we cannot be certain that the effect of changing the activity of E_2 will be the same, regardless of whether that of E_1 has changed or not, and because strictly speaking percentages are not additive in this way: If your salary increases by 30%, and then increases again by 20% the combined effect will not be an overall increase of 50%, but one of 56%, because the 20% is not 20% of the original salary but 20% of the result of increasing it by 30%: $1.3 \times 1.2 = 1.56$. As long as we confine our discussion to small changes, the first objection is probably trivial and the second is certainly trivial: An increase of 5%, followed by a second increase of 5%, produces an overall increase of only slightly more than 10%, close enough to 10% for the difference not to be important.

Increasing all activities by 5%

If there are n enzymes altogether in the system, and if we increase the activity of each of them by 5%, then the overall effect on the flux will be, by an obvious extension of the same argument, $C_1^J + C_2^J + C_3^J + \ldots + C_n^J$%. To make this clear, suppose that we have a tank with a faucet and an outlet, and suppose that we start filling it with water at a constant rate of 10 liters per minute, with the outlet left open. Initially the water will drain out more slowly than it enters, because there will be no head of water to produce the pressure necessary to drive it out as fast as it comes in. This will cause the water level to rise, generating a pressure and causing the water to drain more quickly. If the tank does not overflow first,[10] the water will rise to a level exactly sufficient to produce a pressure on the outlet just enough to force the water out at 10 liters per minute. Let us suppose that this level is equal to 0.2 meter (the exact value is not important). The tank will then be in a steady state, in which the water is entering and exiting at exactly the same rate of 10 liters per minute, and the level inside the tank is exactly constant at 0.2 meter.

All this is easy enough to imagine, and most of us are familiar enough with water tanks to believe without checking experimentally that things will happen in the way I say, but if you are not convinced, it is easy to test with an ordinary basin (Figure 8.3). Leave the faucet running at a constant rate (fast, though not fast enough to flood your house by causing an overflow), and do not block

[10]If the outlet is wide enough to allow the water to flow out at 10 liters per minute or more when the tank is full, then it will not overflow.

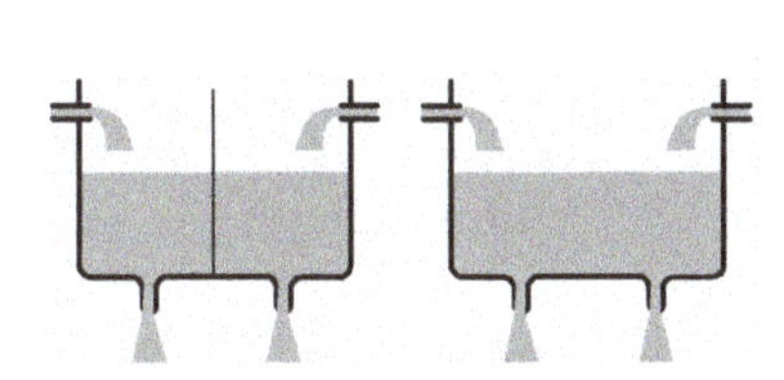

Figure 8.4 Identical steady states in identical tanks. If two identical tanks are studied with identical input flows and identical outlets, they will reach the same steady-state level. Nothing of this will change if the partition separating them is removed, so that they become one tank with two faucets and two outlets.

the exit. You will see that the water level will rise, fast at first and then more slowly, and will stop rising when the pressure is enough to force the water out through the outlet at the rate it is coming in from the faucet. If you repeat the experiment several times, you will find that a consistent flow from the faucet produces a consistent steady-state level of water in the basin, that higher rates correspond to higher levels, and lower rates to lower levels. Moreover, if you block the exit until the level in the basin is well above the steady-state level and then open it, the level will fall until it reaches the same steady state as it approached from the other direction.

Once you are convinced of this, consider what will happen if you do exactly the same experiment with another tank that is identical in every way to the first. It would be surprising if it did not behave in the same way, and that is indeed what we ought to expect (and what you will find if you do the experiment), all the way to giving exactly the same steady-state level of water for a given flow rate, 0.2 meter for 10 liters per minute, according to our hypothesis. Let us now imagine that the two tanks have a design with a vertical wall that allows them to be placed side by side and touching, as in Figure 8.4. Instead of two separate tanks we could have just one tank with a thin partition between two identical halves, each with its own faucet and its own outlet. As long as the partition is in place, neither tank "knows" that the other exists, and each will behave as if the other were not there. So, with just one faucet running, we will have 10 liters flowing per minute and a level of 0.2 meter; with both faucets running at 10 liters per minute, the combined rate will be 20 liters per minute and the level in each of the halves will be 0.2 meter.

What will now happen if we abruptly remove the partition? Why should anything happen? The levels are the same on the two sides, so there is no reason for any net flow of water from one side to the other; both outlets are feeling the same pressure as they did when the partition was in place, and so they will continue removing water at the same rate. The partition is therefore unnecessary. The consequence of all this is that doubling the capacity of the faucets to supply water, and simultaneously doubling the capacity of the outlets to remove it, has the effect of doubling the flow without changing the level. There is nothing special about a factor of two, and if you thought it worthwhile, you could do a more elaborate experiment with, say, 101 tanks and could compare the effects of having all 101 in use with that of having just

Table 8.1 Older terms, with the current equivalents

Symbol	Current term	Kacser and Burns	Heinrich and Rapoport
C_i^J	Flux control coefficient	Sensitivity	Control strength
$C_i^{s_j}$	Concentration control coefficient	Substrate sensitivity	(Not used)

100: In this case, you would find that increasing the capacity of the faucets and outlets by 1% will increase the flow by 1% and leave the level unchanged. The same result applies to any fractional change you care to test: Increasing the capacity of the system by any factor increases the flow by that factor and leaves the water level unchanged.

Shared flux control

The behavior of the corresponding metabolic system is just the same: If the activity of every enzyme in the entire system[11] is increased by 1%, then the flux through the pathway (or the flux through each branch of it if there are more than one) will increase by 1% and the concentrations of all the metabolites will be the same as they were before the enzyme activity was increased. So, it follows that the total we originally wrote as $C_1^J + C_2^J + C_3^J + \ldots + C_n^J\%$ can equally well be written as 1%, or, because we do not need the % signs, $C_1^J + C_2^J + C_3^J + \ldots + C_n^J = 1$. This result is called the *summation relationship* or, if we want to be more precise, because there are other summation relationships that define other metabolic variables like metabolite concentrations, the *flux summation relationship*. The individual C_i^J values were called *sensitivities* by Kacser and Burns when they first presented the relationship (and *control strengths* by Heinrich and Rapoport), but are now more usually called *control coefficients* or, more precisely, *flux control coefficients*. These terms, together with those for the concentration control coefficients that I will consider later on, are listed in Table 8.1.

As mentioned, there are other summation relationships that can be derived in a similar way. For example, in arriving at the result above, we deduced that a 1% increase in the activity of every enzyme would produce not only a 1% change in flux, but also a zero change in the concentration of any metabolite. It follows from this sort of consideration that if we define concentration control coefficients for any particular metabolite, they will sum to zero. These other relationships have in general received less attention than the flux summation relationship, which is now regarded as being fundamental to the understand-

[11]It is important to note that we refer to the entire system: If we just consider part of it, for example the enzymes in one pathway, the analysis will not be valid.

ing of metabolic control.

The simplest way of interpreting it is to suppose that there is a finite amount of control over the flux through a pathway, or through any branch of any pathway, and that this control is shared among all the enzymes of the system. There is, however, a complication that needs to be disposed of before taking this idea too far. We have tacitly assumed that flux control coefficients must be positive, but this is not necessarily true. It is already obvious that concentration control coefficients can be negative; otherwise, the only way they could add up to zero would be if they were all exactly zero, which is hardly believable (and demonstrably not true, if we analyze the behavior in more rigorous detail than is appropriate here). If we know that concentration control coefficients can be negative, why should we suppose differently for flux control coefficients?

Returning to the analogy of the small stream, it seems obvious that any improvement we might make to the flow at any point—removing a rock, digging a deeper channel, and so on—must improve the flow as a whole (though not necessarily by a perceptible amount). This is also true in the metabolic system: As long as we are dealing with an unbranched chain of reactions, any improvement at one step will be an improvement overall (though again, not necessarily by a perceptible amount). Once we admit branches to the pathway, matters become more complicated. Suppose that instead of being straight, our stream divides into two channels as it passes an island. It now becomes easy to believe (and is true) that improving the flow through one of the channels will tend to decrease it in the other. The same is true in a branched metabolic pathway: Increasing the activity of an enzyme in one branch will tend to increase the flux in that branch (the relevant flux control coefficient will be positive), but can decrease it in the other (the corresponding flux control coefficient now being negative).

This complicates the interpretation of the summation relationship, because the idea of sharing is less clear if some of the shares can be negative. We can easily imagine having several individual shares of one if there are enough negative shares elsewhere to give a total of one. Easy to imagine, perhaps, but not so easy to design such cases in the computer or to find examples of them in real studies of metabolism. In the computer, one can define conditions where some of the flux control coefficients have small negative values, but it is more difficult to create conditions that will produce large negative ones. We are forced to replace our nice clean mathematical statement that control must be shared because all the values must add up to exactly one by a less tidy statement: Although negative coefficients are possible, they are not particularly common, and when they occur, they are usually small, so the positive coefficients add up to a number that is not much bigger than one. In Chapter 10, we will see that the liver enzyme hexokinase D is an exception to this statement and, more important, we will be able to understand the reasons *why* it should be an exception.

Although messier than we might like, this statement still provides us with a useful basis for understanding metabolic control. It still allows us to say that control is approximately shared among all the enzymes in a system, that the average flux control coefficient will be about one divided by the number of enzymes in the system, and that unless there are special mechanisms that ensure that all the control is concentrated in one enzyme, there is no particular reason to expect the flux to be proportional to the activity of any given enzyme. The essential idea was well expressed by Italo Calvino in the words he ascribed to Marco Polo:

> Marco Polo describes a bridge, stone by stone. "But which is the stone that supports the arch?" Kublai Khan asks. "This bridge is not supported by one stone or another," Marco Polo answers, "but by the line of the arch that they form."[12]

For detailed analysis, whether theoretical or experimental, we usually take "the system" to consist of a short pathway of three or four enzymes connecting two reservoirs, but it does not have to be as simple as that. At least approximately, we can take the whole organism to be "the system," with one reservoir consisting of food and oxygen, the other consisting of waste products. Large animals like humans neither feed nor excrete continuously, but they do so often enough for there to be long periods when the bloodstream is continuously receiving the digested intake from the last meal, the lungs are continuously taking in oxygen and removing carbon dioxide, and the bladder and bowel are being continuously filled with other waste products. Because of variations in the level of activity—sleeping, eating, walking, running, and so on—the metabolic rate is not constant either, but neither does it vary so enormously that we cannot reasonably ask how its average value might vary in response to the activity of one enzyme. In this case, therefore, the system contains many thousands of different enzymes, and the average flux control coefficient is much less than 0.001, which is effectively zero when we take account of the difficulty of measuring it accurately. It follows that if we choose any enzyme at random and vary its activity by a few percent, then we will in all likelihood detect no change whatsoever in the metabolic rate of the organism.

Saving "excess" production of enzymes

It is only one step from here to start thinking that if changing the activity of an enzyme by a few percent has no detectable effect, then the organism has more of

[12]Marco Polo descrive a Kublai Kan un ponte, pietra per pietra: "Ma qual è la pietra che sostiene il ponte?" chiese Kublai Kan. "Il ponte non è sostenuto da questa o quella pietra," risponde Marco "ma dalla linea dell'arco che esse formano." *Le Città Invisibili*, Italo Calvino (1972) Mondadori, Milan. Translated by William Weaver as *Invisible Cities*, Harcourt Brace Jovanovich, San Diego.

that enzyme than it needs: The enzyme is "in excess." For many enzymes we might well find that we could decrease their activities by as much as one-half without seeing much effect. Surely this is wasteful. Surely the organism could become more "efficient" by making only half the amount of each such enzyme. But we need to think more deeply before reaching any such conclusion. To simplify matters, suppose we are dealing with an enzyme with a flux control coefficient of exactly 0.001 and normally produced in sufficient quantities to constitute 0.1% of the total protein in the organism. Let us now consider how much "wasted effort" we could "save" by synthesizing 1% less of it.[13] Doing this would save 0.001% (1% of 0.1%) of the protein-synthesizing investment of the organism, and it would lower its metabolic output by 0.001%. This latter figure is certainly negligible, because we should never be able to detect so slight a diminution of output. So, apparently we have something for nothing: a saving in investment for no detectable cost. Or have we? The problem is that the "saving" is just as negligible as the "cost," because it has exactly the same value of 0.001% of the total. The organism would no more notice a saving of 0.001% of its investment than it would notice a 0.001% loss of output.

As soon as we start to be more greedy, saving substantial amounts of the "unnecessary" enzyme activity, and decreasing the activities of numerous enzymes rather than just one, the savings become much more significant, but so, unfortunately, do the costs. On the simple proportional model we have been adopting, the costs would just increase in direct proportion to the savings: If we were greedy enough to try to manage with one-third less protein synthesis, for example, we should end up with one-third less output. Even with this oversimplified model, it is obvious that the "saving" is not working out as we hoped: We are paying less and receiving less in return. It would be worse than this, because flux control coefficients do not remain constant when conditions change: Almost invariably a flux control coefficient increases when the activity of the enzyme it refers to decreases. It is not difficult to see in general terms why this should be so: No matter how unimportant an enzyme may be, it will eventually become the weakest link in the chain if we decrease its activity enough; in other words, as its activity approaches zero, its flux control coefficient for the flux through the particular branch where it is located must approach unity.

More generally, it follows from this discussion that the title of this chapter is a nonsense, and that neither we nor any other living organisms are living on a knife edge where any departure from the delicate balance of enzyme activities that natural selection has achieved during many millions of years will be disastrous. On the contrary, most metabolic steady states are remarkably

[13]If you feel that 1% is too little to be worth bothering with, then do not worry: I will consider larger savings almost immediately, but I start by considering small values because they make the system essentially linear—that is, they allow simpler arithmetic.

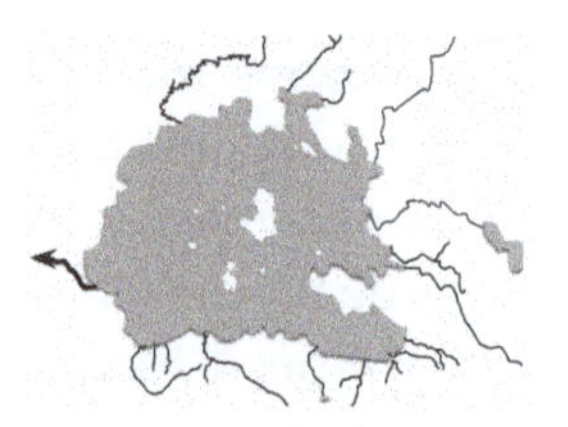

Figure 8.5 Unregulated flow. If there is no interference by humans or beavers, a typical lake (Lago Ranco, in la Región de los Lagos, Chile, in this illustration) has many inputs and exactly one exit. Lakes with a high rate of evaporation, such the Dead Sea, may have no normal exits at all.

robust and can tolerate quite large changes in the activities of many different enzymes. This has many consequences, some of which we will be examining in Chapters 9 and 10, but first we must recognize and explain one respect in which metabolic systems are quite different from flowing systems in the nonliving world, like rivers.

Unregulated flow in rivers

If you examine any lake in the world—I know of no exceptions—you will see that although there will normally be numerous points where water enters, there will be a maximum of one point at which water exits (Figure 8.5). There may be no exits at all (apart from evaporation), as with the Caspian or the Dead Sea, but usually there is exactly one. Lake Erie, for example, receives 80% of its water from the Detroit River; in addition, 11% comes from direct precipitation, and 9% from a large number of other rivers, including the Maumee River, the Cuyahoga River,[14] the Buffalo River, and Grand River. It is drained by just one, the Niagara River, and the flow is unidirectional until the Saint Lawrence estuary is reached.

Controlled flows are different. If we look at how traffic flows in a road network, we find that it divides as often as it unites, and although there may exist major routes onto which the minor ones converge, the points of convergence are also points of divergence. The circulation of blood in the animal body is similar. Even if there are no two-way roads as such, there are separate routes to allow flow in the two directions, and the many points of convergence to the major arteries and veins are matched by a correspondingly large number of points of divergence.

Even though all metabolic reactions are driven by thermodynamic imbalances, just like any other chemical reactions, and in that sense like the gravitational forces that drive the flow of water in rivers, their arrangement nonetheless resembles that of a man-made network like a road system far more than it resembles the drainage network of a river valley. Even though there is a general direction to the entire process—roughly speaking we can say that food plus oxygen is irreversibly converted into carbon dioxide plus water—there are just as many points of divergence as there are points of convergence. This is not how flow systems behave if left to their own devices, and there is no mathematical analysis that leads us to expect it.

[14]"The river that caught on fire."

In contrast to the first half of this chapter, therefore, where we saw that some properties of metabolism mistakenly attributed to natural selection are just the inevitable consequence of mathematical necessity, here we have properties that certainly call for explanation, because they are metabolic properties that do not follow from mathematical necessity, but are much more similar to the way systems designed by engineers behave. This explanation can only come from natural selection, but it is not enough to say that, because we need to recognize what properties have been selected that allow such behavior.

Regulated flow in metabolism

Circulating water remains relevant, but now we must look not at river basins but at unnatural networks like irrigation systems. Although in advanced economies these may involve pumps and external power sources, in the simplest cases the driving force is exactly the same as it is for a river—namely gravity, so that the water must always flow downhill. Nonetheless, by arranging the channels so that they are almost level but separated from the reservoirs and sinks by gates, it is possible to make the water flow along whichever channel and in whichever direction is required at a particular moment. We will see shortly what correspond to gates in metabolic systems, but first we must pause to ask whether they rely solely on "gravity"—corresponding to thermodynamic forces in the chemical case—or whether they incorporate anything resembling pumps.

The answer to this is largely just a question of definition. At one level we can certainly say that every metabolic reaction, at all times and in all circumstances, proceeds in the direction ordained by thermodynamic considerations. True though this is, it requires us to think of reactions in terms of all their components: When we consider a reaction, we must always consider the complete reaction, explicitly taking account of all of its substrates and all of its products. Often this is not convenient: When we consider the conversion of glucose to glucose 6-phosphate, for example, we may prefer to ignore the other half of the reaction. In converting glucose to glucose 6-phosphate, the phosphate group has to come from somewhere, and the hydrogen atom that it replaces has to go somewhere.

These requirements can be satisfied by *coupling* the half-reaction that converts glucose to glucose 6-phosphate to some other half-reaction, as illustrated in Figure 8.6, but we are not restricted to a unique possibility for this second half-reaction. It could be coupled to the conversion of ATP to ADP as shown, as we discussed in Chapter 1, in which case it will proceed forwards if the concentrations of all four components are equal; it could be coupled to the conversion of the inorganic phosphate ion to water,[15] in which case it will proceed backwards; or it could be coupled to phosphorylation of a sugar

[15]Yes, I know this sounds absurd. I will explain what I mean in a moment.

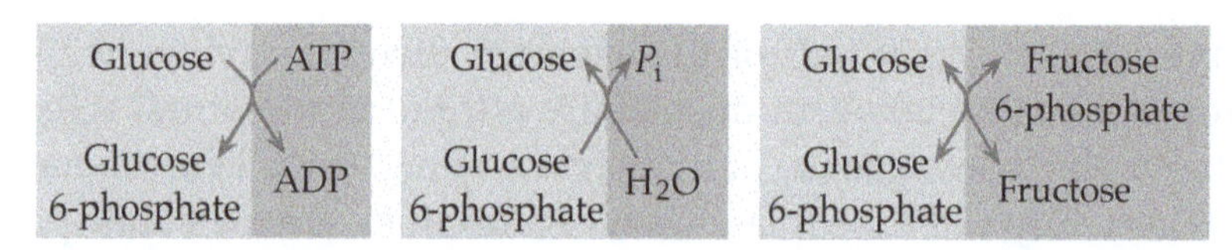

Figure 8.6 Half-reactions. The transformation glucose → glucose 6-phosphate is a *half-reaction*. It cannot proceed unless it is *coupled* to another half-reaction, such as ATP → ADP, in which case it proceeds forwards at ordinary concentrations, or P_i (inorganic phosphate) → H_2O, in which case it proceeds backwards, or fructose 6-phosphate → fructose, in which case it can readily proceed in either direction.

similar to glucose, such as fructose, in which case it will not proceed far from equality of the four concentrations.

The first two of these three choices are of great importance in metabolism. Coupling to ATP is catalyzed by hexokinase,

$$\text{Glucose} + \text{ATP} \rightarrow \text{glucose 6-phosphate} + \text{ADP}$$

and when this enzyme is active, the traffic through glycolysis, one of the principal highways of metabolism, proceeds almost irreversibly from glucose to smaller molecules. Coupling to inorganic phosphate is catalyzed by glucose 6-phosphatase, and when this enzyme is active, the traffic is in the opposite direction:

$$\text{Glucose 6-phosphate} + \text{H}_2\text{O} \rightarrow \text{glucose} + P_i$$

We can regard the half-reactions with ATP and with inorganic phosphate as pumps that decide which direction the half-reaction from glucose to glucose 6-phosphate will go, or we can say that we are talking about two distinct complete reactions, each of which proceeds in the direction ordained by thermodynamics. Regardless of how we express it, the direction of glucose phosphorylation is determined by which of the two enzymes is active: There is always enough water available to ensure that glucose 6-phosphate can be converted virtually completely to glucose and inorganic phosphate if glucose 6-phosphatase is active, and there is enough ATP available in a healthy cell to ensure that glucose will be largely converted to glucose 6-phosphate if hexokinase is active.

You will have noted an apparent absurdity a couple of paragraphs ago, when I talked about inorganic phosphate being converted to water. What can this possibly mean, given that water contains no phosphorus atom, and the phosphate ion (at least in the triply charged form commonly drawn in textbooks) contains no hydrogen? Surely I am not proposing that living systems can realize the ancient dream of the alchemists, the transmutation of one element into another? No, I am simply underlining that when we talk about a half-reaction, the very term implies the existence of another half-reaction that takes care of the necessary bookkeeping that the first one seems to ignore. The objection that the conversion of inorganic phosphate into water is impossible applies with equal force to any other half-reaction: It is no less impossible to convert glucose into glucose 6-phosphate, for the same reason that glucose 6-phosphate contains phosphorus and glucose does not. If we are allowed to talk about converting glucose to glucose 6-phosphate (something that biochemists do all the time), then there is no obvious reason why we should not be allowed to talk about converting inorganic phosphate to water. If doing so sounds rather strange, then so much the better, because it alerts us to the dangers in talking about any other half-reactions as if they were complete reactions. The point is that it *should* sound strange, and so should talking about any half-reaction as if it were a complete reaction.

Using gates to control the flow of water in irrigation systems is in principle quite simple, and has proved to be reasonably easy in practice as well, because many societies throughout history have succeeded in designing such systems and managing them effectively. The corresponding metabolic problem is much more formidable. Physical barriers such as cell walls and membranes are used in metabolism to prevent the free transfer of metabolites to places where they are not wanted, but the metabolic network is so vast and complicated (remember Figure 4.3) that there are many circumstances where the chemical flow needs to be controlled without the use of physical barriers, and this can only be done by varying the catalytic activities of enzymes. Unlike a physical door that can be opened to its fullest extent or completely closed with little expenditure of energy, varying the activity of an enzyme over a wide range is difficult.

Switching enzyme activity on and off

The most effective way of decreasing an enzyme activity to zero is to break the enzyme down to its amino acid components and resynthesizing it when it is needed, but this is too slow for many purposes—when a rabbit sees the fox approaching, it needs to start converting its chemical stores[16] into running energy immediately. There is no question of synthesizing new enzyme molecules to achieve a different metabolic state; it has to make do with changes that can be brought about instantaneously, and this means using inhibitors and activators. Unfortunately (from the point of view of regulatory design), the small molecules that inhibit or activate enzymes do so by binding to them according to the ordinary laws of chemistry, which put severe limits on the amount of change that can be brought about by reasonable changes in the concentration of inhibitor or activator.

Roughly speaking, unless the enzyme has special properties, a change

[16] In the short term, before there is time to mobilize fat stores, these consist most notably of glycogen, as discussed in Chapter 6.

from 90% to 10% of its full activity requires about an 80-fold change in the concentration of a small molecule, whether it is its substrate, an inhibitor, or an activator that modulates it. This is exactly the opposite of what any reasonable designer would want. When we use a light switch, for example, we expect a tiny input of mechanical energy into the switch to take us all the way from completely off to completely on, or vice versa. Ideally, we should like a small change in the concentration of a signal molecule, say a change of a few percent, to switch a pathway on or off, but this is not at all what we get from the usual properties of enzymes and small molecules. An 80-fold change in concentration is a large change, and spanning the range from 10% to 90% activity is a modest interpretation of what we usually mean by *off* and *on*. Matters are made still worse by the considerations we discussed in the first half of this chapter: Very few enzymes have anything approaching complete control of the flux through the pathway where they find themselves, and so even if an inhibitor succeeds in bringing a particular enzyme from 90% to 10% of full activity, the net effect on the flux through the pathway will be less than that.

Responding to signals

This is rather depressing, but there are a number of partial ways around the problem that are available to living organisms. First of all, metabolic signals can act at several different sites: There is no requirement that an inhibitor must inhibit just one enzyme and do nothing else; in reality it can inhibit certain enzymes and activate others, and provided the sites of action are carefully selected, all of the effects can point in the same direction. (The use of the word "selected" here is a deliberate evocation of natural selection: It is quite difficult in this field to avoid anthropomorphic or teleological language that implies the existence of a conscious designer, and the circumlocutions needed to avoid such language are not worthwhile as long as everyone realizes that the long-term effect of natural selection is design without a designer.[17]) Second, more than one small molecule can act on the same enzyme, and even though the effects are not strictly additive, they are certainly to some degree cumulative: It is more effective to inhibit an enzyme and decrease the concentration of its activator simultaneously than it is to do just one of them. Third, the 80-fold change in concentration for a 10% to 90% change in activity that I referred to assumed that the enzyme had "ordinary" properties, but this is not necessary. Enzymes can have structures that allow them to respond more sensitively to changes in concentration, and it appears reasonably easy to decrease the 80-fold range to a range of around fivefold by using the property of *cooperativity* that I will discuss in Chapter 10. Reasonably easy, certainly, but not trivially easy, because most enzymes do not have this property. Presumably there is a price to be

[17]Not everyone accepts this argument, as we will see in Chapter 15.

paid for it: The cooperative enzyme may be less effective as a catalyst than its noncooperative analog, or it may require a more delicate and easily damaged structure. Moreover, extremely few enzymes take cooperativity much further than this—the fivefold range of concentration for a 10% to 90% range of activity—even though it would appear to be extremely useful for regulating fluxes if they did, and this suggests that it is not a particularly easy property to design. It may be just ignorance on my part, but I am not aware that anyone has succeeded in producing cooperativity in an artificial catalyst.

AMP as an enzyme regulator

The fourth way of getting around the difficulty may seem less general, but it is important because so many metabolic reactions involve ATP, and because the status of ATP as the cell currency means that quite small changes in ATP concentration can provide the cell with important clues to changes in metabolic circumstances. I mentioned in Chapter 1 that in addition to ATP and ADP, there was a third member of the adenine nucleotide family, AMP, with just one phosphate group, but apart from mentioning that it participated in fewer metabolic reactions than the other two, I was rather vague about its function. AMP is important in part *because* it participates in few reactions. Because of that, there are few reactions to be perturbed if there are large changes in its concentration. Moreover, its concentration is typically much smaller than those of ATP and ADP, and an equilibrium between the three is maintained in many tissues by the enzyme adenylate kinase (or "myokinase"):

$$\text{ATP} + \text{AMP} \rightleftarrows 2\text{ADP}; \quad [\text{ATP}] \cdot [\text{AMP}]/[\text{ADP}]^2 \approx 0.5$$

At first sight this is rather a pointless reaction: Why maintain ATP and ADP in equilibrium with a third molecule that is present in small amounts and not used for anything much? It is only true to say that it is not used for anything much if we take this to mean that it rarely acts as the substrate or product of a reaction, but there are other things we can use a metabolite for than just as the starting material for making something else. In particular, we can use it as a *signal*.

The equilibrium constant for the myokinase reaction is about 0.5, close enough to 1 for this discussion to be simple: At equilibrium, we can calculate the concentration of AMP roughly by multiplying the ADP concentration by itself and dividing by the ATP concentration. For example, at ATP and ADP concentrations of 10 and 1 millimoles per liter, respectively (chosen as simple numbers to calculate with, but not too grossly far from typical concentrations in the cell), we should estimate an AMP concentration of one-tenth of a millimole per liter. Suppose now that the cellular demand for ATP has become so great that its concentration has fallen to 9 millimoles per liter and that of ADP has risen to 2 millimoles per liter. The concentration of AMP at equilibrium is now four-ninths of a millimole per liter. To be more rigorous, we should have to take into account not only that the equilibrium constant is not exactly one, but also that the extra AMP has to come from a decrease in the total of ATP and ADP, but correcting for these points would not make enough difference for it to be worthwhile being more complicated here.

Notice what this calculation means: A 10% decrease in the ATP concentration has been amplified into a 4.4-fold change in the AMP concentration. It is just a coincidence that the first three letters of "amplified" are "a m p," because AMP was given this symbol long before its amplifying function was recognized (by Hans Krebs, known for many other contributions to biochemistry), but it is a fortunate coincidence that makes it easy to remember one of the main metabolic functions of AMP. Because it is not a substrate or product of many metabolic reactions, a fourfold change in its concentration is unlikely to produce any problems. Instead, it can be used as a signal to enzymes that would "like" to be able to respond to small changes in ATP concentration. Designing an enzyme to recognize a fourfold change in AMP concentration is much easier than designing one to respond to a 10% change in ATP concentration. So, this is what many enzymes do: Wherever logic tells us that it would be useful for regulation for an enzyme to respond to small changes in the concentration of ATP, we often find instead (or in addition) that it responds in the opposite direction to changes in the concentration of AMP; enzymes that "ought" to be activated by ATP are often inhibited by AMP; enzymes that "ought" to be inhibited by ATP are often activated by AMP.[18]

The conclusion from all this is that we do *not* live on a knife edge. We do not need natural selection to explain the observation that much of metabolism can be represented simply as a set of pools of major metabolites at approximately constant concentrations, with chemical flows between them that proceed at rates that over short time scales vary little or not at all. Systems of enzymes catalyzing diverse sets of reactions readily achieve steady states because that is almost an automatic property of such systems. Assigning kinetic properties haphazardly to all the enzymes in a system normally does not produce any exotic properties for the whole system: As Kacser and Burns remarked four decades ago, "Almost any set of enzymes will generate a steady state with all fluxes in operation," with intermediate pools at their "proper levels," and so on. So, desirable as these properties may be, we have no need to invoke natural selection to explain them.

[18]I noted in Chapter 1 that some reactions in plants are driven by the transfer of phosphate from inorganic diphosphate ("pyrophosphate") to inorganic phosphate rather than from ATP to ADP. In this system there is nothing that corresponds to AMP, and no way of responding to signals. The need for signaling may perhaps explain the selective pressure to evolve the more complicated system.

9. *Brown Eyes and Blue*

> I do not think you understand what I mean by the non-blending of certain varieties. It does not refer to fertility; an instance will explain; I crossed the Painted Lady & Purple sweet-peas, which are very differently coloured vars, & got, even out of the same pod, both varieties perfect but none intermediate. Something of this kind I sh$^{d.}$ think must occur at first with your butterflies & the 3 forms of Lythrum; tho' these cases are in appearance so wonderful, I do not know that they are really more so than every female in the world producing *distinct* male & female offspring.
>
> Charles Darwin, letter to A. R. Wallace, February 6, 1866

> On the view that genes act as catalysts and largely through bringing about the production of catalysts of a second order, it is easy to show that increase in the activity of a gene should soon lead to a condition in which even doubling of its immediate effect brings about little or no increase in the ultimate effects.
>
> Sewall Wright, 1929[1]

> Biochemically dominance must be determined by a frightfully complex, and perhaps equally delicate, series of reactions.
>
> R. A. Fisher, Letter to C. G. Darwin, July 16, 1930

YOU MAY BE FAMILIAR with Richard Dawkins' books *The Selfish Gene*[2] and *The Blind Watchmaker*[3], and you may have learned most of what you know about evolution, particularly the evolution of behavior, from them. *The Extended Phenotype*[4] is among the least well-known of his books, and that is a pity, because in many ways it is his finest achievement. Nonetheless, even the most lucid books contain some obscure passages, and you could well have been puzzled by a couple of pages of *The Extended Phenotype* that deal with the theory of modifier genes, which R. A. Fisher proposed in 1930 to explain why the phenotypes of some genes are *dominant*, whereas others are *recessive*, *phenotype* being a technical term for the specific set of observable characteristics that indicate the presence of a particular variant of a gene.

[1]S. Wright (1929) "Fisher's theory of dominance" *American Naturalist* **63**, 274–279.

[2]R. Dawkins (1989) *The Selfish Gene* (2nd edition), Oxford University Press, Oxford.

[3]R. Dawkins (1986) *The Blind Watchmaker*, Longman Scientific and Technical, London.

[4]R. Dawkins (1981) *The Extended Phenotype*, Oxford University Press, Oxford. Although this is directed more to professional biologists than Dawkins' other books, it is readily accessible to the general reader.

These pages are not nearly as obscure and difficult as Fisher's own account of his theory, however, which occupies a chapter of his book *The Genetical Theory of Natural Selection*.[5] Fred Hoyle, a theoretical astronomer that I will discuss in more detail in Chapter 12, recommended this book for its "brilliant obscurity," adding that "after two or three months of investigation it will be found possible to understand some of Fisher's sentences." Even William Hamilton, one of the greatest evolutionary biologists of modern times and a great admirer of Fisher's book, found it heavy going: "Most chapters took me weeks, some months," he wrote.

It may well be true, as Dawkins asserts, that by 1958 the modifier theory was so well accepted, along with Fisher's view that dominance must be an evolved property because it has selective advantage, that he felt no need to justify it when he wrote the second edition of *The Genetical Theory of Natural Selection*. Indeed, the theory was still well accepted in 1981, when *The Extended Phenotype* was written. By chance 1981 was the year in which the matter was clarified once and for all in a landmark paper by Henrik Kacser and Jim Burns, and, as we will see, their explanation differs completely from Fisher's, being instead much closer to the point of view of Sewall Wright, Fisher's great opponent over many years.

RONALD AYLMER FISHER (1890–1962) was a British geneticist and statistician who played a leading role in establishing the classical principles of statistics, which he applied especially to plant breeding. He helped to create *population genetics*, and played a major part in the *Modern Synthesis*, the merging of genetics and natural selection into a coherent modern theory of evolution.

SEWALL GREEN WRIGHT (1889–1988) was an American geneticist known for his influential work on evolutionary theory, and also for his work on *path analysis*, a method for deciding whether a set of data with many variables agrees with a causal model. He remained active until his death at the age of 99, and his last paper appeared in 1988. He wrote his first (unpublished!) book in 1897 on *The Wonders of Nature* at the age of seven.

JOHN BURDON SANDERSON HALDANE (1892–1964) was a British biologist who contributed to many areas of science, including statistics, biochemistry, and genetics, and also to politics, journalism, and popular science. In these last capacities, he wrote articles for the *Daily Worker*, a communist newspaper, and was chairman of its board of management for a while. In the last years of his life he moved to India and became an Indian citizen.

Why is there a problem at all, and why did it take well over a century from the time when Gregor Mendel first described dominance for it to be properly understood? Moreover, why were Fisher, Wright, and J. B. S. Haldane, some of the greatest minds to have influenced the course of genetics, unable to agree about it, and, in the cases of Fisher and Haldane, unable even to come close to the right answer to what seems in retrospect quite a straightforward question?

[5]R. A. Fisher (1930, 1958) *The Genetical Theory of Natural Selection*, a variorum edition edited by H. Bennett (1999), Oxford University Press, Oxford. This is primarily a book for specialists and is not light reading. However, it is not as difficult to read as the comments of Hoyle and Hamilton suggest.

Before trying to answer these questions, we should remind ourselves what it means for the phenotype of a gene to be dominant or recessive. Mendel studied seven different "characters"—green or yellow peas, wrinkly or smooth peas, tall plants or short, and so on, but one is enough to illustrate the idea. First, we must insist on starting with true-breeding strains, which means that if we cross two plants giving green peas, we can guarantee to get another plant giving green peas, whereas if we cross two plants giving yellow peas, we can guarantee to get another plant giving yellow peas. So far so good, but what do we expect if we cross a plant giving green peas with one giving yellow peas? Naively, perhaps, we may expect greenish-yellow peas, but that is not what Mendel found: He found that in the first generation of crosses, all the offspring gave green peas. Even more strikingly, if these first-generation offspring plants were crossed with one another, three-quarters of their offspring gave green peas, but one-quarter gave yellow. Further analysis would show that the descendants giving yellow peas were like their yellow-seed ancestors—true-breeding if crossed with other yellow-seed plants. Of the three-quarters of descendants giving green peas, one-third were true-breeding and the other two-thirds were like the first-generation crosses.

Particulate inheritance

What all this means is that crossing two individuals is not just a matter of mixing their characters in the same way as you might mix the contents of two cans of paint. Instead, Mendel's result shows clearly that inheritance is *particulate*, the particles corresponding to the entities that we now call genes. The simplest interpretation is that a pea plant has two copies of the gene that determines pea color: In the first-generation cross from true-breeding parents giving green and yellow peas, there is a gene for the green phenotype from the green parent and a gene for the yellow phenotype from the yellow parent, but the green phenotype is dominant over the yellow, so that whether a plant has two genes for green color or one its peas are green. In Mendel's time, and until the late 1920s, this was just an observation, and no one tried to explain why some phenotypes were dominant whereas others, the yellow phenotype in this case, were recessive.

Few things in biology are simple, and even when it seems possible to make some tidy generalizations, it does not usually take long for some tiresome exceptions to accumulate. So, before proceeding I should perhaps make it clear that not all genes behave as Mendel observed. First of all, not all organisms are *diploid*, which means that not all have their genes in pairs. Some, like bacteria, are *haploid* and have only one copy of each gene; for them, the concept of dominance has no meaning. Others, such as certain plants, including ones like wheat that are intensively cultivated, are *polyploid*, and have more than two copies; for them, dominance exists but its analysis is more complicated. A few insects, such as bees, are *haplodiploid*, which means that males are haploid and females are diploid: This has important consequences for their behavior, but these have nothing to do with the main themes of this book.

Some organisms, such as the alga *Chlamydomonas reinhardtii*, exist mainly as haploids but occasionally pass through a diploid state: For the moment, I will set this characteristic aside as an unwanted complication, but I will return to it later, because it is highly relevant to this chapter. Most of the organisms likely to interest us, including humans and nearly all other animals, and peas and nearly all other green plants, are diploid, and for them the principles of Mendelian inheritance apply to all simple traits.

GREGOR JOHANN MENDEL (1822–1884) was a monk in the Augustinian Abbey of St. Thomas in Brünn, Moravia (now Brno, Czechia), where he studied inheritance in pea plants until his scientific work largely ended when he became Abbot. The popular image of an unworldly monk isolated from everyday affairs is quite unrealistic. He received a thorough training in mathematics, physics, and biology, first at Palacký University, in Olomouc, and later at the University of Vienna, and he proved to have considerable skills in financial management when he became Abbot. The organ at his funeral was played by the composer Leoš Janáček, who had been a chorister in the monastery.

I hope you noticed the weasel-word "simple" that has slipped in here. What is a simple trait? Obviously, one to which Mendelian inheritance applies! Fortunately we do not have to tolerate such a sloppy circular argument, because the more understanding we gain about what genes code for, the more we can understand that even if one gene does affect just one protein (to a first approximation: I will not worry about exceptions here), more than one protein may be involved in producing an observable character, and, if so, then more than one gene must be involved as well.[6] It is worth spending a moment to reflect on this, because Mendelian inheritance applies well to many cases even if at first sight it may seem to contradict our everyday experience. When this happens, we may be tempted to reject a whole subject because it appears to lead to nonsense, though with more careful explanation it might make good sense.

Eye color

The classic example used in elementary textbooks to illustrate the application of Mendelian inheritance to humans is eye color: The phenotype of brown eyes, it is said, is dominant, whereas that of blue eyes is recessive. True enough, as long as we confine our attention to people whose eyes are bright blue or dark brown; but when I look into a mirror, the eyes that gaze back at me are neither blue nor brown, but the sort of vague greenish-brown color commonly called "hazel," and I cannot easily relate what the more oversimplified textbooks say to my own experience. Unfortunately, many nongeneticists, including some of the sillier leaders of political opinion, are more interested in skin color than eye color, and more interested in both than they are in breeding peas, and when it comes to skin color the effects of

[6]The other side of the coin is that gene effects are *pleiotropic*, a technical term meaning that one gene can affect more than one characteristic. This is easy to explain from Wright's point of view, but creates difficulties for Fisher's.

crosses are much more like mixing paint than elementary notions of Mendelian inheritance would lead you to expect.

Does this mean that Mendelian inheritance is a myth or something that applies well enough to peas but not to species that we care about? No, it means that most of the traits that we can readily observe without instruments, like eye or skin color, are not simple traits. In terms of skin color, it means that there is not just one pigment produced by just one enzyme coded for by just one gene. Moreover, skin color, like most traits, is also affected by environmental, nongenetic conditions, most obviously by exposure to sunlight. Even if we confine attention to purely genetic considerations, several different genes are involved in skin color that interact in ways that are not simple and not fully understood. Moreover, they are not passed on from parent to child in a block, but separately. All this means that skin color is not a good trait to consider if we are interested in finding a clear illustration of the principles of inheritance. Unfortunately, the same complications apply to some of the most important metabolic diseases in humans: Most forms of diabetes, for example, are not determined by single genes but result from the combined effects of several genes.[7] For this reason it is only quite recently that much progress has been made in understanding the genetics of such diseases.

Phenylketonuria and alkaptonuria

It all works much better with simpler cases that are less obvious in ordinary life but still have greater importance for medicine than skin or eye color. The diseases phenylketonuria and alkaptonuria are both concerned with the metabolism of the same amino acids, phenylalanine and tyrosine, and both can be detected by their effects on urine (hence the "-uria" in their names). Phenylketonuria, mentioned briefly in the previous chapter, is a serious disease, and if left untreated it produces severe mental retardation and in many cases death before the age of 25. It is caused by an incapacity to convert phenylalanine into tyrosine (Figure 9.1), but it is not a deficiency disease,[8] because its harmful effects are not caused by a shortage of tyrosine, and cannot be overcome by adding tyrosine to the diet. Instead, they are caused by the toxic effects on the brain of phenylpyruvate, which belongs to a general class of chemical substances known as phenylketones and is produced by the body in its efforts to remove the excess of phenylalanine. The presence of phenylpyruvate in the urine provides a simple method of diagnosis, and the disease is treated by

[7]An added complication is that "diabetes" is not just one disease, and the different forms have different underlying causes. The main division today is between insulin-dependent (type I) and non-insulin-dependent (type II) diabetes mellitus, which are quite different from one another, but the reality is more complicated than that.

[8]It is a *misfolding disease*: The inactive enzyme has all of the amino acids necessary to catalyze the reaction, but the mutation prevents it from being folded into the correct three-dimensional structure.

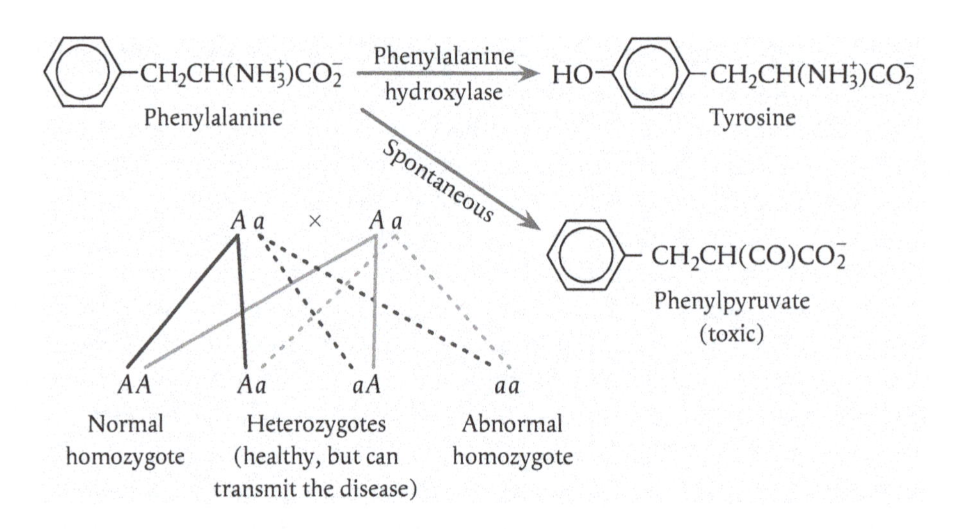

Figure 9.1 A genetic disease. *Phenylketonuria* is caused by a failure to produce the enzyme phenylalanine hydroxylase, which in a normal person is responsible for converting phenylalanine into a different amino acid, tyrosine. When the enzyme is absent, phenylalanine is converted into phenylpyruvate, which is toxic. A normal homozygote has two good copies *AA* of the relevant gene, and a homozygote with the disease has two bad copies *aa*. A heterozygote *Aa* is healthy but can pass on the bad copy to children and thus put them at risk of the disease. The child of two heterozygotes has one chance in four of being a healthy homozygote *AA*, two chances in four of being a heterozygote, and one chance in four of being an abnormal homozygote *aa* with phenylketonuria.

carefully controlling the diet so that it provides no more phenylalanine than is needed for normal health. There is then no surplus to be converted into phenylpyruvate.

Phenylketonuria, therefore, can be treated with a high degree of success, but if left untreated it provides a clear illustration of Mendelian inheritance. One enzyme (phenylalanine 4-monooxygenase) is involved in the conversion of phenylalanine to tyrosine and is coded for by one gene. A normal person has two good copies of this gene, one derived from the mother and the other from the father, but a small proportion of the population are heterozygotes for the phenylketonuria gene, and have only one good copy, the other one coding for something incapable of catalyzing the reaction. These people are also normal and healthy, and if one of them marries a person with two good copies, a homozygote, all of their children will also be normal and healthy. However, if two heterozygotes marry, there is a one-in-four chance that both will pass on the defective copy to any one child; thus, approximately one-quarter of the children of heterozygotes have phenylketonuria. All in all, phenylketonuria behaves just like one of the seven traits that Mendel studied in his pea-crossing experiments.

Alkaptonuria also obeys classical Mendelian laws of inheritance and had an important role in the history of understanding metabolic diseases. It is less serious than phenylketonuria, and it sometimes produces no ill effects at all, but it is easier to detect because it causes the urine of affected people to turn black. This is due to the effect of air and light on homogentisic acid, a substance produced in the metabolism of tyrosine that fails to be metabolized further if the enzyme concerned is missing. It was studied by Archibald Garrod, who realized that it followed Mendelian inheritance, and it became the first known example of what he called *inborn errors of metabolism*.

ARCHIBALD EDWARD GARROD (1857–1936) was the physician who pioneered the study of metabolic diseases and their inheritance, especially in relation to alkaptonuria, which he described in a classic book, *Inborn Errors of Metabolism*. He subsequently described other examples, pentosuria, cystinuria, and albinism. He was born in London, the son of Alfred Baring Garrod, also a physician, who discovered the abnormal levels of uric acid in the blood of patients with gout.

Nowadays we regard Mendelian genetics and evolution by natural selection as inextricably bound up with one another, but it was not always so. Although Darwin and Mendel were contemporaries (and Mendel was aware of Darwin's work), Darwin, in common with the rest of the scientific world of his time, worked in ignorance of Mendel's experiments,[9] and he thought of inheritance in paint-mixing terms. This misconception almost resulted in complete abandonment of natural selection as a credible theory, because in terms of paint mixing it was impossible to answer a criticism raised by an engineer named Fleeming Jenkin in his long review of *The Origin of Species*.[10]

Breeding pea plants under Mendelian rules ...

According to Ernst Mayr, Jenkin's review was "based on all the usual prejudices and misunderstandings of the physical scientists,"[11] but that is rather uncharitable because his argument made sense in the context in which it was made. It can be understood by examining how pea breeding proceeds under Mendelian rules and how it would proceed if it followed paint-mixing rules. Suppose that I am a pea grower with a large number of plants with yellow peas and a single plant with green peas, but I am anxious to have just green peas. To make the experiment much easier to visualize, we will not allow crosses between plants of different generations,

[9]Some authors have suggested that Darwin did know of Mendel's work, and, for example, suggested his name for including in the article on hybridism in *Encyclopaedia Britannica*. For a convincing account of the evidence, arriving at the conclusion that Darwin probably did not know of Mendel's work, see Andrew Sclater (2006) "The extent of Charles Darwin's knowledge of Mendel" *Journal of Biosciences* **31**, 191–193.

[10]Fleeming Jenkin (1867) "The Origin of Species" *The North British Review* **46**, 277–318.

[11]Ernst Mayr (1985) *The Growth of Biological Thought: Diversity, Evolution, and Inheritance*, Harvard University Press, Cambridge, page 512.

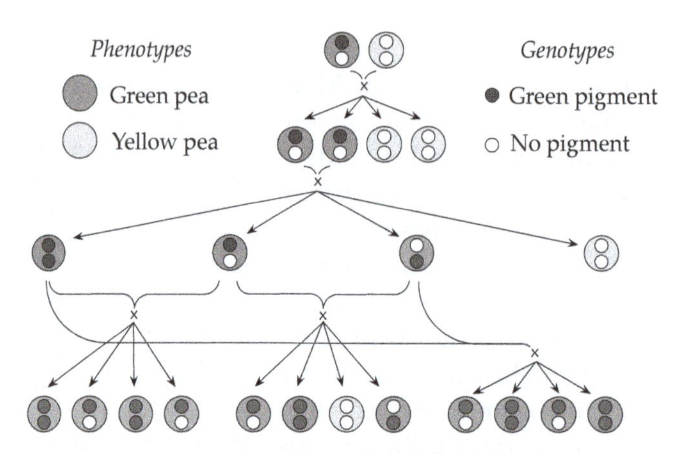

Figure 9.2 Breeding plants according to Mendelian principles. Crossing a heterozygous plant giving green peas with a homozygous plant giving yellow peas will yield 50% of plants giving green peas, all them heterozygotes, and 50% plants giving yellow peas. Discarding the latter and crossing the green-pea plants will yield 25% homozygous green-pea plants and 50% heterozygous green-pea plants. It becomes more complicated in the next generation, because we must consider all possible crosses between the green-pea plants, yielding on average 5/12 homozygous green-pea plants, 6/12 heterozygotes, and 1/12 yellow-pea homozygote. Even if the breeder has no information about the genotypes of the plants, the proportion of homozygote plants with green peas increases in every generation, if all the plants with yellow peas are discarded.

and in particular we will not allow the original plant with green peas to be crossed with plants from later generations.

Suppose that in each generation I can obtain four offspring from any pair of parents (pessimistic, but it allows simple calculations). On the true (Mendelian) model, crossing the original green-pea plant (assumed to be heterozygote) with a yellow-pea plant will then give me two green-pea heterozygotes in the first generation, and two yellow-pea homozygotes (Figure 9.2). Discarding the yellow-pea plants, and crossing the heterozygotes with one another will give, on average, one yellow-pea homozygote in the next generation, two heterozygotes, and one green-pea homozygote. In two generations I have increased the number of plants with green peas 12-fold and the number of heterozygotes fourfold. It becomes complicated to calculate exact numbers after this, because we need to allow for all the possible crosses we can make in each generation. Nonetheless, it is clear that after the first generation I am always crossing green-pea plants with green-pea plants, so the proportion of plants with green peas increases by a factor between three and four each time, and as I discard all plants with yellow peas apart from those needed initially to maintain the population, the proportion of homozygotes among the plants with green peas also increases in each generation. So, even if the detailed calculation is complicated, it is not difficult to see that in a few generations I can produce a population consisting overwhelmingly of plants with green peas. I may continue to produce a small proportion of plants with yellow peas for many generations, because in the absence of modern genetic techniques, there is no way to recognize the heterozygotes until it is too late, but this is a minor nuisance.

Suppose the problem were the opposite: that I started with a population consisting overwhelmingly of plants with green peas, apart from one plant with yellow peas, but I wanted to have just yellow peas. In one sense this may appear more difficult, because the first crossing of yellow-pea plants with green-pea plants will give me no plants with yellow peas at all. After that it becomes easy, because now any green-pea plant with a yellow-pea parent must be a heterozygote.[12] So, in the second generation I produce as many offspring as I can by crossing heterozygotes, and one-quarter of their offspring will have yellow peas. Once I have even a few plants with yellow peas, I can cross them with one another, and the proportion of yellow-pea plants then increases rapidly in each generation.

... and in a paint-mixing world

Selecting for a characteristic that I want is thus relatively easy in the real, Mendelian world, regardless of whether the gene that interests me is dominant or recessive. Matters would be completely different in a paint-mixing world (Figure 9.3). In such a world it would not matter whether I wanted green peas or yellow, so we can return to the original form of the problem where all plants but one had yellow peas, and I preferred green. What could I do? I would start as before, by crossing the lone green-pea plant with a yellow-pea plant, but now I would get not two plants giving green peas and two with yellow, but four giving greenish-yellow peas. If I crossed these with one another, I could obtain other plants with the same characteristics, or if I crossed them with yellow-pea plants, I could get plants with peas more yellow than green, but there is no way I could get a plant with green peas. With completely unnatural selection, the best I could hope for in the long term would be a population with greenish-yellow peas.

Darwin spent a great deal of time in his later life meeting and talking with pigeon breeders, and he was well aware that they could produce an amazing variety of pigeons, in contrast with the gloomy view of their prospects that the paint-mixing model suggests. Dogs provide perhaps a better example for people more familiar with them than with pigeons: In the wild they are as uniform as any other wild species, but unnatural selection can produce varieties as different as the pekingese, the dalmatian, the St. Bernard, or the

[12] It is easy today, because we know the rules of Mendelian inheritance, but even in the past it would have been easy for a breeder to learn from experience that a plant with a yellow-pea parent needed to be discarded.

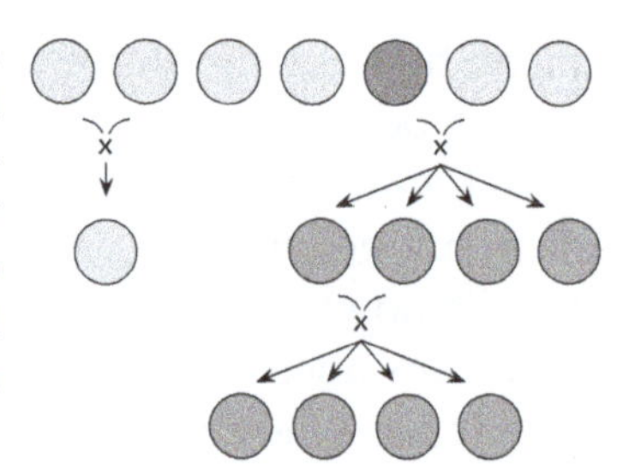

Figure 9.3 Breeding plants in a paint-mixing world. Crossing a green-seeded plant with a yellow-seeded plant can only produce plants with peas of intermediate color, and crossing these with one another can only produce offspring of the same intermediate color. To simplify the analysis, I am assuming that the original green-seeded plant can only be used once. This is an unrealistic restriction for plant breeding.

chihuahua. Richard Dawkins referred in *The Blind Watchmaker* to R. A. Fisher's example of an obvious observation that stares us in the face every time we see a human family: The child of a man and a woman is usually not of intermediate sex, but is either completely boy or completely girl.[13,14] Clearly sex is not a character that can be mixed like paint, but is inherited as a whole.

We do not always see the things that are staring us in the face, and although by Darwin's time breeders of pigeons, dogs, agricultural animals, and plants had acquired a great deal of practical knowledge of how to obtain the results they wanted, this knowledge was not based on a correct theory of inheritance. Until Mendel pointed it out, no one appeared to have noticed that inheritance was particulate.[15] Even then, no one took much notice of Mendel, and when eventually they did, at the beginning of the twentieth century, they mostly took his work not as explaining how natural selection could work, but as making it unnecessary as a mechanism for evolution. As late as 1932, T. H. Morgan, in his influential book *The Scientific Basis of Evolution*,[16] regarded natural selection merely as an effective purifying mechanism for eliminating harmful mutations. This attitude was standard at that time, but it seems so perverse today that it is hard to understand how it came about.

It is also hard to understand why so little attention was paid to Mendel's work during his lifetime, when he had provided the answer to questions of widespread interest and importance. The neglect is often attributed—

[13]As the first quotation at the beginning of this chapter shows, Darwin had already commented explicitly on this in 1866 in a letter to Alfred Russel Wallace.

[14]In these more open times we realize that matters are not as clear-cut as I was led to believe as a child: Intersex individuals are rare in the population as a whole, though much less rare than most of us used to think. One study [M. Blackless, A. Charuvastra, A. Derryck, A. Fausto-Sterling, K. Lauzanne, and E. Lee (1980) "How sexually dimorphic are we? Review and synthesis" *American Journal of Human Biology* **12**, 151–166] estimated that they account for as many as 2% of live births.

[15]Darwin himself tabulated the results of crossing long-styled with short-styled *Primula auricula* in his book *The Different Forms of Flowers on Plants of the same Species* (John Murray, London, 1877), and reported that he counted 25 long-styled offspring and 75 short-styled offspring. The 75:25 ratio leaps out at the eye of the modern reader familiar with Mendelian genetics, but Darwin attached no importance to it.

[16]T. H. Morgan (1932) *The Scientific Basis of Evolution*, W. W. Norton & Company, New York.

sometimes explicitly, but more often by implication—to his supposed isolation from the world in a monastery. This image of an unwordly monk tending his pea plants is misleading, because Mendel was an example—one of the first and most successful—of someone bringing a training in mathematics and physics to bear on a biological problem. He had had an excellent university training in these subjects, first at Olomouc and later at Vienna, where his teachers included the distinguished physicist Christian Doppler.[17] His results were by no means unknown during his lifetime: He corresponded with the great scientists of his day and his published work was widely circulated. The failure of his contemporaries to appreciate his discoveries must be interpreted in terms of lack of understanding, not lack of information.

Shared flux control

Returning to metabolic systems, we saw in discussing them that although stability is a desirable feature, we had no need to call on natural selection to explain it, because just assigning kinetic parameter values at random to all the enzymes in a system would result in a stable steady state much more often than not. We also saw that all enzymes would appear to be present "in excess" if we looked at them in too simple-minded a way. According to the summation relationship discussed in the previous chapter, the control of the flux through a pathway is shared by all of the enzymes in the pathway, and it follows that the average share held by any single enzyme is of the order of one divided by the number of enzymes: If there are 10 enzymes, the average share is one-tenth, and so on. It is slightly more complicated than this, because some of the "shares" can be negative, allowing some of the positive ones to be larger than a simple analysis would suggest, but although this represents a real complication in some circumstances, it is usually unimportant enough to be ignored.

How many enzymes need to be considered depends on how large a system we try to analyze. At the grossest level we can regard the system as the entire organism, and the flux as just the rate of growth. For growth of the whole organism, we must be talking about thousands of enzymes, so the average share of the rate of growth held by any one of them is less than one part in 1000, and this means that we can alter the activity of a randomly chosen enzyme by quite large amounts without any detectable effect on growth at all. There is a limit to this, however: If the particular enzyme activity is absolutely essential to the life of the organism and there is no alternative way of providing the essential function, then eliminating the enzyme entirely will certainly have an effect; the organism will be unable to live. In practice, as I will discuss in the

[17]Doppler is remembered today for explaining the change in tone that you hear when an express train passes close by. He arranged for a brass band to play while seated on a rapidly moving train: Listeners in the station heard the music as sharp while the train was approaching, but as flat when it had passed. This is called the *Doppler effect*, and, applied to light, it has made crucial contributions to our modern understanding of the Universe.

Figure 9.4 Small differences in phenotype. Rectangles 1 and 3 look virtually the same (and quite different from the white rectangle 2 between them), even though the first is 100% black, whereas the third is only 85% black. The difference remains difficult to see when the change occurs gradually across the whole width, as in rectangle 4, but the boundary can be perceived when it is abrupt, as in rectangle 5. Because printing technology does not always faithfully reproduce different shades of darkness, the final version of this figure may not accurately reflect the description. I have several times made experiments with groups of students who have to report the moment when they perceive that an image of a human eye on the screen is no longer brown while it is gradually changing from brown to blue over a period of 30 seconds. They typically notice this when it has become about 60% brown.

last chapter, this happens less often than one might guess, because organisms usually do have alternative ways of achieving any particular function. In organisms where the question has been studied, therefore, organisms as diverse as mice and yeast, only about one out of every five genes is absolutely essential in this sense.

Even if we are talking about a more specific characteristic than the capacity to grow, for example the colors of peas, then the pathway responsible for producing the characteristic, typically a set of reactions leading to a dye, will usually contain several enzymes. So, we should expect to see several enzymes that share most of the control of the rate at which the dye is produced. One of these enzymes may have a flux control coefficient of one-fifth for production of the dye, meaning that if its activity is decreased by 5%, the dye will be produced about 1% (one-fifth of 5%) more slowly. This will certainly not be noticeable without accurate measurements. Extrapolating to larger changes is inaccurate (because in general flux control coefficients become larger as the enzyme activity concerned becomes smaller), but it remains fair to estimate that a decrease to one-half of the normal activity of the enzyme will decrease production of the dye by between 10% and 15%—big enough to be detected quite easily with instruments, but small enough to be missed in a judgement by eye (Figure 9.4). If we decrease its activity to zero, and if there is no alternative way of making the dye, the amount of dye produced must fall to zero as well.

These simple arguments mean that regardless of any genetic considerations, there is a huge difference between decreasing an enzyme activity by one-half and decreasing it to zero: In the one case, the effect may easily pass unnoticed; in the other, some metabolic product will not be made, unless the organism has another way of making it (as often happens). In the one case, peas are near enough to an average green to fall within the expected scatter for any biological variable; in the other case, peas have no green dye at all and will be seen as yellow. This is, in essence, the explanation that Henrik Kacser and

Jim Burns offered in 1981 of why nearly all mutations in diploid organisms are recessive. It can be regarded as a development of Sewall Wright's view that the explanation of dominance needed to be sought in the effects of different gene doses on enzyme activity, which, in addition to the short article quoted at the beginning of this chapter, he later expanded into a longer review.[18] Comparing the upper half of his Figure 7 with the graphical representation of the relationship between genotype and phenotype illustrated in Figure 9.5 for eye color, the resemblance is evident. Incidentally, Wright himself did not claim priority for this insight, but attributed it to a much earlier paper of Lucien Cuénot.[19]

Inheritance in Chlamydomonas reinhardtii

The explanation of dominance and recessivity that I have outlined differs from Fisher's in being easy to understand and so obviously right that any experimental test may seem superfluous, but it is always a good idea to devise an experimental test, even of something that appears obviously right, especially if it goes against the established wisdom in the field accumulated over a half century. Allen Orr did this in 1991,[20] taking advantage of the capacity of *Chlamydomonas reinhardtii* to reproduce occasionally as a diploid organism, even though it spends the overwhelming proportion of its time as haploid. Moreover, unlike some other organisms, such as fungi, its haploid state is truly haploid, with no more than one nucleus per cell, and no more than one copy of each chromosome. Thus, it has no opportunity to experience multiple copies of a gene except during diploid generations. Even if Fisher's explanation of dominance is correct, therefore, *Chlamydomonas* has too few opportunities to benefit from its occasional diploid moments to show any effect of natural selection exerted on diploid cells. On Fisher's hypothesis, mutant genes in *Chlamydomonas* should be recessive much less frequently than they are in species that are always diploid and can experience natural selection of Fisher's modifier genes in every generation. No such effect is observed in *Chlamydomonas*: The great majority of mutations in *Chlamydomonas* are recessive when studied in diploid cells, exactly as found by Fisher in his original study of the fruit fly *Drosophila*, a normal diploid species.

This result leaves no escape for Fisher. If a pattern that requires selection in diploid (or polyploid cells) to work proves to be exactly the same in a species that is nearly always haploid, the explanation cannot be right. As noted by Orr, his observations also dispose of two theories proposed by J. B. S. Haldane,

[18]S. Wright (1934) "Physiological and evolutionary theories of dominance" *American Naturalist* **68**, 24–53.

[19]L. Cuénot (1903) "L'hérédité de la pigmentation chez les souris" *Archives de Zoologie Expérimentale et Générale* 4th series **1**, 33–41.

[20]H. A. Orr (1991) "A test of Fisher's theory of dominance" *Proceedings of the National Academy of Sciences USA* **88**, 11513–11415.

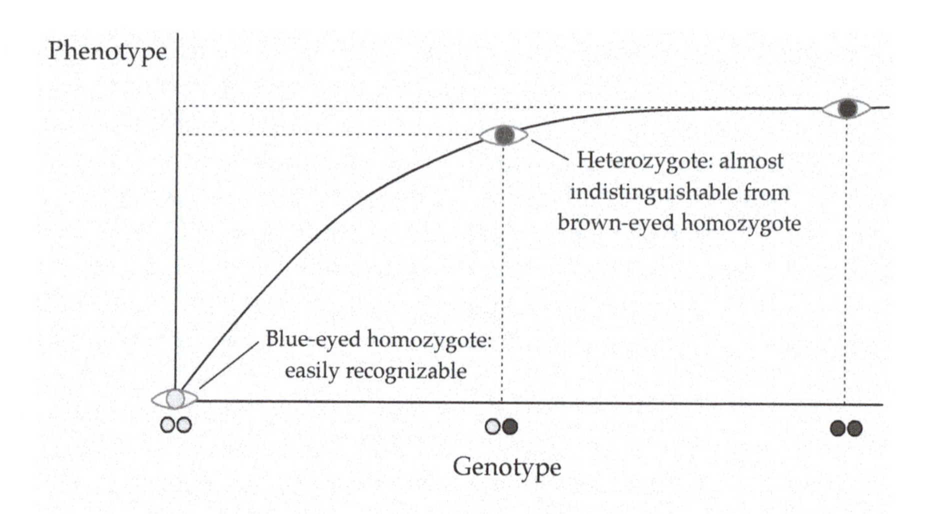

Figure 9.5 Recessive character of blue eyes. This is a graphical illustration of the point made in Figure 9.4. Most enzymes are located on the plateau of the curve relating phenotype to genotype, as in Figure 8.2. Decreasing the gene dose for brown eyes by 50% (or increasing it by any factor) has almost no effect on the perceived color, but decreasing it by 100% has an obvious effect.

one of which is similar to Fisher's, though (for me at least) much easier to understand. This theory supposes that natural selection has a tendency to compensate for the harmful effects of mutations by replacing wild-type alleles that produce "just enough" enzyme with ones that produce "too much." Now, as we saw in the previous chapter, the whole idea of having "too much" of most enzymes is a misconception (though an excusable one for Haldane, who was writing more than 80 years ago): Just about any mixture of enzymes catalyzing a series of linked reactions will result in a state in which there is apparently more than enough of each enzyme. As Orr noted, there is a different reason for rejecting this explanation of dominance. Like Fisher's, it requires selection in heterozygotes, and therefore cannot explain why mutant genes are just as likely to be recessive in the rare heterozygotes of a principally haploid species like *Chlamydomonas* as they are in species that are always diploid. This result parallels the theme of the whole book and illustrates that even as great a geneticist as Fisher can be guilty of finding an adaptation in a phenomenon that requires no evolutionary explanation.

Before leaving Fisher it is worth mentioning that he calculated that Mendel's reported results were too good to be true, in the sense that they agreed with his theory beyond reasonable expectation:[21] If we find, as Darwin did in his experiments on *Primula auricula*, that among 100 offspring 25 were long-styled and 75 short-styled, we can attribute the exact 1:3 ratio to a lucky chance

[21]R. A. Fisher (1936) "Has Mendel's work been rediscovered?" *Annals of Science* **1**, 115–137.

(especially since Darwin appears to have had no particular expectation of what the ratio ought to be), but when Mendel reported that 253 plants gave 7324 seeds, of which 5474 were round and 1850 wrinkled, the ratio of 2.96:1 seems suspiciously good, and when similar (or better) agreement is found in other cases, it seems hard to escape the conclusion that the results were too good to be true. Fisher was a great admirer of Mendel and it does not appear from reading his paper that he was accusing him of fraud, though he was certainly shocked by what he thought he had found. It is more likely that in the middle of the nineteenth century Mendel was not conscious of all the sources of bias that have been discovered and analyzed subsequently, and, in addition, Daniel Hartl and Daniel Fairbanks[22] conclude from a careful analysis that Fisher may have misunderstood what Mendel actually measured in two crucial experiments, and that his "allegation of deliberate falsification can finally be put to rest, because on closer analysis it has proved to be unsupported by convincing evidence."

Humans: a unique species?

As we have seen, mutant genes are normally recessive, but humans seem to be the exception, remarkably enough: An abormally high proportion of mutant genes in humans are reported to be dominant. How can this be explained without returning to a world view in which the human species is unique unto itself and quite separate from animals, plants, fungi, and bacteria? The most plausible interpretation comes from the tendency for human phenotypes to be analyzed far more precisely than is ever done with animals or plants. The idea of a "symptomless disease" might seem absurd, but it is applied in all seriousness to conditions like maturity-onset diabetes of the young ("MODY"), in which small differences in blood-sugar levels allow a prediction that diabetes may develop later. If Mendel had measured the exact amounts of pigment in peas from different plants, he would have been able to detect differences that are not obvious to the unaided eye.[23]

[22] D. L. Hartl and D. J. Fairbanks (2007) "Mud sticks: on the alleged falsification of Mendel's data" *Genetics* **175**, 975–979.

[23] A. Cornish-Bowden and V. Nanjundiah (2006) "The basis of dominance," pages 1–16 in *The Biology of Genetic Dominance*, edited by R. A. Veitia, Landes Bioscience, Georgetown, Texas.

10. *An Economy that Works*

> With an increase in the number of bronze-workers articles of bronze may become so cheap that the bronze-worker has to retire from the field. And so again with ironfounders.
>
> Xenophon (–354)[1]

> The whole quantity of industry annually employed in order to bring any commodity to market, naturally suits itself in this manner to the effectual demand. It naturally aims at bringing always that precise quantity thither which may be sufficient to supply, and no more than supply, that demand.
>
> Adam Smith (1776)[2]

XENOPHON IS MOSTLY REMEMBERED TODAY as the author of the *Anabasis*, where he told the story of the 10,000 Greek soldiers ("The Ten Thousand") from Cyrus' army that he led back home after the unsuccessful war with Artaxerxes. However, he has other calls on our attention, the most relevant to this book being that he was the first known to have expressed the idea that the demand for a commodity is related to its market value. This formed the basis, more than 2000 years later, of Adam Smith's first serious analysis of economic phenomena. Xenophon's remark that if there are too many bronze-workers the price of their work falls, causing some to go out of business and helping to restore the value of the work of those that remain. This is effectively the first expression of what we now call the law of supply and demand.

How well do economic laws work?

However, no observer of modern economic systems can fail to notice that economic theory does not seem to work very well.[3] This is in part due to the tendency of governments to think that the laws do not apply to them and that they can manipulate them for their own purposes. We will examine other possible explanations at the end of the chapter. Meanwhile,

[1]Xenophon (–354) Πόροι ἢ περὶ Προσόδων (*Ways and Means*), section IV, translated by H. G. Dakyns (1897) in *Complete Works of Xenophon*, Macmillan, London; English text available online at `https://tinyurl.com/ke9ct9h`.

[2]Adam Smith (1776) *An Inquiry into the Nature and Causes of the Wealth of Nations*. An edition of 1977 (a facsimile of one of 1904) is edited by Edwin Cannan, University of Chicago Press, Chicago. Other editions are available from other publishers, and it can be found online at `http://www.online-literature.com/adam_smith/wealth_nations/`.

[3]Maybe I should qualify that. No *scientist* observing economic systems could fail to notice it, but some economists apparently think that climate science must be judged by its adherence to economic theory. Writing in the *Washington Times* (March 15, 2015), Stephen Moore asked, "How can a movement be driven by science when its very agenda violates basic laws of economics?"

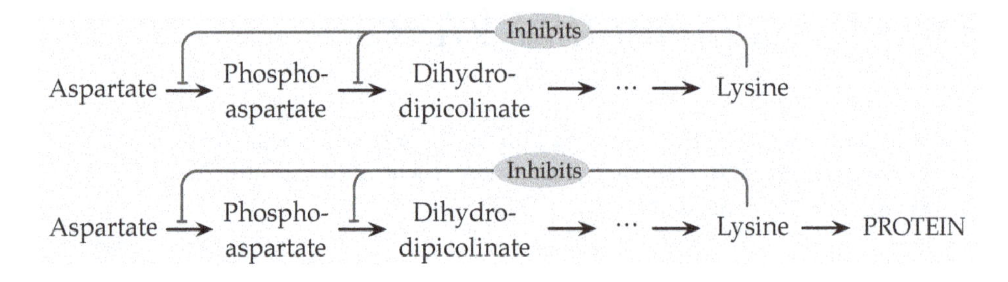

Figure 10.1 Biosynthesis of lysine. *Top:* In textbooks, metabolic pathways are typically drawn with no indication that the end product (lysine in this example) is synthesized in order to be used. Lysine is needed for protein synthesis (*bottom*), and if this essential step is not taken into account, it is impossible to rationalize the regulation of the pathway.

we need to look at a domain in which the laws of supply and demand work supremely well—namely, the organization of metabolism in a healthy organism. In efforts to explore the relationships between economics and metabolism, Jannie Hofmeyr and I proposed a theory of supply and demand in metabolism as part of a modern theory of metabolic regulation.[4] This forms the basis of this chapter.

The metabolic factory needs to synthesize a vast array of different products to satisfy all of its activities, and each of these is made in just the right quantities at just the speed needed. In crises, such as may follow a physical injury or infection by a microbe, sudden unexpected demands for particular metabolites may exceed the capacity of the organism to satisfy them, and in extreme cases this may result in death, but I am speaking here of periods of health. In any case, organisms typically respond far more efficiently to crises than human societies do.

Biosynthetic pathways

The regulation of much of metabolism according to supply and demand is obscured in most biochemistry textbooks by omitting the crucial demand component of a metabolic pathway as it is usually drawn; as a result, this essential aspect of the organization passes unnoticed by most biochemistry students. Thus, "everyone knows"—every biochemist, anyway—that a metabolite such as lysine is made in order to be used as a building block for protein synthesis. But a textbook illustration of lysine biosynthesis in bacteria such as *Escherichia coli* will typically stop at lysine (Figure 10.1). Worse still, it is typically called an "end product," or even sometimes an "ultimate end product." This may not matter much if the aim is just to illustrate the chemical reactions that are used to transform aspartate into lysine. After all, no one expects an account of the

[4]J.-H. S. Hofmeyr and A. Cornish-Bowden (2000) "Regulating the cellular economy of supply and demand" *FEBS Letters* **476**, 47–51.

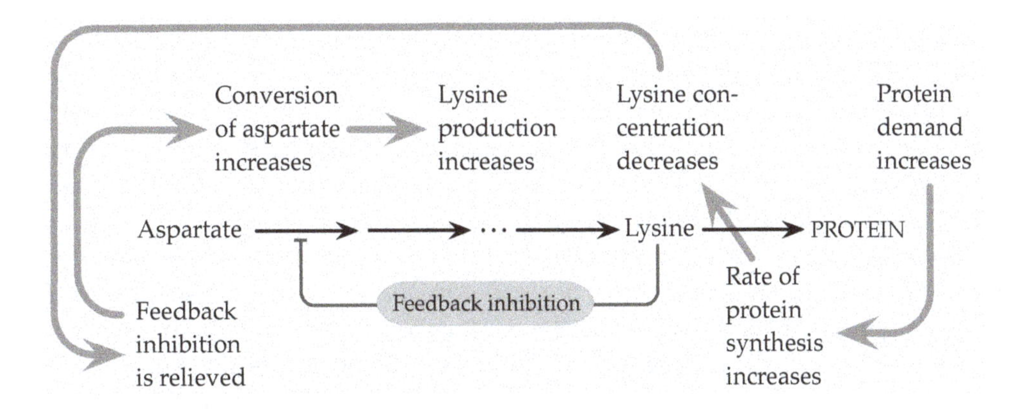

Figure 10.2 Regulation of lysine biosynthesis by feedback inhibition. The explanatory notes should be read in the order indicated by the thick gray arrows, starting with "Protein demand increases" and ending at "Lysine production increases." Notice that what is regulated is the concentration of lysine, not the flux.

processes used in a factory that makes shoes to include a description of how the shoes are taken away to be sold and worn by consumers—"everyone knows" that this is what happens to shoes after they have been made. Nonetheless, an account of the economics of shoe factories would be regarded as seriously deficient if it dealt only with the methods of production, entirely ignoring the demand for the finished product.

So it is with biosynthetic pathways. As long as we are talking just about the chemical transformations, it is of no great account if we ignore what happens afterwards, but if we also ignore demand when we ask how the pathway is regulated, we make a serious error, because our description then omits the key to the whole business. Unfortunately, most accounts of metabolic regulation have done just that since biochemical feedback was discovered in the 1950s,[5] and although the principles are not difficult, they are frequently presented so badly that many years later they are still poorly understood.

Feedback regulation

If a bacterial cell needs more lysine, say, for protein synthesis, it simply uses more. This causes a transient fall in the concentration of lysine, which is sensed by enzymes at the beginning of the pathway that transforms aspartate into lysine,

[5]Why so late, given that the kinetic properties of enzyme-catalyzed reactions had been worked out more than a generation earlier, and the principles of feedback in engineering had also been known for a long time? However, the deviations from expected behavior could only be recognized when enough examples of the expected behavior were known [M. L. Cárdenas (2013) "Michaelis and Menten and the long road to the discovery of cooperativity" *FEBS Letters* **587**, 2767–2771], and the early development of enzyme kinetics was achieved on the basis of a small number of enzymes. Moreover, it was done at a time when few metabolic pathways were known.

making them more active. Conversely, when the cell needs less lysine it just uses less, causing a transient increase in the concentration of lysine, which has the opposite effect of making the same enzymes at the beginning of the pathway less active. The way the mechanism works is illustrated in Figure 10.2. It is called *feedback inhibition*, or sometimes *allosteric inhibition*, for a reason that we will come to shortly.

This regulatory system works just like the negative-feedback systems used by engineers to control the output of regulatory devices, such as the thermostat in a refrigerator. In the latter case, the explanation in an engineering textbook will make it quite clear that what is being regulated is not the flow of heat but the temperature. A thermostat does not prevent you from leaving the refrigerator door open, but it can respond to the increased heat flow if you do leave it open by working harder to keep the temperature constant. By contrast, although biochemistry textbooks will often describe the regulation of lysine synthesis in much the same way as I have just done, they will often confuse the interpretation by implying that it is the metabolic flux that is being regulated, whereas the regulatory design is a design for regulating concentrations, not fluxes.

Before continuing we need to ask whether we are entitled to attribute the existence of feedback regulation in metabolism to natural selection. Maybe it is just an inevitable consequence of the organization of metabolism into series of linked reactions, like some other properties mistakenly attributed to natural selection that we discussed in Chapters 8 and 9. To decide this, we should compare metabolism with other flow systems found in nature or engineering. In natural river systems, for example, there is no feedback and no law of supply and demand. The existence of a drought around the lower reaches of a river does not cause the tributaries at the higher levels to flow any faster; the existence of a flood does not make the snow melt more slowly in the mountains. On the contrary, a river is a water delivery system that is completely indifferent to the "needs" of the plants and animals that live at the lower levels.

What about artificial water supply systems? Maybe just enclosing the flow from reservoir to final user in watertight pipes is sufficient to ensure that the laws of supply and demand will apply. As illustrated in Figure 10.3, we can imagine a pipe leading from a reservoir situated in the hills 500 meters above the user, open at the level of the reservoir but terminated by a faucet at the level of the user. Opening the faucet will make the water flow; closing it will make it stop. Not only that, but we can obtain a less-than-maximal flow by opening the faucet only partially.

Surely we have here a satisfactory flow system that obeys the laws of supply and demand and requires only the crudest of "designs"? No. Although such a system might appear to work as a design on paper (if we refrain from

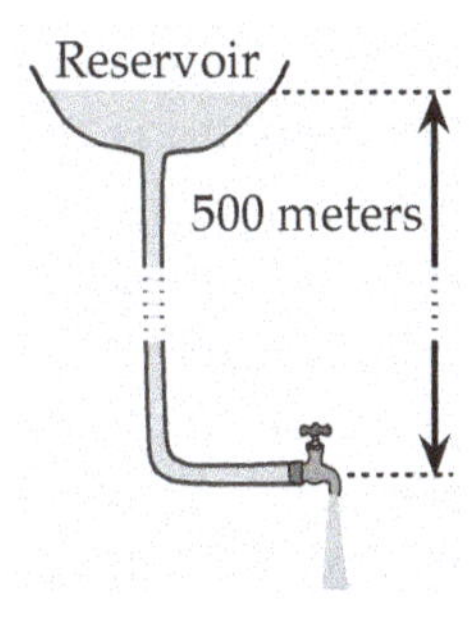

Figure 10.3 An unworkable regulation system. Simply connecting the outlet of a water supply to the reservoir far above it with a watertight pipe and changing the state of the faucet to control the flow would *fail* as a method of regulation, as explained in the text.

asking any questions about the pressure inside the pipe), it would work very badly in practice, if at all, as any civil engineer would realize immediately. A head of water 500 meters in height implies a pressure of about 50 times atmospheric pressure at the level of the faucet when it is closed and there is no flow of water (this is about 10 times the pressure in an unopened bottle of champagne). To resist this, the pipes would need to be extremely strong. Moreover, as we can hardly conceive of a single unjointed pipe leading all the way from the reservoir in the mountains to the faucet at the other end, there would need to be separate portions of pipe with joints between them, and at the point of highest pressure there would need to be a joint between the end of the pipe and the faucet. All these joints would need to be extremely strong and completely watertight, and any repairs that would imply opening the system would be impossible to undertake because of the powerful jets of water that would spurt out at any point of opening.

Worse still, the pressure at each joint would not be a constant static pressure, but it would fall drastically each time the faucet was opened and the water started flowing, only to increase again when the faucet was closed. So, the joints would need to withstand the changes in pressure. The faucet itself would be almost impossible to manipulate because it would be under 50 times atmospheric pressure when closed: This is the pressure that bears down on an area of one square centimeter carrying a weight of 50 kilograms (or an area of one square inch carrying a weight of more than a quarter of a ton); if you prefer, it is the pressure that a submarine feels at a depth of 500 meters.[6]

So, although a sealed tube connecting a reservoir to a faucet might work quite effectively on paper as a way of regulating the flow of water, it would fail in practice because the only way in which the flow would be regulated at the point of exit from the reservoir would be by the back pressure in the tube, and, as we have seen, this would vary so much, and go to such large values, that

[6]Remarkably, however, it is also the pressure inside the capsid of a virus such as the bacteriophage φ29, and the motor that packages the DNA in the capsid is strong enough to work against it [D. E. Smith, S. J. Tans, S. B. Smith, S. Grimes, D. L. Anderson, and C. Bustamante (2001) "The bacteriophage φ29 portal motor can package DNA against a large internal force" *Nature* **413**, 748–752].

it would result in intolerable stresses on the whole system. In practice, civil engineers solve this problem by ensuring that the final point of demand is not the only place where the flow is regulated. There needs at least to be a feedback loop from the point of demand to the point where the supply is initiated. In other words, the exit from the reservoir needs to receive information about the demand that comes by a route that avoids the need for the information to be transmitted solely by the pressures in the pipe. If there are several different points of demand (as there normally will be in any water supply system), there needs to be a separate regulatory device at each exit from a point where the supply pipe divides into two or more smaller ones: You do not want to be deprived of the possibility of taking a shower because your neighbors are watering their garden. This can happen in real water (or gas, or electricity, or whatever) supply systems, but it is a sign of a design that is insufficient to meet the demand. In a well designed and well maintained system, your water supply does not depend on how much water your neighbors are using.

Notice that although the simple sealed tube with a faucet at the end requires (in principle) no special design, and has properties that follow automatically from its structure, the same is not true of systems with feedback loops. There is nothing automatic or inevitable about a signal that passes around all the plumbing to inform the regulators at the reservoir about changes in demand: This must be a design feature deliberately installed by the engineer.

Nearly all of this analysis applies equally well to metabolic regulation. The sort of system illustrated in Figure 10.3 is no more possible in metabolism than it is in water supplies. You may object that that sort of regulation does work when the range of pressure is small, for example for regulating the flow of water from a domestic storage tank, and in fact there are a few metabolic pathways that are regulated by the build up of intermediate concentrations, but it is only possible for exceptionally short pathways, such as the biosynthesis of the amino acid serine in mammals, with only three steps.[7] Even then the pathway in bacteria is regulated by feedback inhibition of the first enzyme.

The only important difference between the regulation of water supplies and the regulation of metabolism is that the water system that supplies your home was planned and designed by real engineers with conscious intentions, whereas the mechanisms that regulate metabolic systems were not planned at all and their "design" is just the result of the selection of designs that work better, with rejection of the ones that work less well. The explanation here is just the same as the one that Darwin and his successors apply to all suggestions of design in biology, and there is no need to belabor the arguments here. Suffice it to say that metabolic regulation, like any other biological adaptation,

[7]D. A. Fell and K. Snell (1998) "Control analysis of mammalian serine biosynthesis: Feedback inhibition on the final step" *Biochemical Journal* **257**, 97–101.

has arrived at a state that looks as if it had been consciously planned by an engineer.

Once the basic features of metabolic regulation were known, study of many examples led to the generalization that each metabolic pathway is regulated by "feedback inhibition at the first committed step." What does this mean? Feedback inhibition means that an enzyme acting early in a pathway is inhibited, or made less active, by accumulation of a metabolite that occurs late in the pathway, and the first committed step is the first reaction after a branch point that is not readily reversible. Essentially this is just like what we discussed in the context of the water supply: Signals that measure the demand for end product loop around the series of chemical reactions to act directly on the supply at an early enough point to avoid large variations in metabolite concentrations between the two.

We cannot carry out an experimental test of what would happen if feedback inhibition did not exist at all in a living organism, because the organism would not be able to support the stresses that this would produce, even if we knew how to make all the genetic modifications necessary to create such an organism. The best we can say is that if gene manipulation is used to suppress the feedback inhibition in just one pathway in a bacterium, this can create difficulties for the cell: Poor growth, leakage of metabolites through the membrane, and so on, which can be interpreted as resulting from inadequately controlled variations in concentrations. Sometimes, it must be admitted, the effects are less dramatic than one might guess, but this probably just reflects the incompleteness of our knowledge and understanding. For example, in a yeast strain in which the enzyme phosphofructokinase lacks the regulatory controls believed to be important in the normal strain, metabolic fluxes and growth are normal, but some concentrations of intermediates are abnormally high, and the mutant strain responds to changes in conditions more sluggishly than the normal strain.

On the other hand, in the computer it is easy to model what would happen in an organism with no feedback controls at all, and it turns out that a pathway without feedback can vary its flux quite satisfactorily in response to the demand, but it does so at the expense of huge variations in metabolite concentrations. In other words, it behaves just like the water supply in a sealed tube without feedback, as illustrated in Figure 10.3.

For a water supply we can, if we are interested, examine the history of its construction and confirm that the regulatory controls that it contains were put there deliberately and are not just haphazard consequences of its structure. Can we be equally sure that the corresponding features in metabolic regulation are the result of natural selection and do not follow automatically from the properties of all enzymes? After all, enzyme inhibition is a common enough phenomenon that can be exerted by many small molecules without the need to

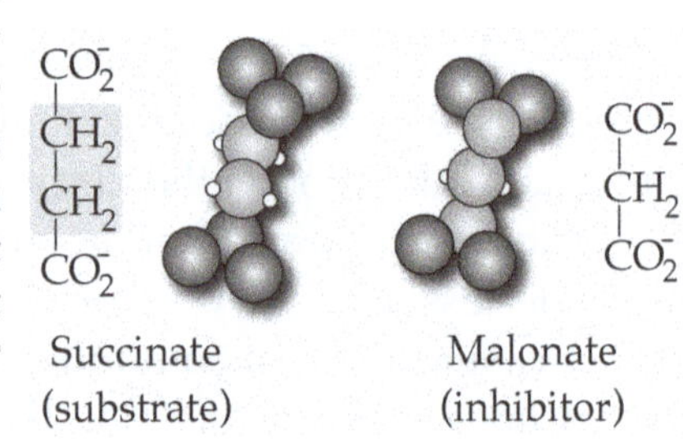

Figure 10.4 Substrate and inhibitor. The reaction catalyzed by succinate dehydrogenase alters the $-CH_2-CH_2-$ grouping (shown with a gray background) in succinate. Malonate is similar in size and structure, but it does not have this grouping: It interacts well with the groups on the enzyme that allow succinate to bind, and so it can be bound to the site of reaction; but it cannot undergo the reaction, so it is an inhibitor rather than a substrate.

postulate a biological function. One answer might be that although feedback inhibition is common, the opposite phenomenon of feedback activation is rare almost to the point of nonexistence. This would not be altogether convincing, because *all* forms of activation (including feed-forward activation, which could fulfill a plausible physiological function in some circumstances) are rare compared with inhibition. On the other hand, feed-forward inhibition, when an enzyme late in a pathway is inhibited by an metabolite that occurs early in the pathway, is also extremely rare, but if feedback inhibition were a haphazard property we should expect feed-forward inhibition to be just as common.

Moreover, if we can claim that feedback inhibition occurs in metabolism because it fulfills a useful stabilizing role, we can also claim that one of the reasons why feedback activation is rare is that if it occurred it would be dangerously destabilizing. Feedback activation occurs in everyday life when a microphone is placed within range of a loudspeaker that it feeds, and the howling noise that we hear in these circumstances is a symptom of a system out of control.

After mention of feedback inhibition and activation, as well as feed-forward inhibition, it will be evident that there is a fourth possibility: feed-forward activation. This is also rare, but, significantly, when it does occur it is sometimes in pathways that are not demand-driven. We will meet an example of this shortly when we consider glycogen production in the liver.

Classical inhibition

In addition, feedback inhibition has two other characteristics that would be unlikely to arise haphazardly: In longer pathways, there is often little structural resemblance between the end product that inhibits and the substrate of the inhibited enzyme; and the inhibition is usually cooperative. Let us consider what these two properties mean. When inhibition arises for no obvious biological reason, it is usually for an obvious enough chemical reason: The substrate and inhibitor are similar enough in terms of chemical structure that the inhibitor can bind to the same site on the enzyme as the one where the substrate binds, but because it lacks some feature necessary for the chemical reaction, it does not react, but just binds and inhibits the enzyme. A classical example is provided

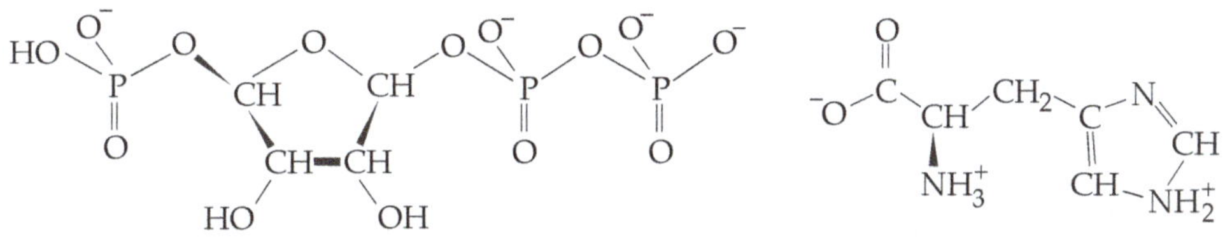

Phosphoribosyl diphosphate Histidine

Figure 10.5 Allosteric inhibition. The first step in the biosynthesis of histidine uses phosphoribosyl diphosphate as substrate, which is completely different in structure from histidine. Nonetheless, the reaction is inhibited by histidine: This makes sense in physiological terms, but cannot be explained in the same way as the inhibition of succinate dehydrogenase by malonate illustrated in Figure 10.4.

by the enzyme succinate dehydrogenase, which uses succinate as its substrate, but is inhibited by malonate (Figure 10.4). Succinate and malonate have almost the same chemical structures, so either is likely to bind to a site intended for the other, but malonate lacks the particular carbon–carbon bond that is transformed in the reaction catalyzed by the enzyme, so it does not react.

Allosteric inhibition

By contrast, a feedback inhibitor may be quite different in structure from the substrate of the inhibited enzyme. For example, the amino acid histidine is made in some bacteria from a sugar derivative, phosphoribosyl diphosphate. This is a far from obvious transformation, several chemical steps being needed to connect them, because histidine and phosphoribosyl diphosphate are structurally very different (Figure 10.5), and there is no chemical or structural reason for histidine to bind to a site intended for phosphoribosyl diphosphate. Yet the enzyme that transforms phosphoribosyl diphosphate is indeed inhibited by histidine. Moreover, the two do not bind to the same site: The inhibitory effect can be destroyed by poisoning the enzyme with mercury, but this poisoning does not eliminate either the catalytic activity or the capacity of histidine to bind; it just destroys the connection between the site where histidine binds and the site where the reaction takes place. This type of inhibition by a molecule that does not resemble the substrate and does not bind to the active site is called *allosteric inhibition* (from a rather rough translation into Greek of the idea of a different shape) and is characteristic of metabolic regulation. Allosteric inhibition is quite common in circumstances where it has an easily identifiable biological function and virtually unknown otherwise. Thus, it can only be a design feature produced by natural selection and not a random property of the enzymes affected.

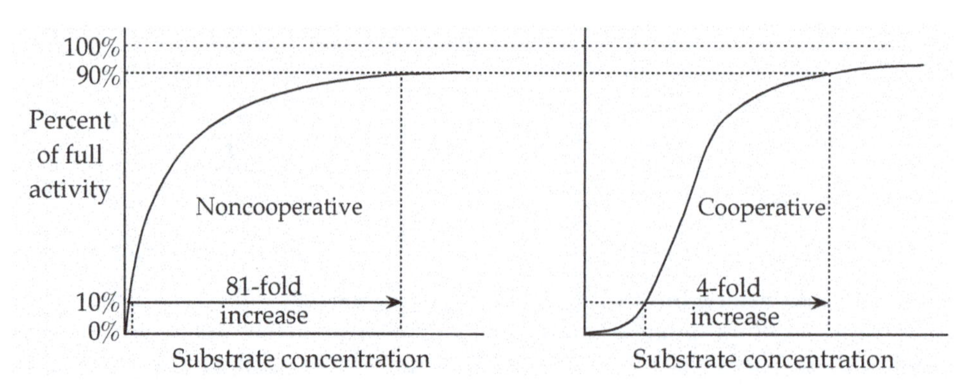

Figure 10.6 Cooperative binding of a substrate to an enzyme. In a simple enzyme with no regulatory properties (*left*), the activity depends on the substrate concentration according to a curve that requires an 81-fold increase in concentration to pass from 10% to 90% of full activity. Some enzymes use a property known as *cooperativity* (*right*) to convert this into an S-shaped curve, so that this ratio is substantially decreased, to three- or fourfold.

Cooperativity The other property that can only be reasonably interpreted as a design feature is *cooperativity*. This describes a capacity of some enzymes to respond more sensitively to biologically significant inhibition than most enzymes do to inhibition in general. For properties that arise just from the inherent structure of enzymes, the dependence of the rate on the concentration of substrate or inhibitor that modulates it is rather sluggish. Typically, to increase the rate of a reaction from 10% to 90% of the maximum by varying the substrate concentration, the latter needs to increase 81-fold (Figure 10.6). For inhibitors the relationship is more complicated, but essentially the same idea applies: To decrease the rate from 90% to 10% of what it would be in the absence of the inhibitor, the inhibitor concentration needs to be increased by a large factor, not necessarily exactly 81-fold but of that order of magnitude.

This is clearly unsatisfactory as the response to a signal. Imagine driving a car that could not go more slowly than 10% of its maximum velocity, even with the brakes fully on, and could not go faster than 90% of its maximum, even with the accelerator fully depressed. It would not work. You might be willing to put up with a car that could never go faster than 90% of its theoretical maximum (most of us do drive cars that cannot go nearly as fast as the maximum indicated on the speedometer), but a car that could not stop, because it could not go at less than 10% of the maximum, would be intolerable. Likewise in metabolism, enzymes that need to respond sensitively to signals do so cooperatively: This just means that they are more sensitive than they would be if they just had the standard off-the-shelf properties that any enzyme has.

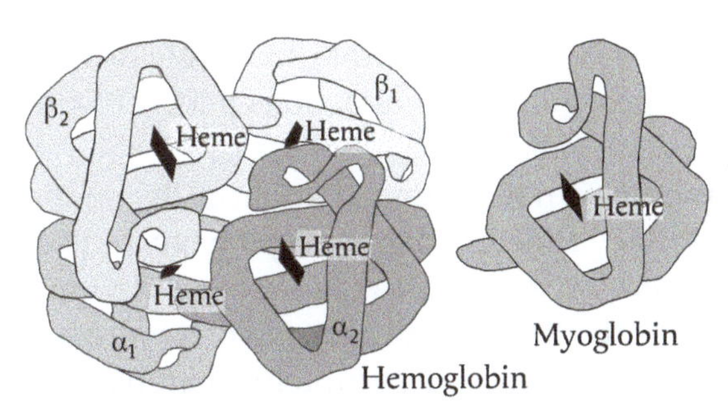

Figure 10.7 Hemoglobin and myoglobin. Hemoglobin has two α and two β subunits, which are quite similar to one another, and resembles a tetramer of myoglobin: compare the structure of myoglobin with that of the α_2 subunit of hemoglobin. Oxygen can bind to each of the heme groups.

Hemoglobin and myoglobin

The two oxygen-binding proteins that we met already in Chapters 2 and 3, hemoglobin and myoglobin, illustrate the difference well. They are not enzymes, because they do not catalyze a reaction, but hemoglobin contributed so much to our understanding of cooperativity that it was often called an "honorary enzyme," because it shares the properties of cooperative enzymes that are relevant to the discussion. In particular, filling hemoglobin from 10% to 90% of capacity with oxygen requires much less than an 81-fold increase in the oxygen pressure. The actual ratio varies somewhat with the conditions of measurement, but is of the order of fivefold. Because hemoglobin has to be able to take up oxygen in response to relatively small variations in oxygen pressure in the lungs, and release it in response to relatively small variations in oxygen pressure in the tissues where it is used, this sensitivity fulfills an obvious biological need.

By contrast, myoglobin is used for storing oxygen in muscles. To do this effectively, it needs to bind oxygen more tightly than hemoglobin does, so that it can strip off the oxygen from the hemoglobin that arrives, but it has no need of cooperativity because it releases oxygen according to demand over a wide range of working conditions. Accordingly, therefore, we should not be surprised that oxygen binds more tightly to myoglobin than it does to hemoglobin and does so without cooperativity.

These two proteins also illustrate quite nicely the structural features that allow proteins to achieve cooperativity (Figure 10.7). Hemoglobin has four oxygen-binding sites on each molecule, and the term cooperativity comes from the idea that these sites "cooperate" with one another: Once one binding site has taken up an oxygen molecule, interactions with the other sites make it easier for them to do so as well. By contrast, myoglobin has just one oxygen-binding site on each molecule and has no possibility of a corresponding cooperation between sites. So, we have one protein with cooperativity and multiple binding sites, and another with just one site and no cooperativity, and together they illustrate the normal combination of structural characteristics and binding properties. Enzymes with just one binding site cannot display cooperativity between sites (and in the rare monomeric enzymes that seem to show cooperativity, it has to be explained in more complicated ways than simple cooperation between sites). Enzymes with multiple sites may or may not display cooperativity, because the possibility of interactions between sites does not mean that such interactions must exist.

Like allosteric inhibition, cooperativity is not scattered haphazardly among the known enzymes without regard to any function that it might have, but occurs almost exclusively in enzymes where a regulatory function seems obvious, for example in an enzyme that catalyzes the first committed step of a pathway. Even though there appears to be no structural feature of proteins that makes it necessary that an enzyme with an allosteric response to a particular metabolite must also have cooperative kinetics, or vice versa, they do in practice occur so often in the same enzymes that the two terms have often been regarded as synonymous. It would be thus perverse to regard the regulatory properties of enzymes as anything but an adaptation produced by natural selection.

Detoxification

Although, as I have suggested, the most common need satisfied by enzyme regulation is to ensure that the laws of supply and demand are obeyed as painlessly as possible—with minimal variation in metabolite concentrations—it would be wrong to suggest that this is the only need or that all pathways are necessarily designed in the same sort of way. Not all pathways exist to produce useful metabolites for the cell; some exist to get rid of harmful substances that come from toxins in the food or are released by infective agents. Expecting these pathways, collectively known as *detoxification pathways*, to obey the same laws as biosynthetic pathways would be as absurd as requiring the sewage works in a city to operate under exactly the same rules as those that govern the supply of fresh water.

As far as I know, no one has tried to develop a general regulatory theory for detoxification, but I should be surprised if ideas of supply and demand proved to be as prominent in it as they are in biosynthetic pathways. On the contrary, we should expect the major considerations to be getting rid of the harmful substance as fast as possible, while making sure that anything it was converted into was either less harmful or at least could be removed rapidly. Detoxification pathways must deal with toxins that may have never been seen before during evolution, and so we must expect them to make mistakes sometimes, especially with chemical poisons not found in the natural world. For example, a routine transformation may sometimes convert something mildly unpleasant into something much worse. Although in general it may be a good bet to start working on a water-insoluble hydrocarbon from the petrol industry by oxidizing it to something that can be dissolved in water, in some specific cases this may not be a good idea at all, and some of the more potent cancer-producing agents are produced by the body itself in its efforts to detoxify foreign substances.

We have seen various examples already of harmful substances produced by metabolism itself. In phenylketonuria, described in the previous chapter,

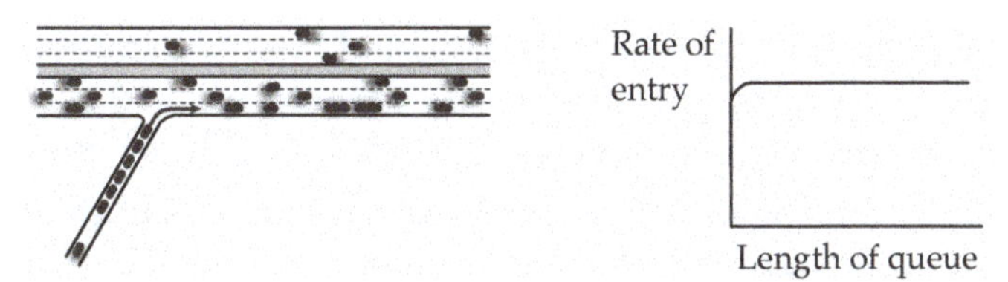

Figure 10.8 Constant flux at saturation. The rate at which vehicles can enter a busy highway from a side road depends mainly on the frequency of gaps in the traffic and is essentially independent of the length of the queue unless it is very short. Enzymes like pepsin operate under these conditions in the stomach, thereby assuring an almost constant supply of digestion products.

the major harm is done by a molecule produced when there is an excess of the amino acid phenylalanine, which is not harmful at all when present in normal amounts. We also encountered the enzyme catalase (in Chapter 1), which exists to destroy the dangerous chemical hydrogen peroxide that arises as an unavoidable side effect of using oxygen in the first stages of dealing with a wide range of unwanted molecules. Superoxide dismutase, which we will meet in the next chapter, has a similar role, in destroying an even nastier substance that arises as a side effect of reactions that use oxygen.

Even with a toxic substance like ethanol (ordinary alcohol), which is certainly not a new product from the chemical industry but has been present on an evolutionary timescale, it is chemically convenient to begin the detoxification by converting it into something worse—namely acetaldehyde. In such a case, the regulatory design has to ensure that the acetaldehyde is removed as fast as it is produced. In this instance, medical practice has taken advantage of the toxicity of acetaldehyde by using *Antabuse*, or *disulfiram*, as a treatment for alcoholism. This substance has little toxicity of its own, or little effect of any kind when there is no acetaldehyde in the system, but it blocks the enzyme that catalyzes the removal of acetaldehyde. If there is no ethanol in the system, this does nothing, but in the presence of ethanol, the relatively mild symptoms of ethanol poisoning are replaced by the much more unpleasant symptoms of acetaldehyde poisoning—a throbbing headache, difficulty in breathing, nausea, copious vomiting, and sweating are the least of the problems. Patients administered Antabuse are normally strongly advised not to touch alcohol, or even to use products like after-shave lotion that have ethanol in them, as long as the effect of the Antabuse lasts.

Digesting food

We must also expect considerations of supply and demand to need some modification in the early stages of food processing. Herbivorous animals like cows and sheep eat almost continuously, partly from necessity, because the food they eat has so little nutritive value that they have no other way of getting enough, but also because they have no other more pressing things to do. Almost continuous feeding is also convenient for very young animals like neonatal rats, even if their diet is richer in proteins than that of sheep. Having to eat nearly all the time would be a nuisance for adults of many species, including our own (though a visit to a modern amusement park may suggest otherwise), and impossible for large carnivores like lions that tend to have large quantities of food available only at infrequent moments. In these cases there is no advantage in being able to digest food fast. On the contrary, it satisfies the requirements better if digestion of one meal takes a large fraction of the time available until the next meal. As a result, enzymes like pepsin, responsible for the first stage of digesting proteins in the human stomach, do not have any sophisticated feedback regulation, but just operate most of the time at saturation, like the traffic flow illustrated in Figure 10.8. This is like a queue of vehicles at a traffic intersection: Only when the queue is very short does the number crossing the intersection each minute vary much with the number waiting; in busy periods the rate is essentially independent of the length of the queue. The effect of this saturation of the protein-digesting enzymes is that the bloodstream of a meat-eating animal continues to receive a steady supply of amino acids—the products of protein digestion—long after the last ingestion of protein.

Although carbohydrate eaten as starch is digested rather slowly, like protein, animals can also obtain quite large amounts in the form of small molecules like sucrose, glucose, and fructose. Sucrose is ordinary sugar as bought from a supermarket, and also occurs as such in some sweet plants. Glucose and fructose are natural components of sweet food like fruits and honey. Ingestion of any of these small-molecule sugars produces a rapid and potentially disastrous increase in the amount of glucose circulating in the blood. But even if natural selection has not yet had enough time to respond to the existence of supermarkets and cheap sugar (discussed in detail in Chapter 11), it has had plenty of time to adjust to the availability of wild fruit and honey,[8] and in a healthy animal, disaster is avoided by rapid conversion of the excess glucose into glycogen, which was the subject of Chapter 6. This is stored temporarily in the liver, being subsequently released and converted back into glucose when the need arises.

This latter conversion back to glucose must certainly be demand-driven, like the biosynthetic processes I discussed earlier, but it is equally certain that the initial conversion of glucose into glycogen cannot be. This is because the liver has no "demand" for glycogen, and although it uses a little glucose for its own purposes, the amount is small compared with the amount it processes. On

[8]Honey has been available since time immemorial, but in a peculiar way: Setting aside modern bee-keeping and distribution practices, it has not been available at all most of the time, but at certain moments appears in large amounts. What this means is that metabolism needs to have evolved to be able to cope with a large sudden intake of sugar when honey is found, albeit not as large as what is possible today.

the contrary, conversion of glucose into glycogen in the liver has to be supply-driven, so we may expect it to be an exception to some of the generalizations made above, and so it proves. Hexokinase D, the enzyme that catalyzes the first chemical step, is exceptional in a number of ways. Unlike other hexokinases that serve the physiological needs of organs like brain and muscle, it is unresponsive to the concentration of glucose 6-phosphate, the product of its reaction; in other words, it ignores the demand for product. In contrast, it responds cooperatively to the concentration of its substrate, glucose. Not only that, but even though glucose 6-phosphate does not inhibit hexokinase D, it does activate the later steps of the conversion of glucose to glycogen in the liver. It is not by chance that this unusual example of feed-forward activation occurs as a mechanism in an unusual example of a supply-driven pathway. Perhaps most striking of all, hexokinase D is as close as we can find to a genuine rate-limiting enzyme, something that I referred to as being largely a myth when I discussed it in Chapter 8.

Overexpressing the rate-limiting enzyme

Enough of exceptions: There is little doubt that most of the processes that occur in metabolism satisfy the laws of supply and demand, which means that they respond to changes in demand and ignore small changes in supply. Moreover, there is abundant evidence that natural selection brought them to that state, but this has not prevented a large amount of wishful thinking. Failure to take account of the real regulatory design of metabolism is a major reason for the failure of the biotechnology industry over recent decades. Since genetic manipulation became possible at the end of the 1970s, a vast amount of money has been wasted—or "invested," if you prefer a more conventional word—on the search for the mythical rate-limiting step in each pathway that would allow commercially valuable metabolic products to be overproduced at will in suitable microorganisms.

Ethanol, for example, has many industrial uses in addition to its presence in beverages, and is produced on a huge scale. About half a million tons of glutamate are produced each year, and other amino acids like lysine and methionine owe their tremendous industrial importance to their use by farmers to supplement corn as a feed for cattle, which is cheap but does not contain enough of these essential components. Even carbon dioxide, which may seem to be just waste if we think of yeast fermentation only as a way of making still wine (and mainly as waste in making beer or sparkling wine), becomes the whole point of yeast fermentation in bread-making. These and other substances are produced on such a large scale in the world's industry that even a small increase in productivity translates into many millions of dollars per year. The idea is that once you identify the enzyme that catalyzes the rate-limiting step, you can overexpress the enzyme so that its activity in the living organism is higher than it would normally be; then the organism will make more product and everyone will be happy.

The major problem with this approach is that it does not work. In Chapter 9 we have seen some good theoretical reasons for believing that it cannot work, but let us also look at some experimental evidence. The crucial experiments were done in Germany in the 1980s, early enough in the story for vast amounts of money to have been saved if the results had been taken seriously. One of the things that "everyone knows"—everyone who has followed a standard course of biochemistry, in this case—is that phosphofructokinase controls glycolysis. We have met glycolysis already in Chapter 4 as a central metabolic pathway in virtually all organisms, essential for converting sugars into energy, and phosphofructokinase is an enzyme that catalyzes a reaction of glycolysis. Its activity responds to an impressive variety of metabolic signals that indicate the immediate requirements of the organism. This behavior is certainly crucial for regulating glycolysis, but it is translated in an excessively simple-minded way into the naive idea that phosphofructokinase is the rate-limiting enzyme and that increasing its activity will inevitably result in a higher glycolytic rate. When the activity of phosphofructokinase in fermenting yeast was increased by a factor of 3.5, however, there was no detectable effect on the amount of alcohol produced.[9]

Before we try to understand this I must emphasize that this is an experimental result; it is not just the wild musings of an armchair biochemist. By now it has been repeated, not only in yeast but in other quite different organisms, such as potatoes, and overexpressing phosphofructokinase in potatoes has just the same effect as it has in yeast: It has no effect. Nor does it have any effect in other organisms where similar experiments have been done.

To understand these results, we have to recognize that the main value of metabolic regulation of biosynthetic pathways is that it allows metabolites to be made at rates that satisfy the demand for them, without huge variations in the concentrations of intermediates. This last qualification is so important that I will say it again: Adjusting rates is quite easy and requires little sophistication in the design. The difficult part is keeping metabolite concentrations virtually constant when the rates change, and this is needed if the effects are to be prevented from leaking into areas of metabolism where they are not wanted.

When we refer to demand we are talking about the organism's own demand, not demands made by an external agent such as a biotechnologist. This seems such an obvious point that it ought not to be worth making, but because it is one that often seems to have escaped the biotechnology industry altogether it is perhaps not as obvious as you might think. An organism such as *Saccharomyces cerevisiae*, the yeast used in baking bread, brewing beer, and

[9]I. Schaaff, J. Heinisch, and F. K. Zimmermann (1989) "Overproduction of glycolytic enzymes in yeast" *Yeast* **5**, 285–290. There is no special significance in the value 3.5: It just happens to be the value Schaaff and colleagues used.

wine-making, has been perfecting its metabolic organization for a long time. Brewing is known to have been used in Egypt at least 4000 years ago, and thus for many more *Saccharomyces cerevisiae* generations than there have been human generations. It is the height of arrogance to think that a little tinkering with the expression levels of its genes will allow us to force as much glucose through its fermentative system as we would like. Its regulatory mechanisms have evolved for the benefit of *Saccharomyces cerevisiae* and nothing else. The yeasts used in the baking and fermentation industries have undergone a considerable amount of unnatural selection over the centuries, but this has always, at least until the past few years, involved selecting natural variants capable of growth under the conditions preferred by the technologist, not tinkering with the basic structure of the organism in the hope of obtaining quite unnatural properties.

As a general rule, therefore, if you try to overcome by brute force the regulatory controls built into any organism, you will fail. On the other hand, with a little understanding of what these controls are designed to achieve, you may hope to achieve your aims by subterfuge. Once it is understood that most biosynthetic pathways are designed to satisfy the demand for end-product and to be as unaffected as possible by changes in supply and in the work capacity of the enzymes involved in the biosynthesis, it should become obvious that manipulating either the supply of starting material or the activities of the enzymes will have no effect other than to stimulate the regulatory controls to resist. It is as futile as trying to stop the import of cocaine by interfering with growers or traffickers in Colombia: As long as the demand is there, it will be satisfied.

Engineering a leak

By contrast, if you could fool an organism by manipulating the demand for a desirable end product, you might persuade the regulatory mechanisms to help rather than hinder. An interesting example comes from the monosodium glutamate ("MSG") industry. This flavoring agent, responsible for the "Chinese restaurant syndrome," a form of glutamate poisoning, was produced during the 1950s by growing the bacterium *Corynebacterium glutamicum* (at that time called *Micrococcus glutamicus*) on pure cane sugar as a carbon source, a cheap source of sugar imported in large quantities from Cuba. The bacteria were deliberately deprived of biotin, a vitamin essential for making cell membranes, with the result that their membranes leaked glutamate, which could then be harvested by the food company.[10]

At the end of the 1950s, the cheap source of pure sugar was no longer

[10]A. Ault (2004) "The monosodium glutamate story: the commercial production of MSG and other amino acids," *Journal of Chemical Education* **81**, 347–355; A. L. Demain (2010) "History of industrial biotechnology," pages 17–76 in *Industrial Biotechnology: Sustainable Growth and Economic Success*, edited by W. Soetaert and E. J. Vandamme, Wiley-VCH, Weinheim.

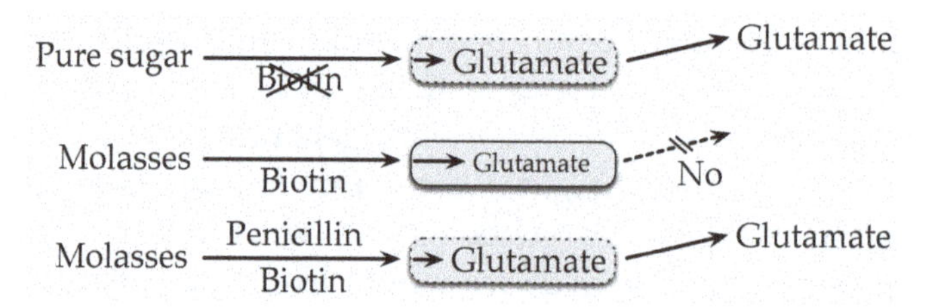

Figure 10.9 Engineering a leak. When grown on pure cane sugar with no added biotin, *Corynebacterium glutamicum* produces imperfect membranes through which the industrially important product glutamate can leak into the medium. Growth on molasses, a cheaper source of sugar that contains biotin as an impurity, produces cells that do not leak and synthesize only as much glutamate as they need for their own metabolism. The same cells, when grown in the presence of an antibiotic that interferes with membrane synthesis, such as penicillin, again leak glutamate.

available and molasses had to be used instead. Although this was sufficiently cheap, it was much less pure and contained large amounts of biotin, which would have been far too expensive to remove. The bacteria no longer made low-quality membranes, which no longer leaked, so glutamate ceased being lost to the growth medium. To overcome the problem, a different way of making the membranes leaky was needed, and this was achieved by growing the bacteria in the presence of penicillin (Figure 10.9), which acts by interfering with the building of membranes.

The point of all this is that as long as the bacteria grew at all, they produced enough glutamate for their own purposes, principally for making new proteins. If a large part of the glutamate made was being lost through their leaking membranes and ending up in soup for human consumption, they just made some more. That is what their regulatory mechanisms had evolved to achieve, and that is what they did. More generally, it follows that one of the most effective ways of causing a microorganism to produce more of a desirable metabolite will be to engineer a leak.

I do not know why the engineers starved their bacteria of biotin in the first place. Maybe it was just because it was cheaper and they found that it worked. But whatever the reason, it is an example that is worth following. Engineering a leak, or finding some other way of artificially stimulating the demand for the desired product is likely to work as a general approach. Trying to force more metabolites through a pathway by increasing the supply or the activities of the pathway enzymes is just a way of triggering the regulatory controls that have been naturally selected to ensure that it will not work.

I started this chapter by noting that the classical laws of economics appear to work rather badly when applied to the domain for which they were devised, and indeed the title of the chapter is intended to reflect this idea. I continued by arguing that in the field of metabolism, in great contrast, they work extremely well. This may seem to present a problem: What difference between real economies and the economy of the living cell could justify believing that the same laws work badly in the one and well in the other? I could reply that the living cell is the product of many millions of years of evolution, and that even if not strictly a designed system, it shows all the characteristics of a designed system, whereas human economies are not designed and have not existed for long enough to have evolved to a state in which they appear to be designed. This is not a very satisfying answer, because some national economies, especially during the twentieth century, were thoroughly planned, and to some degree therefore designed, but there is little to suggest that they worked any better than market economies; most, indeed, would argue that they performed much worse.

Complexity

I think the answer is that real economies are complex and metabolism is not. Enzymes and the other components of the cell are governed by physical laws, and human behavior, which determines how real economies behave, is not. In metabolism, moreover, atoms are conserved, though they are moved around and transformed in different ways, but corresponding quantities in economies, such as money and the things it can buy, are not. We are still faced with a difficulty: What could be more complicated than a living cell, with its thousands of chemical reactions proceeding simultaneously, so how can I claim that it is not complex? In everyday conversation, we often use the terms "complicated" and "complex" interchangeably, and ordinary dictionaries encourage this by giving each word as one of the definitions of the other, but in modern science they are not at all equivalent. A system is complicated if it contains many components, all of which need to be understood for the whole system to be understood. To be complex, a system needs more than that: It needs in some sense to be more than the sum of its parts, to display *emergence*, a manifestation of interactions between the different components that could never be deduced by examining them one at a time.[11]

Metabolic systems do not entirely lack such interactions. Some of the different enzymes do interact with one another and with components of the cell architecture that ensure that enzyme molecules are not evenly distributed around the compartments they find themselves in. Nonetheless, computer models of cell metabolism that ignore these complications manage to reflect fairly accurately, sometimes even quantitatively, the real behavior of the cell. This means that even when complications exist, they act far more to stabilize than to destabilize the cell. The sort of positive feedback effects that make

[11]I must add that not everyone agrees with the distinction I am using in this book. Enrique Meléndez-Hevia in particular considers that metabolism is complex but not complicated. I cannot discuss in detail why I disagree, so I will just say that I believe the distinction that I am making is widely accepted.

the behavior of real economies difficult to predict, and contribute to problems such as runaway inflation, barely exist in metabolism. As we have seen in this chapter, negative feedback is common in metabolism, but positive feedback is rare almost to the point of nonexistence.

Where complex properties occur in living systems, as for example in chaotic cardiac arrhythmias, they are characteristic of pathological failure of the normal controls—in this case, in some kinds of heart disease—not of health. Even at the level of an individual enzyme an example of chaotic behavior has been known for many years, the enzyme peroxidase from horseradishes,[12] but we have no idea how its properties benefit the horseradish, and they contribute nothing to our present understanding of how metabolic systems are controlled in general. The sort of positive feedback loops that account for the complex behavior of economic systems, political relations, and indeed some aspects of biology such as the extravagant overdevelopment of the peacock's tail, are conspicuously absent from the major pathways in the metabolic economy. There is no reason, therefore, to see any contradiction in the claim that classical economic theory works much better in metabolism than it does in the domain in which it was developed.

Not everyone likes this analysis. Complexity is a fashionable subject, and the idea that metabolism is complex has considerable appeal, especially if you can see a Nobel Prize beckoning. Given the real complicatedness of the subject, it becomes easy to slide into calling things complex when they are just complicated. Moreover, the noun "complicatedness" is such a mouthful that even people who are careful not to describe things as complex when they are just complicated will often use the word "complexity" as the noun for both. One should resist, however, because the subject is sufficiently difficult as it is without inventing difficulties that do not exist. In any case, even if much of metabolic regulation can be explained and analyzed without invoking complexity, some of it cannot, at least for the present, because there is genuine complexity in the organization of living organisms as well. Not all of the metabolic properties of a living organism can be understood in terms of the properties of its components, though not everyone agrees about the reasons. Some argue that it just reflects our current lack of adequate tools and knowledge, others that there is a fundamental barrier that can never be breached no matter how complete our knowledge becomes, and no matter how powerful our computers become. When the time comes to grapple seriously with this, it will not be helpful to find that we cannot use the natural word to describe it because it has been misappropriated to describe something else.

[12]L. F. Olsen and H. Degn (1977) "Chaos in an enzyme reaction" *Nature* **267**, 177–178.

11. *Failures of Natural Selection*

> Osteoarthritis is by far the most common type of arthritis and is one of the leading chronic diseases in the United States, affecting an estimated 27 million individuals 25 years of age and older.
>
> Lori L. Alexander (2013)[1]

> Over the past three decades childhood obesity rates in this country have tripled, and today nearly one in three children in America are overweight or obese.
>
> Michelle Obama (2010)[2]

IN CHAPTER 8 WE SAW an example of a property that is a mathematical necessity and is thus unaffected by natural selection or any other evolutionary process, and in Chapter 7 we discussed observations that are sometimes attributed to natural selection but which are just mathematical tricks. Now we will discuss two contrasting problems that natural selection could in principle have solved, but in practice has not, so we need to ask why not. Both of them are important for human health, and both have been analyzed by Enrique Meléndez-Hevia, like so much in this book. The first is linked to degenerative diseases like osteoarthritis, and the other is obesity—not a disease in itself, but an important contributor to different diseases, in particular type II diabetes mellitus, one of the major increasing health problems of the twenty-first century.

What is an animal?

If you were trying to name the defining characteristic of animals, you might say that they are eukaryotic, or that they are multicellular, motile, sexual, or heterotrophic. The first and last of these terms may need some explanation. A eukaryotic organism is one with cells that store their genetic material in a nucleus; a heterotrophic organism is one that cannot survive on purely nonbiological food, but depends on chemicals produced by other organisms. All of these are indeed characteristics of animals, but they all fail as definitions because they fail to exclude organisms that are clearly not animals. Plants, fungi, and protozoans are also eukaryotes. Plants and many fungi are also multicellular. Some protozoans and bacteria are motile.[3] Some plants are sexual, and although the great majority of animals are sexual, there are exceptions: The bdelloid rotifers are a large group of microscopic animals found in fresh water all over the world, and all are female, reproducing asexually for 85 million years; other

[1] L. L. Alexander (2013) *Osteoarthritis* CME Resource/NetCE, Sacramento, California.

[2] M. L. R. Obama (2010) Speech in the White House, Washington DC, February 9, 2010.

[3] Chemotaxis, mentioned in Chapter 5, is an example of bacterial motility.

	Animals	Plants	Fungi	Protozoans	Eubacteria	Archaea
Eukaryotic?	Yes	Yes	Yes	Yes	No	No
Multicellular?	Yes	Yes	Some	No	No	No
Motile?	Yes	No	No	Yes	Some	Some
Sexual?	Most	Some	No	Some	No	No
Heterotrophic?	Yes	Some	Yes	Some	Some	Some
Contains collagen?	Yes	No	Some	Some	No	No

animals, such as aphids, produce asexually during part of their lives. As well as all animals, a few carnivorous plants,[4] all fungi, and some protozoans and bacteria, are heterotrophic. These characteristics are summarized in Table 11.1.

We are thus left with a feeling of knowing what an animal is, and being able to recognize one when we see one, but of being unable to define what it is. There is a less obvious characteristic that seemed until recently to have no exceptions:[5] All animals synthesize the protein collagen, and it was present in the first animals, about 580 million years ago, remaining an essential feature of them ever since. It is a necessary component of bone, skin, muscle, heart, cartilage, and ligaments, none of which could exist without it. The bottom line of the above table therefore provides a test for being an animal that has few known exceptions.

Collagen

Collagen is an untypical protein, with the structure illustrated in Figure 11.1, which resembles that of no other apart from elastin, which allows tissues such as skin to be elastic, able to return to their original shape after stretching, squeezing, or bending. In general, proteins have amino acid sequences with no obvious regularity, so that they appear

[4]As with so many definitions in biology, there is some ambiguity. Plants like the Venus flytrap are certainly carnivorous in the wild, but can be grown artificially without any insect food. On the other hand, many plants not considered to be heterotrophic grow better if they are provided with animal products such as bonemeal, bloodmeal, or horse manure.

[5]Until recently I did indeed think that there were no exceptions, but it seems that this is not an example of a biological generalization that has no exceptions. The genome of the choanoflagellate *Monosiga brevicollis* turns out to contain at least two genes coding for collagen-like proteins [N. King and 35 others (2008) "The genome of the choanoflagellate *Monosiga brevicollis* and the origin of the metazoans" *Nature* **451**, 783–788]. This is a particularly interesting exception, because the choanoflagellates, a small group of free-living unicellular organisms, are generally regarded as the closest known relatives of the animals. More worryingly, collagen has also been found in fungi [M. Celerin, J. M. Ray, N. J. Schisler, A. W. Day, W. G. Stetler-Stevenson, and D. E. Laudenbach (1996) "Fungal fimbriae are composed of collagen" *The EMBO Journal* **15**, 4445–4453]. It appears that an unambiguous definition of an animal has yet to be found.

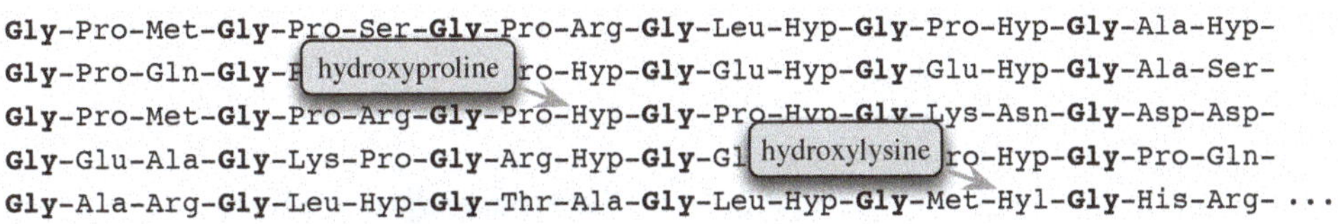

Gly-Pro-Met-**Gly**-Pro-Ser-**Gly**-Pro-Arg-**Gly**-Leu-Hyp-**Gly**-Pro-Hyp-**Gly**-Ala-Hyp-
Gly-Pro-Gln-**Gly**-P [hydroxyproline] ro-Hyp-**Gly**-Glu-Hyp-**Gly**-Glu-Hyp-**Gly**-Ala-Ser-
Gly-Pro-Met-**Gly**-Pro-Arg-**Gly**-Pro-Hyp-**Gly**-Pro-Hyp-**Gly**-Lys-Asn-**Gly**-Asp-Asp-
Gly-Glu-Ala-**Gly**-Lys-Pro-**Gly**-Arg-Hyp-**Gly**-Gl [hydroxylysine] ro-Hyp-**Gly**-Pro-Gln-
Gly-Ala-Arg-**Gly**-Leu-Hyp-**Gly**-Thr-Ala-**Gly**-Leu-Hyp-**Gly**-Met-Hyl-**Gly**-His-Arg- . . .

Figure 11.1 Amino acid sequence of collagen. Collagen has a very unusual structure, easily visible in this fragment, which shows only the first 90 residues out of a sequence of more than 1000, for one of the numerous forms of collagen. Note the following in particular: Every third residue is glycine (**Gly**); many of the others are proline (Pro) or hydroxyproline (Hyp), and some are hydroxylysine (Hyl). These last two occur in very few other proteins and are not encoded in the DNA: They are formed by modifying proline and lysine residues after the synthesis of procollagen, the precursor of collagen.

to be random at first sight.[6] Collagen, on the other hand, has an obvious regularity: Every third residue is glycine, and the glycine is often, but not always, followed by proline; the other residues "look" as random as they do in any other protein (Figure 11.1). This regular structure allows collagen to exist as a fibrous protein, in which three polypeptide chains are intertwined to form a triple helix, producing a fiber of great strength.

The great mechanical strength of collagen fibers allows all of the tissues mentioned earlier to achieve and maintain their structures. Because of this, collagen has acquired considerable economic importance, as cosmetic companies have successfully persuaded many customers that collagen creams, collagen dietary supplements, or collagen injections[7] will be useful for health. Maybe, but I have not seen any serious evidence that any of these treatments have any useful effect. It is the collagen you make yourself, and gets installed by the proper biochemical mechanisms in the places where it is used, that you need, not extraneous collagen that you inject or rub into yourself. Dietary collagen may have a small useful effect, because digestion will convert part of it into glycine that can be used for making your own collagen, but that is a very expensive way of increasing your glycine intake. Injection seems as likely to be effective for maintaining the structure as laying steel cables along an aging and deteriorating suspension bridge would be.

Glycine

About one-third of the residues in collagen are glycine, and glycine accounts for about one-quarter of the total mass of the protein—less than one-third because glycine is the lightest of the amino acids

[6]It is important to emphasize that regardless of how random a natural amino acid sequence may *look*, it is not *in fact* random.

[7]To quote one advertisement, "collagen injections are minimally invasive procedures aimed at skin rejuvenation without the risk, recovery time and cost of major surgery." All that is true enough, as long as we recognize that "aimed at" says nothing about whether such treatments work, and that "without the cost of major surgery" does not imply negligible cost.

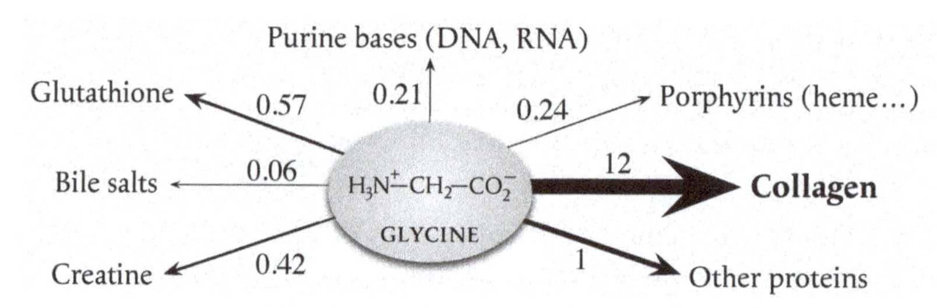

Figure 11.2 Structure and uses of glycine. Glycine has the simplest structure of all the amino acids. By far its major use is for the synthesis of collagen, but it has at least six other functions that are necessary for health. The numbers next to the arrows indicate the estimated number of grams of glycine needed each day by an adult human of about 70 kilograms. Except for collagen, which would require such a thick arrow that it would overwhelm the diagram, and for bile salts, for which the arrow would be invisible, the thickness of each arrow is proportional to the daily requirement.

(Figure 11.2).[8] The importance of this becomes more evident when we realize that collagen is by far the most abundant protein in the human body, about one-quarter of the total, and synthesizing it implies a substantial consumption of glycine. That in turn did not seem to be a great problem, for two reasons: Until fairly recently, bone and other tissues containing large amounts of collagen were believed to turn over so slowly—over years—that the large amount of collagen in them did not imply a high rate of synthesis; secondly, animals, including ourselves, have an enzyme, glycine hydroxymethyltransferase, that can synthesize glycine from serine, an abundantly available precursor. Both of these arguments fail, for different reasons.

It is now known that collagen-rich tissues are replaced far more rapidly than had been thought. The average lifetime of bone is about 40 days, somewhat more for skin and ligaments, and significantly less for heart, muscle, cartilage, and lung. In all tissues, we are concerned with weeks, not years, and this implies a high rate of collagen synthesis. This in turn represents a glycine consumption of around 10 grams per day, taking account of all the uses of glycine. Although collagen synthesis is by far the most substantial consumer, it is not the only one, and glycine is also needed for the synthesis of other products: proteins, porphyrins (the essential non-protein component of hemoglobin), purines (essential components of DNA), creatine (needed for muscle function), and glutathione (essential for resisting the toxic effects of oxygen).

Does not the existence of glycine hydroxymethyltransferase and an

[8]It is not by chance that glycine and glucose have similar names, despite being chemically quite different from one another, because both derive their names from the Greek word γλυκός, which means "sweet." Everyone knows that many sugars, including glucose, are sweet, some, like fructose and sucrose, being much sweeter than glucose. Less well-known is that glycine is also sweet, sweeter than glucose and tasting somewhat like sucrose.

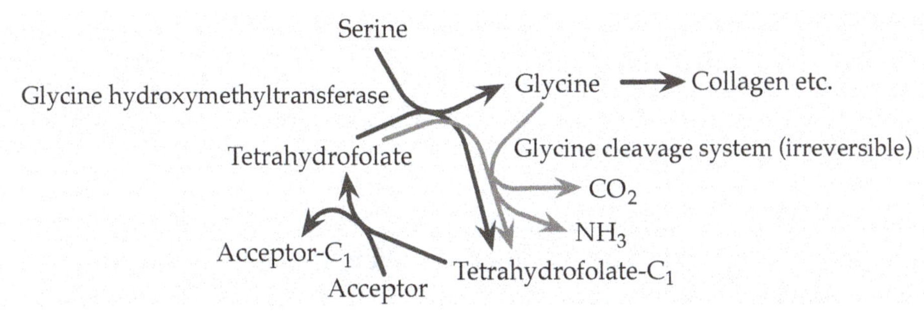

Figure 11.3 Conversion of serine to glycine by the reaction catalyzed by glycine hydroxymethyltransferase. The glycine produced in the reaction is used for synthesizing collagen and other proteins, and for satisfying other biochemical needs. The other product, tetrahydrofolate-C_1, is used in various biosynthetic processes, and each time a C_1 unit is transferred to an acceptor molecule, the substrate tetrahydrofolate is regenerated. Surplus glycine can be converted to tetrahydrofolate-C_1 in an *irreversible* process, the glycine cleavage system, but no known process can convert surplus C_1 units to glycine.

abundant supply of serine, together with the glycine that can be obtained by digesting proteins in the diet, mean that an animal can obtain as much glycine as it needs, and thus synthesize as much collagen as it needs? Unfortunately, no, because of a rather subtle feature of the glycine hydroxymethyltransferase reaction[9] that needs to be carefully explained. Serine contains one carbon atom more than glycine, and the reaction that converts it into glycine transfers the additional C_1 unit to a molecule called tetrahydrofolate, from which it is transferred to other molecules (regenerating tetrahydrofolate) as a way of adding carbon atoms to them (Figure 11.3). The details of how this is done are not important for the present context: It is sufficient to note that C_1 units are needed for growth and are thus more important for growing animals than for mature ones. If an animal's need for C_1 units is greater than its need for glycine, as will usually be the case for growing animals and, for reasons I will come back to, for small animals in general, then surplus glycine can be converted into C_1 units and transferred to tetrahydrofolate, but that reaction is irreversible and no reaction is known that can bring about the reverse process of converting unwanted C_1 units into glycine. All this is illustrated in Figure 11.3, in which it should be clear that every time a glycine molecule is made, a C_1 unit must be transferred to tetrahydrofolate.

Because the argument is rather technical, it may be clearer if we apply the same ideas to a more everyday example, the production of bread and beer. Both depend on the fermentation of yeast and both produce alcohol, though in the case of bread the amount of alcohol left in the finished product is so

[9]E. Meléndez-Hevia and P. de Paz-Lugo (2008) "Branch-point stoichiometry can generate weak links in metabolism: the case of glycine biosynthesis" *Journal of Biosciences* **33**, 771–780.

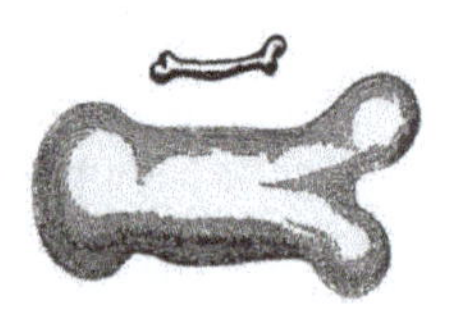

Figure 11.4 Allometric scaling. Galileo's drawing in *Two New Sciences* (1638) greatly exaggerates the effect, and unless the animals differ enormously in size, the difference in their bones is much less than suggested here. Not even a shrew and an elephant would show as great a difference as this.

small that I do not suppose anyone has become drunk as a consequence of eating too much bread. There is another difference between bread and beer: The production of beer leaves more yeast than there was at the beginning, which can be used for other processes, but when bread is baked the yeast is killed. Let us imagine that no source of baker's yeast was known other than as a by-product of brewing. This would mean that the amount of bread that could be made would be strictly dependent on the amount of beer produced, and a religious movement that banned the brewing of beer and other alcoholic beverages would also produce as a side effect a ban on the baking of bread.

We saw in Chapter 10 that biosynthetic pathways are typically regulated by the demand for end product, but the synthesis of glycine is regulated by the demand for C_1 units and not by the demand for glycine. (In our hypothetical situation, the production of bread would be limited by the demand for beer.) This pattern of regulation presumably evolved in bacteria before the appearance of animals, and before, therefore, the need for collagen. It still works perfectly well in all of the organisms that do not require collagen, because in all of these, insufficiency of C_1 units is far more likely to be a problem than the shortage of glycine.

Allometric scaling

Even in animals it only becomes a serious problem for large, adult, terrestrial animals. The difference between rapidly growing animals (which consume large amounts of C_1 units) and adults should be obvious, but the other two points require some explanation. In Chapter 2, we noted that a domestic cat is built differently from a lion and a mouse differently from a rat. These are examples of a principle known as *allometric scaling*, which was first described and illustrated by Galileo, and his drawing is illustrated in Figure 11.4 (with a more realistic one in Figure 11.5). The point can be understood by considering the length, surface area, and volume of any object, for example a cube with a length of 1 cm for each edge. The area of each face is $(1\ \text{cm})^2$, or $1\ \text{cm}^2$, and because there are six faces the total surface area is $6\ \text{cm}^2$. The volume is $(1\ \text{cm})^3$, or $1\ \text{cm}^3$. Suppose now that we scale it up by a linear factor of two, to produce a cube with a length of 2 cm for each edge. The surface area is now $6 \times 2^2\ \text{cm}^2 = 24\ \text{cm}^2$, and the volume is $2^3\ \text{cm}^3 = 8\ \text{cm}^3$.

Thus, increasing the linear dimension by a factor of two does *not* mean increasing other dimensions by the same factor: The surface area increases by a factor of four, and the volume (and mass) by a factor of eight. This sort of differential scaling is important for engineering structures as well as for

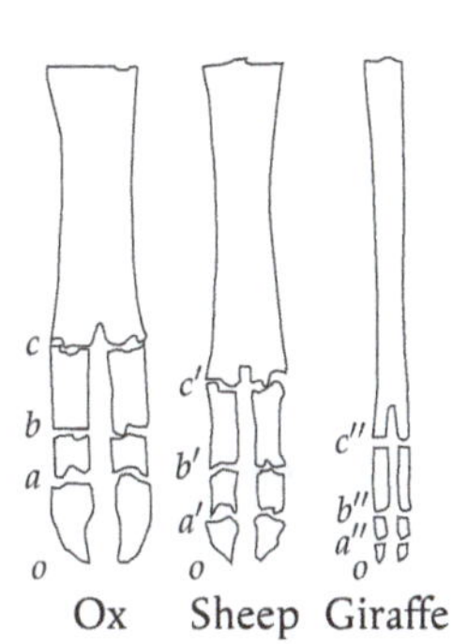

Figure 11.5 More realistic allometric scaling. The illustration of the metacarpal bones from ox, sheep, and giraffe is redrawn from Figure 506 of D'Arcy Thompson's book *On Growth and Form*, discussed later in Chapter 14 (© Cambridge University Press; reproduced with permission). It conveys the allometric relationship much more realistically than Galileo's drawing in Figure 11.4. The three bones are rescaled so as to give the same height for each. The giraffe bone may seem surprisingly slender compared with the other two, but a full discussion of this would be inappropriate here.

biological structures: A bridge that is perfectly satisfactory for crossing a river 10 meters wide cannot be rebuilt to span a river 100 meters wide according to exactly the same design with all lengths increased tenfold. Such a scaled-up bridge would not be 10 times as heavy, but 1000 times as heavy, and would not be able to support its own weight, let alone any traffic across the river. So it is with biological structures: A silk thread so thin as to be virtually invisible can easily support the weight of a spider, but a scaled-up thread could not support a rat, let alone a man. A flea can jump to 300 times its own height, but an elephant cannot jump at all, despite being vastly stronger than a flea.

Applying these ideas to collagen requirements is somewhat more complicated, because the powers involved are not simple integers and are less obvious. The amount of collagen needed to support the body increases with the 1.1 power of the weight, whereas the metabolic capacity increases with the 0.75 power. This was discovered by Max Kleiber and is called *Kleiber's law*. It is illustrated in Figure 11.6 for a wide range of mammals from shrews to whales.[10] In a 70-kilogram human, about 5% of the body weight is collagen, equivalent to about 1 kilogram of glycine. Such a man is about 230 times heavier than a 300-gram rat, and thus needs $230^{1.1} = 400$ times as much collagen, but the metabolic rate is only $230^{0.75} = 60$ times greater. It is thus about seven times as difficult for a man to produce enough collagen as it is for a rat.

GALILEO (1564–1642) is best known for his work in astronomy and for his disputes with the Inquisition, but he also introduced the concept of *allometry*, whereby different characters of otherwise similar animal species scale differently. His own drawing in Figure 11.4 greatly exaggerates the effect.

This explains why it is more difficult for a large animal to make enough collagen, but why do we need the qualification *terrestrial*? Why should the

[10]In *The Ancestor's Tale* (Weidenfeld & Nicolson, London, 2005), Richard Dawkins shows a graph spanning a range of organisms from small bacteria to large mammals, and asserts that "the truly astonishing thing about Kleiber's law is that it holds good from the smallest bacterium to the largest whale. That's about 20 orders of magnitude." That is an exaggeration, however, because the data for microorganisms, cold-blooded organisms, and warm-blooded organisms lie on three different lines, and the metabolic rate of a whale is about 200 times greater than it would be if one just extended the line for bacteria to reach the size of a whale.

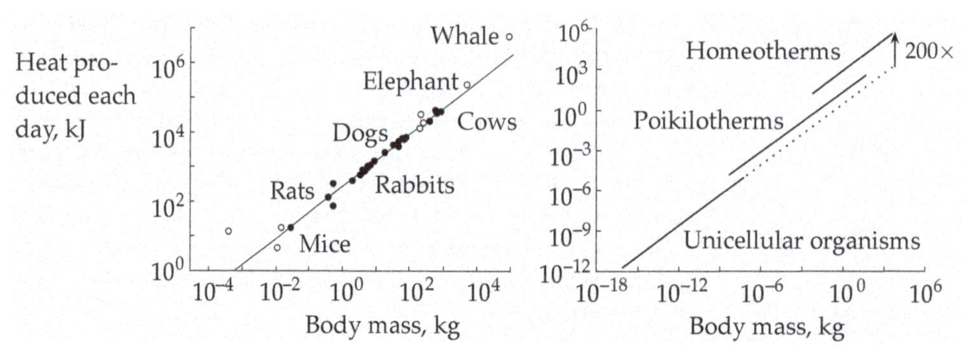

Figure 11.6 Kleiber's law. The metabolic rate of an animal varies with the 0.75 power of its body mass (*left*) over a wide range of mass. The points are plotted from a table given by Max Kleiber (1947) "Body size and metabolic rate" *Physiological Reviews* **27**, 511–541. Open circles represent data for animals measured under conditions that Kleiber did not consider comparable, but apart from some extra scatter, they agree well with the line calculated from the filled circles. Not all species are labeled. More extensive data (not available to Kleiber) allow results for a wider range of body masses to be compared (*right*), from the smallest bacteria to the largest mammals. Results for homeotherms (warm-blooded animals), poikilotherms (cold-blooded organisms), and unicellular organisms lie on three different lines.

problems not be even more severe for a sperm whale than they are for humans? This is because terrestrial animals must be strong enough to support their own weight, whereas aquatic animals do not have to be, because they are supported by the buoyancy of the water in which they live. As a result, animals like sperm whales need much less collagen in relation to their size than the allometric relationship for terrestrial animals would suggest and can reach sizes that would be impossible for terrestrial animals.

How much glycine is available?

It is, unfortunately, complicated to determine with certainty whether the amount of glycine available to an adult human is enough to satisfy the need for collagen synthesis. There are several different sources of glycine: Glycine hydroxylmethyltransferase (Figure 11.3) and the diet are the most important, but they are not the only ones. In addition, glycine is needed for several different biochemical functions, not just for the synthesis of collagen. These are both difficult to quantify accurately, and there is a third complication that renders the problem even more difficult to solve: the *procollagen cycle*. In common with some other proteins, such as the hormone insulin and enzymes of digestion such as the stomach enzyme pepsin, collagen is not initially synthesized as the finished product, but as *procollagen*, a molecule that undergoes various chemical changes before it can be assembled into the triple helix of collagen: Some of its proline and lysine residues are changed into hydroxyproline and hydroxylysine, two amino acids that are not part of the "standard 20" encoded by the DNA and are not found in other proteins.

The chemical modifications need to be done in a precisely correct way to produce a satisfactory collagen fiber, and in practice errors occur. The triple helix must therefore undergo a thorough process of quality control, and any defective products that are not correctly folded are degraded and resynthesized. The proportion of collagen molecules that are thus degraded varies from tissue to tissue, but in some cases it is huge—about 90% in bone; it is even higher in the heart, but that is less important for the total because the heart consumes relatively little collagen. If all the free amino acids produced by the degradation of defective molecules could be reused, the procollagen cycle would not affect the calculation. There are two difficulties with this, of which the first is that the amount of recycling is unknown and difficult to measure, and the second is that such data as exist indicate that the proportion is unlikely to be higher than 95%. The activity of the procollagen cycle is so great, however, that even a wastage of 5% of the glycine will add up to several grams per day in an adult person.[11]

Independent evidence that adult humans suffer from a glycine deficiency comes from medical observations. Various authors over recent decades have suggested that glycine should be defined as *conditionally essential*, which means that despite its availability from the diet and despite the existence of a reaction to synthesize it, the amount available may not be enough. Collagen-related diseases are common in older people, and although not usually life-threatening, they have a major impact on the quality of life. Strikingly, although degenerative diseases are not normally found in large wild animals, osteoarthritis and osteoporosis can be observed in such animals as elephants, for example. Moreover, study of ancient human skeletons showed that these diseases are far older than human culture and are thus unlikely to be caused by modern diet and behavior.

Obesity

Before discussing why natural selection appears to have failed to solve the problem of insufficient collagen production—on the face of it not an obviously difficult problem, given that glycine is the simplest of the amino acids and elaborate chemistry is not needed for making it—I will discuss a different problem that natural selection has not solved, the obesity that follows fom excessive intake of carbohydrates, with all the health problems that accompany obesity. Although once largely confined to a few countries, it is on its way to becoming universal, and countries like France, in which obesity was once rare, have seen their levels doubling in the past 25 years (Figure 11.7). The high place of Mexico on the list is not

[11]E. Meléndez-Hevia, P. de Paz-Lugo, A. Cornish-Bowden, and M. L. Cárdenas (2009) "A weak link in metabolism: the metabolic capacity for glycine biosynthesis does not satisfy the need for collagen synthesis" *Journal of Biosciences* **34**, 853–872.

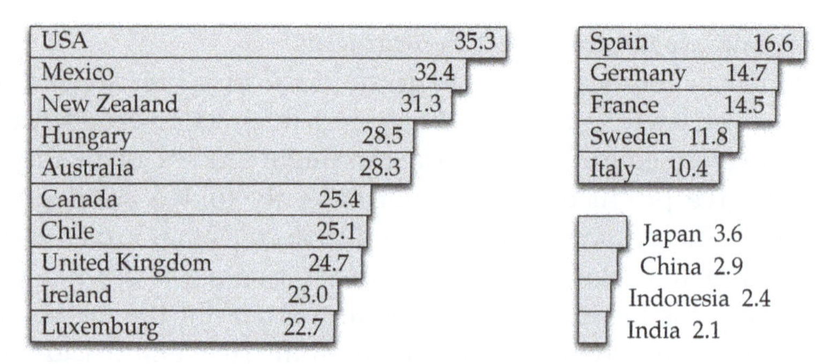

Figure 11.7 Obesity in the world. The ten countries with the highest levels of obesity are listed, together with the four lowest and five others. The numbers are percentages of obese people over the age of 15. The figure is based on data for 2012 of the Organization for Economic Co-operation and Development (OECD): `http://tinyurl.com/m3a7vfq`

surprising, given its proximity to the USA,[12] but some of the others may appear surprising. In particular, Hungary and Chile exemplify the dangers that await countries that see rapid increases in prosperity over the time of a generation.

Dietary carbohydrate

High consumption of carbohydrates is a consequence of agriculture. Before agriculture, that is to say before 10,000 years ago, a short time on an evolutionary scale, our remote ancestors had few sources of carbohydrates. They were essentially limited to ripe fruit, but only when it was in season, and wild honey, which was never any easier to find and harvest than it is today.[13] Widespread obesity has appeared on an even shorter evolutionary scale, within the past half century, so it can hardly be attributed to carbohydrates as such. Cheap refined sugar is a product of modern industry, however, as are cheap sweet beverages. These are indirect products of agriculture, but products nonetheless.

Why should carbohydrate, whether refined sugar or starch, be harmful?[14] Glucose is an essential constituent of blood, because it is the fuel used for energy production, but a need for *some* glucose does not imply a need for *unlimited* glucose. To understand why it might be harmful, we need to examine how glucose is used in two central metabolic pathways, glycolysis and the tricarboxylate cycle, often called the *Krebs cycle*, because it was discovered by

[12] *¡Pobre México! ¡tan lejos de Dios y tan cerca de los Estados Unidos!* (Poor Mexico: so far from God and so close to the United States!): attributed to Porfirio Díaz (President of Mexico until the revolution of 1911).

[13] Remember also that modern supermarket fruit has been artificially selected to be larger and sweeter than its equivalent in the wild.

[14] Like so much in this book, the discussion of obesity and carbohydrate intake in this section (other than textbook information like the reactions of the tricarboxylate cycle) is based on ideas of Enrique Meléndez-Hevia, in this case ones presented at a meeting that he organized in Puerto de la Cruz, Spain, in 2007.

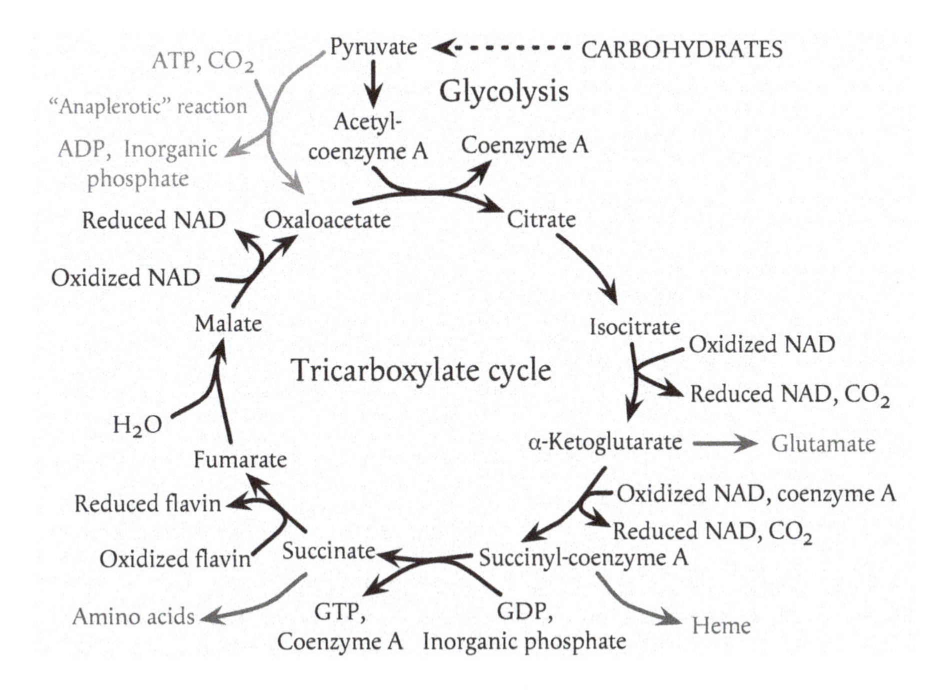

Figure 11.8 Glycolysis and the tricarboxylate cycle. These two pathways bring about the oxidation of carbohydrates to produce energy in the form of ATP. Glycolysis converts carbohydrates to pyruvate, a highly reactive molecule that is introduced into the tricarboxylate cycle by coenzyme A, where it reacts with oxaloacetate. In one turn of the cycle, the pyruvate is completely consumed, but the oxaloacetate is regenerated. Although the principal function of the cycle is to produce ATP, this is produced in a separate pathway, the *electron transport chain*, which uses molecular oxygen to reoxidize reduced NAD. The reactions shown in dark gray are processes that remove intermediates to synthesize other metabolites, and the *anaplerotic reaction* shown in light gray has the function of replacing these, so that the cycle is not brought to a halt.

Hans Krebs.[15] These are illustrated in Figure 11.8.

Treating the tricarboxylate cycle just as a cycle, and considering only the reactions shown in black in Figure 11.8, the intermediates, from citrate to oxaloacetate, are *catalysts*, because they are not consumed but regenerated in each turn of the cycle. Matters are not as simple as that, because we are ignoring the important fact that the cycle is needed not just to allow the combustion of carbohydrates and the production of ATP by the respiratory chain, but also acts as a source of starting materials of various *other* metabolic pathways. Succinate, for example, is used for the production of various amino acids, and other intermediates are similarly used in other pathways. If a catalyst is removed

[15]Most modern textbooks call it the *tricarboxylic acid cycle* or the *citric acid cycle*, but these names are misleading, because the cycle involves the negatively charged anions, not the acids.

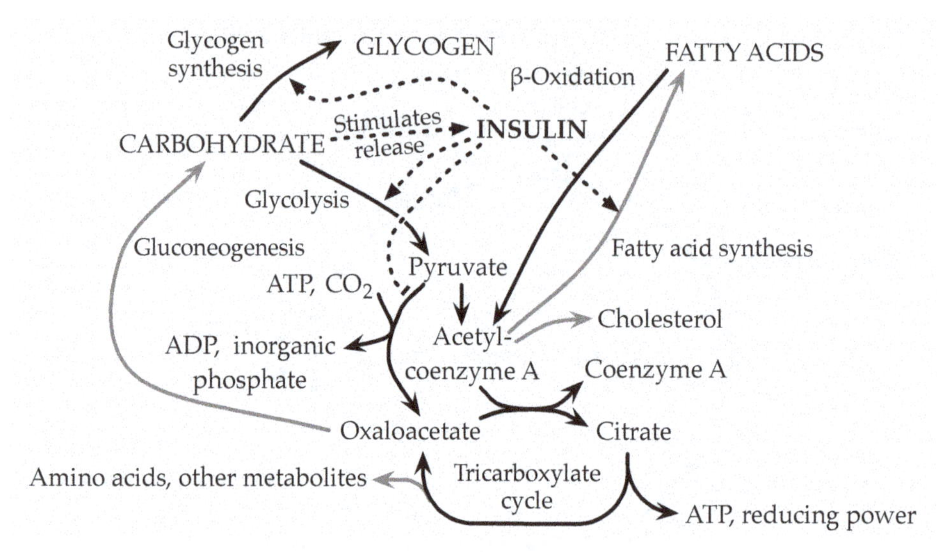

Figure 11.9 Regulation of the tricarboxylate cycle. The reactions of Figure 11.8 are shown as a summary to better display the regulatory interactions, and to show that carbohydrates are not the only source of molecules for entry into the cycle; fatty acids can also enter it by *β-oxidation*. The fuels can also be taken from the cycle: Carbohydrates can be regenerated from oxaloacetate by *gluconeogenesis*, and fatty acids can be regenerated by *fatty acid biosynthesis*. The hormone *insulin* is the principal regulator of the various processes: It inhibits the anaplerotic production of oxaloacetate and stimulates glycolysis, glycogen synthesis, and fatty acid biosynthesis. Insulin itself is always present, but its concentration is increased as a response to high glucose levels. This is a rather complicated diagram, and one needs to distinguish between chemical transformations, represented by continuous arrows, and regulatory effects, represented by broken arrows. Processes that tend to decrease cycle activity are shown with gray arrows.

from a system and used for other purposes, less and less is left to fulfill its catalytic function, and this use of intermediates for other functions suggests that the cycle must slow down and eventually stop.

That this does not normally happen is due to an additional reaction, called an *anaplerotic reaction*, which just means a "refilling reaction," that allows pyruvate to be converted directly to oxaloacetate without passing through coenzyme A. This anaplerotic reaction requires ATP, so it is an energy-consuming reaction. In normal circumstances it supplies just enough oxaloacetate to replace the cycle intermediates consumed by biosynthetic side reactions.

To understand all this, we need to look at the summary of the metabolism that is shown in Figure 11.9, which shows not only that carbohydrate can enter the tricarboxylate cycle, but it can also be regenerated from the cycle by *gluconeogenesis*. In addition, carbohydrate is not the only fuel involved, because fatty acids—loosely speaking just fats—can also act as a fuel, by *β-oxidation*, or can be produced from the cycle by *fatty acid biosynthesis*. Just by inspection of Figure

11.9, it would appear that carbohydrates can be converted to fatty acids and fatty acids to carbohydrate, but it is not so simple. Carbohydrates can indeed be converted to fatty acids, so the final result of excessive carbohydrate intake from the diet is deposit in the form of fat, but, despite appearances, the reverse is *not possible*: No net synthesis of carbohydrates from fats is possible. We need the word "net" here, because all the reactions are present, and individual molecules of fatty acids can be converted to carbohydrate molecules, but the total balance is zero. The reasons for this are complicated, and I will not go into them here: We just need to accept as a fact that in humans there is no net conversion of fat into carbohydrate.

This has a serious consequence. It means that once fat has been deposited, it can only be mobilized by exercise: It can be used as a fuel to drive the tricarboxylate cycle and hence to produce ATP for driving the exercise, but it cannot be converted to carbohydrate and thus used to replenish stores of glycogen. The only way to reduce fat deposits, therefore, is to convert them to energy by muscular activity—that is, by exercise, but that will not happen if there is an abundant supply of energy in the form of carbohydrate and if people take little exercise.

Regulation of the tricarboxylate cycle

We now need to consider how all this is regulated, because it will clearly not be beneficial to have all the different processes occurring simultaneously. In primitive times, people lived on the edge of starvation (like many wild animals today), and could only survive periods of hunger by mobilizing their fat stores, and once those were completely consumed, they died. In those circumstances, fat stores were an essential protection against lack of food, so it was clearly beneficial to build them up when the opportunity arose. Eating a substantial amount of carbohydrate, typically in the form of ripe fruit or honey, would provide such an opportunity, and the primary effect of the presence of large amounts of carbohydrate is to stimulate the release of the hormone *insulin*. This stimulates *glycogen synthesis*, causing the glycogen reserves to be filled, but there is a limit to the amount of glycogen that can be stored, and when that is reached, insulin has other effects on the processes of Figure 11.9: It inhibits the anaplerotic conversion of pyruvate to oxaloacetate, thereby inhibiting the replenishment of the cycle intermediates when they are removed as precursors for biosynthesis; in addition, it promotes the biosynthesis of fatty acids and cholesterol. Both of these are valuable properties for people at risk of starvation, but they are not at all good properties for people with abundant cheap high-carbohydrate food and at risk of overeating. Inhibition of the anaplerotic reaction could be problematic in a low-protein diet, though possibly in earlier times the level of insulin would not have been high enough to inhibit it sufficiently.

It is hard to escape the conclusion that the great increase in availability of high-carbohydrate food that followed the invention of agriculture is responsible for obesity. That still raises some questions: Agriculture has been known for about 10,000 years, a trivial time in evolutionary terms, but still a long time in terms of human history, so why is it only now that we are seeing a huge increase in obesity, if the cause is 10,000 years old? I have already alluded to one reason: Yes, starch has been available for thousands of years, but refined sugar is much more recent, cheap refined sugar is even more recent, and more recent still are abundant, cheap, sweet beverages. Refined sugar is not necessarily worse for the health than starch, apart from two points: As discussed in Chapter 6, sugar has an immediate effect on the osmotic pressure of the blood that starch does not have; more relevant to this chapter, an excessive intake of starch triggers the instinct to stop eating, but this instinct is much weaker in the case of refined sugar, so that it is possible to eat a harmful amount of it before one is unable to continue. The other big change in the past 50 years is in the amount of exercise people take, and we have seen that exercise is the only way of burning off fat stores. Here we need to consider not only deliberate exercise, but also involuntary exercise such as *shivering*, which has been greatly decreased in modern times by artificial methods of maintaining a comfortable ambient temperature indoors.[16]

The reason why agriculture is now practiced over almost the entire world is not primarily that it is healthier than hunting and scavenging, or that it is less work, or that it makes people happier. On the contrary, studies of modern hunter-gatherers indicate that they survive on fewer hours of work than their farming neighbors, that they are just as healthy (or healthier), and they are not particularly attracted by the possibility of farming. The major advantage of farming—if we can call it an advantage—is that it can support much denser populations than hunting and scavenging. After agriculture appeared in the Middle East, it took about 4000 years of gradual spreading to reach Western Europe, at an average speed of about 1 kilometer per year—far slower than a garden snail can move. That is clearly not consistent with the idea that hunter-gatherers were so impressed with the merits of farming when they saw their neighbors doing it that they immediately imitated them, but it is consistent with the idea that the population increase of the agriculturalists gradually

HANS ADOLF KREBS (1900–1981) was born in Germany and did his first work there as assistant to Otto Warburg, but his appointment in Freiburg-im-Breisgau was ended by the Nazi government in 1933, and he moved to Cambridge. He was subsequently Professor of Biochemistry in Sheffield, where he created what became the strongest biochemistry department in England, and later at Oxford. In addition to the tricarboxylate cycle, he made numerous other contributions to metabolic biochemistry and was awarded the Nobel Prize in Physiology or Medicine in 1953.

[16]Newborn babies are not limited to shivering as a natural means of maintaining an adequate body temperature, because they have substantial amounts of *brown adipose tissue*, a form of fat that allows the direct metabolic generation of heat.

squeezed out the hunter-gatherers.

Why has natural selection failed?

In this chapter we have seen two major problems of human health that natural selection seems to have been unable to eliminate, even though both in principle would seem to be susceptible to selection. How can we explain this? The second is probably the easier to deal with. If the human species survives for another million years, it is certainly possible that it will have evolved to the point of being able to support a high-carbohydrate diet without becoming obese. Even such a simple process as selection for individuals who do not *like* sweet food and drinks, or, like cats, have lost the receptors that allow them to taste sweetness, would go a long way toward an improvement. However, 10,000 years is negligible on an evolutionary scale, and 50 years is completely imperceptible.

The problem of collagen-related diseases is very different, because 580 million years is by no means negligible on an evolutionary scale. There has been no shortage of time, so why are today's animals still suffering from a problem that has existed for so long? The problem was almost certainly negligible at the beginning, because the first animals were small and aquatic, and required relatively much less collagen than large terrestrial animals, probably no more than could be supplied as a side effect of the production of C_1 units, and so 580 million years overestimates the time available. Moreover, rapidly growing animals need large amounts of C_1 units and can accordingly synthesize as much glycine as they need for collagen synthesis. Collagen-related diseases in large animals are not normally life threatening and have a negligible effect on reproductive capacity, and natural selection is in general indifferent to the quality of life of individuals that are no longer reproducing and no longer essential for the well being of their offspring.[17]

Two other considerations are pertinent. Metabolic innovation is easily observed in bacteria, and not unknown in plants, but animals in general show almost no capacity to evolve new metabolic pathways. They can lose ones their ancestors had, as the primates and some other animals have lost the capacity to synthesize vitamin C, so it needs to be present in the diet (whereas most mammals have no need for dietary vitamin C), but they cannot invent new ones. The second point is that glycine is the simplest of the amino acids, and the first organisms could probably obtain as much as they needed from their environment, so bacteria had no need to evolve regulatory mechanisms to ensure an adequate supply.

[17]Humans are unusual, even among primates, in having a long post-reproductive life. For most animals, the post-reproductive lifetime is short to the point of being negligible, caring for aged individuals being a specifically human characteristic, as noted in Chapter 7.

12. *A Small Corner of the Universe*

> Consideration of the conditions prevailing in bisexual organisms shows that ... the chance of an organism leaving at least one offspring of his own sex has a calculable value of about 5/8. Let the reader imagine this simple condition were true of his own species, and attempt to calculate the prior probability that a hundred generations of his ancestry in the direct male line should each have left at least one son. The odds against such a contingency as it would have appeared to his hundredth ancestor (about the time of King Solomon) would require for their expression forty-four figures of the decimal notation; yet this improbable event has certainly happened.
>
> R. A. Fisher (1954)[1]

> L'an mil neuf cens nonante neuf sept mois
> Du ciel viendra un grand Roy d'effrayeur
> Ressusciter le grand Roy d'Angolmois
> Avant après Mars régner par bonheur.[2]
>
> Nostradamus, Quatrain 72, 10th Cycle

OVER THE YEARS I have made only rare and short-lived attempts to maintain a diary—probably a good thing, because the fragments that survive reveal a life largely lacking in excitement and drama. The most sustained attempt occupied the first six weeks of 1959, with the result that I know with some exactness how and when I came to hear the name of Fred Hoyle, one of the most important cosmologists of the twentieth century, but also notorious for his excursions into biological evolution. On January 22, 1959, I recorded that my religion teacher had been reading a lecture by Fred Hoyle and that I disagreed strongly with it about relativity. It seems a curious thing to be reading in a religious knowledge class: The intention was to contrast Hoyle's godless approach to science with a more spiritual one we had heard about the previous week in a recording by another cosmologist, Bernard Lovell, famous at the time but now, unlike Hoyle, largely forgotten.

[1]R. A. Fisher (1954) in *Evolution as a Process*, edited by J. Huxley, A. C. Hardy, and E. B. Ford, George Allen & Unwin, London.

[2]"In the seventh month of the year 1999, a great King of Terror will come from the sky to revive the great King of Angoulème before and after Mars to reign in happiness." I was tempted to leave the quotation in French to avoid arguments about what it means, but there are so many fanciful and absurd interpretations on the web that I thought I should provide a straightforward one. Fortunately, the first line, the only part that matters for this chapter, is unambiguous.

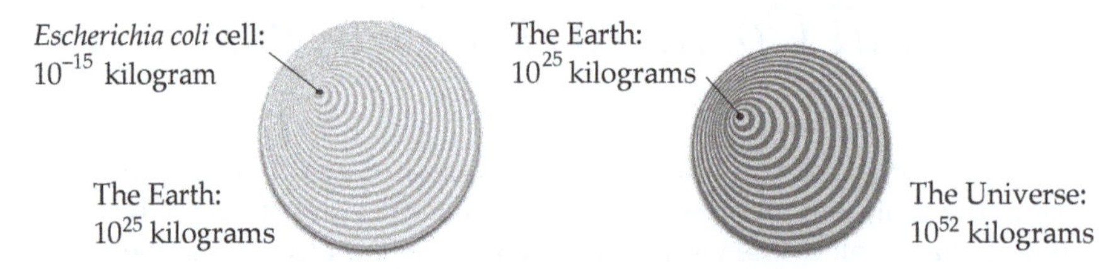

Figure 12.1 Relative sizes of a bacterial cell, the Earth, and the Universe. The Universe is only 10^{27} times larger than the Earth, whereas the Earth is 10^{40} times larger than an *Escherichia coli* cell. The drawings are not to scale, but each shaded circle represents a tenfold increase in size.

Although I remember the occasion independently of the diary, my memory is different from what I wrote, in that I can remember feeling contemptuous of Hoyle's suggestion that somewhere in the Universe there was a team that could beat Australia at cricket, but I have no recollection of disagreeing with him about relativity; indeed, I am astonished that at that time I thought myself qualified to hold an opinion about relativity, especially one in strong disagreement with that of a leading cosmologist. The remark about cricket teams elsewhere in the Universe seems at first sight too absurd to be worth discussing, but it makes sense if the Universe is infinitely large. In an infinitely large Universe, every possible event, no matter how improbable, is occurring, not just once but infinitely often.

Emergence of life

The specific improbable event that interested Hoyle throughout much of his life was the appearance of life and its subsequent evolution into the diversity of life that we know today. He argued that both the initial appearance and the subsequent evolution were too improbable to have occurred in such a small place as the Earth, that the evolution of complex life forms from simpler ones on Earth was absurd, and that we needed the vastly larger space of the entire Universe to explain it. The problem with this argument is that the Universe as it is described by most astronomers is not nearly large enough to be much help. We are so conscious of how much bigger it is than ourselves that we easily forget how small it really is.

Size of the Universe

Although Hoyle himself sometimes referred (without explanation) to a "spatially infinite Universe, a Universe that ranges far beyond the largest telescopes," the usual estimate is that it contains about 10^{79} atoms (one followed by 79 zeroes), a trivial number compared with what we need if we are to be convinced of the necessity to assume an origin and evolution of life elsewhere. 10^{79} atoms may seem like quite a lot, but let us consider it in relation to more everyday objects. Because most of the 10^{79} atoms are hydrogen atoms, it corresponds to a mass of the order of 10^{52} kilograms, about 10^{27} times heavier than the Earth, which weighs about 10^{25} kilograms. By contrast, the common gut bacteria *Escherichia coli* have cells with a volume of about 10^{-15} liter, and hence a mass (taking the density to be about that of water) of about 10^{-15} kilogram. Thus, the Earth is about 10^{40} times bigger than a typical bacterial cell and is vastly bigger on the scale of such a cell than the Universe is on the scale of the Earth, as illustrated schematically in Figure 12.1. Taking a different example, an adult human weighs of the order of 100 kilograms (most of us weigh less than that,[3] but no matter, we are talking in terms of orders of magnitude here to make the arithmetic easy), and the Sun weighs about 10^{30} kilograms, so it is about the same size in relation to the Universe as a human being is in relation to the Earth. The Sun and the other stars are much too hot for life to have originated, evolved, or survived there, and the dark matter that is believed to account for most of the mass of the Universe is much too cold (with an average temperature a few degrees above absolute zero). The Earth, therefore, represents a larger proportion of the inhabitable part of the Universe than these calculations suggest, and anyway a far larger proportion than most of us would guess without doing this simple arithmetic.

A different way to appreciate that the Universe is smaller than we tend to assume is to ask how many stars it contains, compared with how many microbes the Earth alone contains. As Maureen O'Malley[4] points out, the answer is surprising: There are about 100 million times more microbes on Earth than there are stars in the Universe.

The origin of life: a vastly improbable event?

We are at present far from understanding the origin of life. It seems to require vastly improbable events: On the one hand, efficient catalysts made of protein, and on the other hand, an efficient system for storing the information needed to make them, and it is hard to see how one could arise without the other. In Chapter 14 I will discuss how self-organizing systems could arise, but here I will concentrate on the aspect of probability, because it is sometimes argued that the founding event was so improbable that the Earth is too small and young for it to be plausible that it happened here. What I will argue is that the Universe is neither big enough nor old enough for the differences in size and age to affect the argument.

What does it matter if the Universe is only 10^{27} times heavier than the Earth, rather than, say, 10^{270} or 10^{2700} times heavier? Surely even 10^{27} is a large enough factor to cope with the improbabilities inherent in the appearance and evolution of proteins? Unfortunately for this argument, a factor like 10^{27} is utterly trivial in relation to the sort of numbers that Hoyle and others calculate in their efforts to prove that the biologists' view of the origin and evolution

[3]If the trends discussed in Chapter 11 are not brought to a halt, this may cease to be true!

[4]M. A. O'Malley (2014) *Philosophy of Microbiology*, Cambridge University Press, Cambridge.

	1	2	3	4	5	6	7	8	9	10	11	12	
Human	Gly-	Asp-	Val-	Glu-	Lys-	Gly-	Lys-	Lys-	Ile-	Phe-	Ile-	Met-	...
Some birds	Gly-	Asp-	Ile-	Glu-	Lys-	Gly-	Lys-	Lys-	Ile-	Phe-	Val-	Gln-	...
Tuna	Gly-	Asp-	Val-	Ala-	Lys-	Gly-	Lys-	Lys-	Ile-	Phe-	Val-	Gln-	...
Aspergillus niger	Gly-	Asp-	Ala	-	Lys-	Gly-	Ala-	Lys-	Leu-	Phe-	Gln-	Thr-	...

Figure 12.2 Some cytochrome c sequences. The standard three-letter code (Gly = glycine, and so on) is used, and only the first 12 residues are shown. *Aspergillus niger* (a mold), the most distantly related species, has a residue missing at position 4: Such gaps are rare in cytochrome c, but are very common in some alignments of protein sequences, and complicate analysis.

of life is untenable. Their argument asks us to take a "typical" protein with a sequence of 100 amino acids, such as the protein cytochrome c, which does have about 100 amino acids in many species, and is found in all the animals, plants, and fungi that have been checked. In the human, for example, it begins as shown in Figure 12.2. Suppose that the probability of having Gly (glycine) in the first position would be about one-twentieth, or 0.05, if the sequence were arranged at random, and that of having Asp (aspartate) at position 2 would likewise be 0.05, that of having Val (valine) at position 3 would again be 0.05, and so on. (These are suppositions rather than necessary facts, because they involve assuming that all the possibilities are equally likely. In fact they are not, because some amino acids are used much more often in proteins than others, and if this is taken into account, the true probabilities come to about 0.07 at each position. The correction makes little difference to the argument, and I will ignore it.)

If we further suppose that whatever amino acid we have in one position has no effect on the probability of what we will find adjacent to it, then the probability of having both glycine at position 1 and aspartate at position 2 is 0.05×0.05, or $0.05^2 = 0.0025$, the probability of having Gly-Asp-Val as the first three is $0.05 \times 0.05 \times 0.05$, or $0.05^3 = 0.000{,}125$, which is more convenient to write as 1.25×10^{-4}, and so on. By the time we have counted to 10 amino acids, we have $0.05 \times 0.05 \times 0.05 \times 0.05 \times 0.05 \times 0.05 \times 0.05 \times 0.05 \times 0.05 \times 0.05 = 0.05^{10} = 9.8 \times 10^{-14}$ as the probability that a random sequence will begin Gly-Asp-Val-Glu-Lys-Gly-Lys-Lys-Ile-Phe... (and the same probability for any other 10-amino-acid sequence that we might consider). This is already very small, but this is just the beginning. By the time we get to 100 amino acids, we will have $(9.8 \times 10^{-14})^{10}$, or 10^{-130}. Human cytochrome c actually has 104 amino acids, not 100, and the extra four amino acids bring the probability down to 10^{-135}.

This is an optimistic estimate of our chances of producing human cytochrome c by a random shuffling of amino acids, because it assumes that there is some way of doing the shuffling and stitching all the amino acids together in the first place. That is in itself highly improbable, and it is just the beginning of the difficulties: A typical enzyme has many more than 100 amino acids, and we need more than one of them; a single enzyme molecule is not a living organism! Even if we optimistically guess that we could have a viable organism with as few as 100 different proteins (remember from Chapter 3 that *Mycoplasma genitalium*, the simplest known modern organism, has around five times as many), with an average of 100 amino acids each, and we skate over the difficulties of putting the right amounts of these proteins together at the right place and protected from the environment by a suitable cell wall, we still find ourselves with a probability of about $10^{-13,500}$, an almost inconceivably small number.

If we pursue this argument, we arrive almost inevitably at the conclusion that we can "prove" that it is impossible for life to have originated on Earth, and although that may seem like an argument that it originated somewhere else, that is not sustainable because, as we have seen, "somewhere else" is only 10^{27} times bigger than the Earth, and improving our chances by a factor of 10^{27} is barely worth bothering with if we start with odds of one in $10^{13,500}$ against us. How comforted would anyone feel to learn that some enterprise had a chance of one in $10^{13,473}$ of success after thinking it was only one in $10^{13,500}$? In any case, 10^{27} is an optimistic estimate of how much it helps to regard the whole Universe as available for the origin of life to have occurred. Not only is most of the Universe much too hot or too cold for life, but in addition, the hypothesis that life originated elsewhere introduces a new difficulty that does not arise if life on Earth originated on Earth: We need a mechanism to transfer existing life forms to the Earth in a viable form from wherever they originated, so whatever we calculate as the probability that life originated anywhere must be multiplied by the low probability that such a mechanism exists.

FRED HOYLE (1915–2001) was one of the most productive and distinguished cosmologists of the twentieth century, best known for his rejection of the "Big Bang" (a deliberately contemptuous term that he introduced) theory of the origin of the Universe. He was born in Yorkshire and spent most of his professional life at the University of Cambridge. His views on the origin of life have found little favor with biologists, but his standing in his own field is deservably high.

Age of the Universe

There is one aspect of this analysis that we have not yet touched upon. Just as we tend to assume that the Universe is vastly bigger than the Earth unless we make the appropriate numerical comparison, so we also tend to think of the Universe as being vastly older than the Earth. If the Universe is, say, $10^{13,500}$ times older than the Earth, then maybe there has been so much time available for extremely improbable events to have occurred elsewhere in the Universe that there is no longer any difficulty with postulating them. The problem with this argument is that the Universe appears not to be nearly old enough. Although at the time when Hoyle first began to advance his probability arguments there was some support among other cosmologists for his theory of an infinitely old Universe, this appears to have largely dissipated in the past few decades, and the current view favors a Universe that is perhaps three or four times older than the Earth—about 14 billion years for the Universe and about four billion years for the Earth. It would be presumptuous for a biochemist to enter into this argument; suffice it to say there is no support for an infinitely old Universe in the writings of authorities of the standing of Stephen Hawking,[5] Steven Weinberg,[6] and Roger Penrose.[7] Although Paul Steinhardt and Neil Turok[8] do suggest that the Universe may be infinitely old, they see the Big Bang not as the beginning but as a cataclysmic moment in a history of cycles, with no beginning and no end. Nothing from one cycle survives into the next, so the living world that we know has still had only 14 billion years to develop, and their model is different from the infinitely old Universe that Hoyle was proposing in the 1950s.

Thus, not only can we apparently prove that life could not have originated on Earth, we can prove almost as easily that it could not have originated anywhere. Yet this conclusion must be wrong: We are here to discuss it, and we know that life does exist on Earth, so any proof otherwise must be flawed. It seems to me that there are only two possible ways out of the difficulty: Either a divine intelligence planted life on Earth (and possibly elsewhere) in a deliberate act, or the probability calculations outlined above are incorrect. No biologist would accept the counsel of despair implicit in invoking a divine intelligence,[9] and recalling that back in 1959 my teacher of religious knowledge used him as an example of a godless scientist, I do not believe that Hoyle would have done so either; even theologians nowadays appear unenthusiastic about a creator who wound up the clock, defined the laws of physics, and then left it running for 14 billion years without interference.

Only one possible sequence for each function?

This leaves the second possibility, and virtually all modern biologists would agree that the explanation is that the probability calculations are at best inappropriate, and probably plain wrong as well. We cannot just leave it at that, however, and we need to give some plausible reasons for thinking the calculations may be wrong. Let us begin by noticing the implicit assumption that the particular 104-amino-acid sequence that cytochrome c has in the human is the only possible one it could have if it is to

[5]Stephen Hawking (1998) *A Brief History of Time*, Bantam, New York.

[6]Steven Weinberg (1992) *Dreams of a Final Theory*, Pantheon Books, New York.

[7]Roger Penrose (1989) *The Emperor's New Mind*, Oxford University Press, Oxford.

[8]Paul J. Steinhardt and Neil Turok (2007) *Endless Universe: beyond the Big Bang*, Doubleday, New York.

[9]Well, there are maybe one or two: See Chapter 15.

work. If, on the other hand, the particular sequence is one of, say, 10^{120} that would work just as well, then the chance of getting a working cytochrome c by shuffling amino acids ceases to be a hopelessly improbable endeavor. We have no justification for putting the number as high as that, but we certainly have reason to believe that it is bigger than one, that the sequence we have is not the only possible one. First of all, although not only humans but also those other mammals that have been studied have cytochrome c sequences that begin as shown in Figure 12.2, the mammals do vary from one another in the remaining 94 locations, and some variants found in other organisms are also given in the figure. In position 3, the amino acid Ile (isoleucine) is rather similar in its properties to Val (valine), and replacing valine by isoleucine is usually regarded as a "conservative" change. On the other hand, in position 4, Ala (alanine) is rather different from Glu (glutamate), so this is not a conservative change.

Examination of the whole range of known cytochrome c sequences reveals at least some variation at the majority of loci, and they are not all of the same length: Even among higher animals, there is some variation, so that frogs and fish have 103 amino acids instead of 104; most plants have somewhat longer sequences. All of these proteins have similar chemical and physical properties, so it is certainly not the case that only a unique sequence can do the job. On the other hand, there is no suggestion that a vast number of different sequences could do it either. Among other kinds of proteins, there are some that are more conservative than cytochrome c, such as histone H4, a protein associated with the genetic material, DNA, which shows almost no variation over the entire animal and plant kingdoms.[10] There are also many that are much more variable, with barely detectable sequence similarity in comparisons between distantly related animals.

Unfortunately, none of this tells us whether even the most conservative structure, such as that of histone H4, is the only possible one, or whether it is just that having once chosen one of many possible structures, organisms are now "locked" into it, because although others still exist, they cannot be reached by any reasonable series of mutations. Consider a bee living on a small island 20 kilometers from the mainland. Such a bee might well conclude from its forays of a few kilometers from its hive that the island where it lived was the only place in the entire Universe where life was possible, but it would be not only wrong but grossly wrong: A huge number of other places exist where

[10]We need to be careful in this sort of context not to confuse identity of amino-acid sequences with identity of genes. The histone H4 *protein* is almost identical across animals and plants, but the *genes* are thoroughly randomized, to the extent that redundancy in the genetic code allows, and even the genes of a single species show variation (see Figure 1.6). There are very few significant blunders in Richard Dawkins' book *The Blind Watchmaker* (Longman Scientific and Technical, London, 1986), but the confusion between genes and the encoded protein is one.

bees can live, but they just happen to be inaccessible.

In Chapter 3 I discussed some of the knowledge that we have of how much variability in protein sequences organisms can stand. Further suggestions that a wide variety of sequences might be capable of fulfilling the role of any given protein come from studies of lysozyme, an enzyme that is widely distributed and acts to break down the cell walls of certain bacteria. It was famous for a while as a forerunner of penicillin, because it was discovered by Alexander Fleming before he discovered penicillin. Noticing its capacity to destroy some bacteria, he hoped it might have therapeutic value as an antibiotic. It did not prove useful for that, but its discovery may still have prepared his mind for the idea that much better antibiotics, such as penicillin, might exist. For the present discussion, the point is that lysozyme's bacteriolytic activity can be mimicked quite well with random copolymers of amino acids. For example, a "protein" with a sequence consisting of the two amino acids glutamate and phenylalanine in a random order has about 3% of the bacteriolytic activity of lysozyme.[11]

There are some difficulties with this result, because lysozyme is not a particularly good bacteriocide and is inactive against the more dangerous bacteria. It is found in places that have a low risk of bacterial infection, like bone marrow, and it is not found in some places where its bacteriocidal properties would be more useful. So, it is possible that its function is something different that has not yet been identified. The ability of random proteins to mimic its bacteriolytic activity would then be irrelevant. In an emergency you could use the pages of this book to kindle a fire, and they would do the job quite well; you might then find that some dry leaves would make an adequate alternative, but that would tell you nothing about the fitness of the book for its primary purpose.

Two other enzymes offer a different sort of evidence. Carbonic anhydrase catalyzes the dissociation of carbon dioxide from the state that exists when it is dissolved in water. I mentioned it in Chapter 1 as an example of an enzyme that catalyzes a reaction that occurs fast with no catalyst at all, noting that there are some contexts, for example for releasing carbon dioxide gas in the lungs for exhalation, where completing the reaction in a few seconds is not fast *enough*. Remember that we (humans) breathe about 12 times per minute, and so the average time available for the carbon dioxide bound to the hemoglobin that arrives in the human lung to be exchanged for oxygen is about five seconds. The rate at which dissolved carbon dioxide can be dehydrated if there is no catalyst is thus insufficient. Carbonic anhydrase is found in many different organisms apart from mammals, and it exists in three known classes with no detectable structural similarity. In other words, there are three entirely different

[11] V. K. Naithani and M. M. Dhar (1967) "Synthetic substitute lysozymes" *Biochemical and Biophysical Research Communications* **29**, 368–372.

kinds of proteins with no resemblance between them if you just compare their sequences, but which catalyze the same process and fulfill the same biological function.

Another enzyme, superoxide dismutase, is essential for preventing some highly toxic side effects of the use of oxygen in the energy-harnessing apparatus of many organisms and is found in two entirely different classes with no detectable structural resemblance. As a digression, incidental so far as this chapter is concerned but an interesting example of the sort of things that can happen in evolution, the ponyfish (*Leiognathus splendens*) is a luminous fish that emits light by means of a light organ in which lives a light-emitting bacterium called *Photobacter leiognathi*. Both fish and bacterium have superoxide dismutases, and that of the fish is a typical member of the class that other vertebrates have, but the bacterium has two forms of the enzyme, not only the form expected in bacteria but also a second that is typically fishlike. It is quite unlike the enzymes found in other bacteria, but at the same time not so similar to the enzyme found in its host that we could explain its presence by contamination of the bacterial samples with fish tissues.[12] It seems hard to escape the conclusion that a "horizontal transfer" of DNA took place during the long period of association between the two species, and at some moment a bacterial ancestor incorporated some fish DNA in its genome.[13]

Comparisons between the sequences of enzymes fulfilling the same functions in different species give at least a minimum estimate of how many related sequences are capable of doing any given task. For most proteins (histone H4 being an outstanding exception), we find at least one or two differences between the sequences found even in closely related organisms, and so the number of sequences possible for an enzyme that occurs in all species should be at least as large as the number of species, making millions of sequences even if we admit only species that exist on Earth today, but many more if we include extinct species. Within a single species we often find multiple *isoenzymes*, different proteins that catalyze the same reaction. In the human, for example, there are four isoenzymes of hexokinase, which are similar enough to one another in sequence for us to be sure that they have all derived from a common ancestral sequence.[14] What this means is that there are not just

[12]J. P. Martin, Jr. and I. Fridovich (1981)"Evidence for a natural gene transfer from the ponyfish to its bioluminescent bacterial symbiont *Photobacter leiognathi*. The close relationship between bacteriocuprein and the copper-zinc superoxide dismutase of teleost fishes" *Journal of Biological Chemistry* **256**, 6080–6089.

[13]Nowadays the idea of horizontal gene transfer is widely accepted as real, but it was still controversial when Martin and Fridovich proposed it for this example.

[14]Despite the obvious similarity in sequence, there are important differences in kinetic properties, and, perhaps surprisingly, three of them have about double the molecular size of the fourth. This is explained by the fact that the three large molecules are "dimerlike," the two halves of

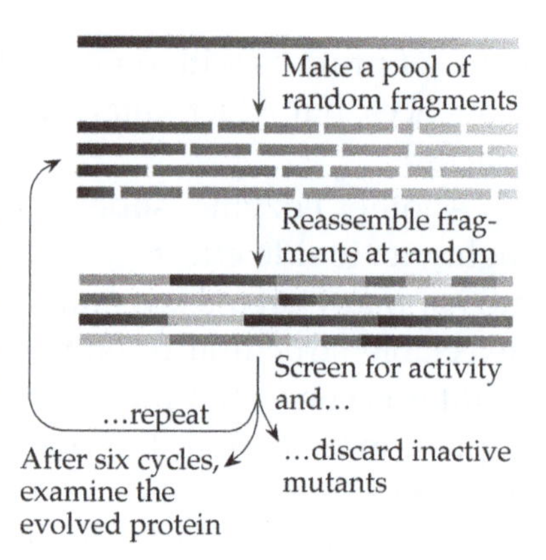
Make a pool of random fragments
Reassemble fragments at random
Screen for activity and...
...repeat
After six cycles, examine the evolved protein
...discard inactive mutants

Figure 12.3 DNA shuffling. Starting with the gene for β-galactosidase, an enzyme that catalyzes the hydrolysis of derivatives of the sugar galactose called β-galactosides, and with little activity toward β-fucosides, similar derivatives of fucose, a different sugar, Stemmer and colleagues used DNA shuffling, as illustrated, to obtain an enzyme with 66-fold higher β-fucosidase activity.

four human enzymes capable of catalyzing the hexokinase reaction, but many more than four, because at least one in every generation of the intermediate sequences that existed during evolution of the four descendants from their common ancestor must have been a functional hexokinase.[15] Although it is not impossible that one of the four, or maybe even two or three, but not all four, lost activity completely during some generations and recovered it in later ones, that is not a particularly plausible hypothesis, and anyway it does not contradict the conclusion that a great many different functional hexokinases must have existed during the evolutionary process.

More information comes from experiments that have been done to "shuffle" the DNA in particular genes, for example by Willem Stemmer and colleagues,[16] as illustrated schematically in Figure 12.3. The idea is to take a set of genes from different strains of a bacterium that code for forms of an enzyme that are all different, but which are all capable of catalyzing the same reaction. If the genes are broken into fragments, mixed together, and then reassembled in a way that allows each reassembled gene to contain a random collection of fragments from different sources, the result is a large collection of genes coding for related but different protein sequences. When these genes are expressed as proteins, many of them turn out to be nonfunctional, but, more interestingly, large numbers of novel proteins do act as catalysts, some of them better than the natural ones. So, the number of related sequences capable of fulfilling any

each molecule closely resembling one another: See M. L. Cárdenas, A. Cornish-Bowden, and T. Ureta (1998) "Evolution and regulatory role of the hexokinases" *Biochimica et Biophysica Acta* **1401**, 242–264.

[15]How do I know? If a functional hexokinase is necessary in order to live to reproductive age, then I must have one. So must both of my parents. So must all four of my grandparents, and so on, indefinitely. But there is no requirement for all of them to have the *same* hexokinase, only that they must all have had one that was functional.

[16]J.-H. Zhang, G. Dawes, and W. P. C. Stemmer (1997) "Directed evolution of a fucosidase from a galactosidase by DNA shuffling and screening" *Proceedings of the National Academy of Sciences of the USA* **94**, 4504–4509.

given functions could be of the order of millions or more.

Unfortunately, these experiments tell us nothing about how many different unrelated sequences would do the job, and we have almost no way to generalize from the observations with carbonic anhydrase and superoxide dismutase to decide how much structural tolerance there may have been in the primitive enzymes that existed at the origin of life. We can be quite certain, however, that there was more tolerance than there is for any organism today, because the first living organism had no competitors. A rival to *Escherichia coli* appearing today in the human gut would rapidly disappear if it needed a day to reproduce itself, because in that day an *Escherichia coli* cell (with a doubling time of about 20 minutes) could have been replaced by more than 10 million copies of itself. No matter if in some respects the rival had a more efficient metabolism than *Escherichia coli*; no matter if each cell secreted a powerful toxin capable of killing 1000 *Escherichia coli* cells; it would still be outgrown and would never be detected.

These concerns would not have mattered to the first organism; there was no race to reproduce because there were no competitors. So, we can guess that the first organisms could survive with many fewer enzymes than any modern organism needs, and that the individual enzymes could have been far less efficient at fulfilling their functions than their modern counterparts, but beyond that rather vague statement, we cannot easily go. I think that few biologists would claim that the origin of life is a solved problem, or even a problem that will be solved to everyone's satisfaction in the next century or so. But few would go to the other extreme either and agree with the creationists that the probabilities against the origin of life are so enormous that it will never be solved without invoking a divine creator, or, as Hoyle and his supporters would prefer, seeding of the Earth with bacteria grown elsewhere in the Universe.

Mikhail Volkenstein approached the question from a quite different point of view in his book *Physical Approaches to Biological Evolution*,[17] asking how many different sequences could lead to the same three-dimensional structure of a protein. Without worrying about the technical details of his argument, the point is that he arrives at a huge number, about 5×10^{122} for a protein of 100 amino acids.

Frequency of improbable events

Extremely unlikely events happen every day, and we can find examples of things that have certainly happened but which have calculated probabilities so small that we could easily be tempted to regard them as impossible. Consider the quotation from R. A. Fisher at the beginning of this chapter. He assesses

[17] Mikhail V. Volkenstein (1994) *Physical Approaches to Biological Evolution*, Springer-Verlag, Berlin.

the probability that a man living at the time of King Solomon would have a descendant in the direct male line living 100 generations later as being of the order of one in 10^{44}, though, as he adds, this improbable event has certainly happened. In case this last point is not perhaps completely obvious, let us spell it out. Any adult man living today has or had a father, so that father had at least one son who survived to maturity. His father likewise had at least one son who survived to maturity, and so on backwards for 100 generations.

Why stop at 100? Humans are now thought to have become differentiated from chimpanzees and gorillas around three to five million years ago, so let us imagine a protohuman sitting by his camp fire on an African savannah at the time of Lucy, around three million years ago, and pondering the likelihood that he would have a distinguished cosmologist by the name of Fred Hoyle in the twentieth century as his descendant through the direct male line. If he was as good at estimating chances as Fisher was, he would arrive at an almost inconceivably small number, given that at least 100,000 generations must have passed in three million years. He would probably conclude that the thing was impossible. Yet, if not for him, then for at least one of his contemporary protohumans, this improbable event has certainly happened. Fisher's point here is that we must refrain from getting carried away by improbabilities viewed after the event. Calculations of the probability of future and unknown events can be useful and help separate good bridge players from the rest, but *a posteriori* probability calculations are rarely illuminating.

I claimed a moment ago that extremely unlikely events happen every day. How could I possibly justify such an extravagant claim? Before doing so, I will put it even more extravagantly: Every day, some person on Earth has an experience that has odds of worse than one in a billion against its occurrence at that time and in that place. Surprising though it may seem, there is no question that this is true. There are many more than a billion people in the world (there are more than that in India alone), and even if we make the conservative estimate that each of them has just one "experience" each day that is in some sense particular to that person, that still makes at least a billion separate experiences each day. Of these, 99% will fail to be "highly significant" in the sense commonly used for statistical tests of experimental observations. This means that 99% will fail to satisfy tests to detect observations outside the range of what is expected 99% of the time (assuming that the tests have been appropriately designed and correctly done, conditions that are often, unfortunately, unfulfilled).

This leaves 1% of highly significant experiences, or 10 million of them every single day, even on a conservative estimate of how many experiences a person has each day. By an obvious extension, we see that every day more than 1000 people have experiences that are expected only once in a million trials, and on an average day one person will have a one-in-a-billion experience. Given that there are quite a lot more than a billion people in the world, and each of them has more than one experience each day, it is surely fair to multiply by at least 10, so that more than 10 people have one-in-a-billion experiences each day, or one person has a one-in-ten-billion experience. But even this is too cautious, because most of us live much longer than a day. In the biblical threescore years and ten, there are more than 25,000 days, so it will be fair to say that in the course of a typical lifetime someone somewhere will experience something that is not expected to occur more often than once in 2.5×10^{15} trials.

To put this in a more familiar perspective, if everyone on Earth was dealt an average of one bridge hand per day, all of them with proper shuffling and no cheating, about once every two years someone would be dealt all 13 hearts. Or, for those that remember the Rubik's cube craze of the early 1980s, if everyone on Earth spent at least three hours every day making random moves on shuffled Rubik's cubes, making about one move every second (if you do remember the early 1980s, you may feel that it did indeed seem like that at the height of the craze), we should expect a report of chance success about every century. This is a low frequency, to be sure, but it is much more than the "impossible" odds of one in 4×10^{19} might lead you to expect.

Coming back to evolution, and bearing in mind that something of the order of a thousand million years were probably available in the prebiotic period before anything interesting happened at the origin of life, and bearing in mind that we are not talking about a mere billion conscious people having experiences, but a much larger number of localities on Earth where haphazard "experiments" were being done by mixing chemicals in arbitrary ways determined by variations in the weather, tides, volcanic and meteorite activity, and so on, we can imagine that huge numbers of combinations were tried out over a vast expanse of time. Even if the overwhelming majority of these experiments yielded no interesting results, some extremely unlikely things could happen in the time available. So, there is really very little reason to suppose that it could not have happened on Earth.

The solar eclipse of 1999

It remains to explain why this chapter is prefaced with a quotation from Nostradamus. It is there to contradict another claim made by Hoyle in one of the last articles that he wrote to justify his view of the impossibility of an independent origin of life on Earth.[18] As I mentioned, he made numerous references to "bee-dancing" to characterize the fuzzy thinking of biologists, which he contrasted with the "real science" practiced by people like himself, and exemplified by the accurate prediction of the total eclipse of the Sun on August 11, 1999. This was an unfortunate example to have chosen,[19] because,

[18]F. Hoyle and C. Wickramasinghe (1999) "Cosmic life: evolution and chance" *The Biochemist* **21**(6), 11–18.

[19]The choice of the term "bee-dancing" as an insult was also unfortunate, because the use of

not only could the eclipse have been predicted many centuries before the tools of "real science" were developed, but it was in fact predicted by Nostradamus about a century before Isaac Newton and around four centuries before Albert Einstein.

There are two objections you might want to make about this "prediction." First, it refers to the seventh month of 1999, but August is not the seventh month. I could dismiss this as a minor inaccuracy, but it is not even that, because according to the Julian calendar in use at the time of Nostradamus, the eclipse was indeed in the seventh month. Second, you may object that although the date is clear, the rest of the quatrain is vague. A reference to the sky, certainly, but what have a King of Terror and the King of Angoulème to do with eclipses? We can answer this with another question: What event in July or August 1999, apart from the eclipse, could a rational person conceivably regard as being accurately predictable in the sixteenth century?[20] Clearly nothing, so if it is a prediction at all, it can only refer to the eclipse. Moreover, although most of Nostradamus' dates are as vague as the rest of his text, the few that definitely refer to dates in what was then the future can all be associated with astronomical events that would have been predictable by the methods available to him.[21] In the circumstances, it would be perverse to interpret it as a reference to anything but the eclipse of August 11, 1999.

One difficulty with disagreeing with physicists, even about biology, is that there is a hierarchy in science whereby lesser scientists are supposed to accept the opinions of physicists as if they were holy writ. This applies as much to popular books and magazines as it does to serious scientific journals. A book questioning the validity of natural selection stands a good chance of being published if it comes from the pen of a physicist or an astronomer, and of being taken seriously by at least some of its readers. A biologist who wrote a book questioning Copernicus' heliocentric view of the solar system would be unlikely to find a publisher. If such a book did appear, it would make a laughing stock of the author and might well provoke articles in serious newspapers about the lamentably low level of culture among biologists.

Nonetheless, physicists have been wrong before in their arguments with biologists and will doubtless be wrong again. Hoyle did not claim originality

body movements by honeybees to signal the location of a good source of nectar has now been proved beyond any possibility of doubt.

[20]There is no shortage of irrational people to interpret it differently, and you can find various bizarre interpretations on the web, including several by people who ignore the historical fact that Angolmois is an old name for the region of Angoulème and prefer to think that the quatrain refers to Genghis Khan (on account of a fanciful resemblance between "Angolmois" and "Mongol": Hey, it has "ngol" and "mo" in it, so it must refer to the Mongols).

[21]For example, he made predictions for "the beginning of 1609" and for October 1727, both of which correspond to eclipses of the Sun.

for his view that life could not have originated on Earth, but traced it back to Lord Kelvin (at the time known as William Thomson) in the nineteenth century, who asserted that "we all confidently believe that there are ... many worlds of life besides our own." This is the same Lord Kelvin who argued that Charles Darwin could not possibly be right in his theory of natural selection, and was equally confident that the Earth could not be more than 400 million years old, and might even be as young as 25 million years old, and thus much too young for natural selection to have operated in the way Darwin needed. No physicist today thinks the Earth is only 25 million years old! Kelvin is perhaps too easy a target (though he was Hoyle's choice, not mine), and among his other confident assertions we may note his views that "heavier-than-air flying machines are impossible" and that "radio has no future."

All of this raises the question of why some physicists appear to take such a naive view of biology, even though they clearly are not naive about their own subject. In his interesting book *Darwin's Dangerous Idea*,[22] Daniel Dennett suggests that what physicists miss in biological theory is a set of laws that allow the properties and states of systems to be expressed in simple (or not so simple) mathematical equations. But natural selection is not a law in this sense, rather it describes a procedure, what computer scientists would call an algorithm, that directs how systems change over long periods of time. Approached with understanding, natural selection allows a vast array of individual biological observations to be brought together in one rational framework. As Theodosius Dobzhansky said in the article that I quoted earlier,[23]

> Seen in the light of evolution, biology is, perhaps, intellectually the most satisfying and inspiring science. Without that light it becomes a pile of sundry facts, some of them interesting or curious, but making no meaningful picture as a whole.
>
> This is not to imply that we know everything that can and should be known about biology and about evolution. Any competent biologist is aware of a multitude of problems yet unresolved and of questions unanswered.

These words, written in 1973, remain pertinent today.

[22]Daniel C. Dennett (1995) *Darwin's Dangerous Idea*, Penguin, London.

[23]T. Dobzhansky (1973) "Nothing in biology makes sense except in the light of evolution" *American Biology Teacher* **35**, 125-129.

13. *Aspects of Cancer*

Before the Flood America was not separated from the other parts of the Earth, and there were no islands.

François Placet (1668)[1]

When a great genius appears in the world you may know him by this sign; that the dunces are all in confederacy against him.

Jonathan Swift (1708)[2]

Most of the arguments supporting the drift hypothesis are shown to be based on erroneous conceptions. The evidence definitely favors the theory of stable continents.

George Gaylord Simpson (1943)[3]

Tumors destroy man in a unique and appalling way, as flesh of his own flesh, which has somehow been rendered proliferative, rampant, predatory, and ungovernable. They are the most concrete and formidable of human maladies, yet despite more than 70 years of experimental study they remain the least understood.

Peyton Rous (1967)[4]

THE JIGSAW-LIKE FIT between Africa and South America has been obvious to everyone who has looked at a world map since sufficiently accurate maps became available. Francis Bacon is sometimes said to have commented on it in 1620 (though some authors say that this is an overinterpretation of what he wrote). François Placet made the point somewhat more clearly in 1688, as did many others long before 1912, when Alfred Wegener proposed his theory of continental drift to explain it. Nonetheless, until the 1960s the standard view of geophysicists was that the fit was pure coincidence, and they heaped on Wegener's ideas the kind of scorn that is today reserved by virologists for the notion that human immunodeficiency virus may not be responsible for the collection of diseases known as AIDS.

[1]Avant le Deluge, l'Amerique n'estoit point séparée des autres parties du Monde, & il n'avoit point d'Isles: François Placet (1668) *La corruption du grand et petit monde,* Alliot & Alliot, Paris.

[2]Jonathan Swift (1708) *The Battle of the Books and Other Short Pieces,* available in the Project Gutenberg at `http://tinyurl.com/pweyqtt`.

[3]G. G. Simpson (1943) "Mammals and the nature of continents" *Americal Journal of Science* **241**, 1–31.

[4]P. Rous (1967) "Nobel Prize Lecture: the challenge to man of the neoplastic cell" *Cancer Research* **27**, 1919–1924.

The standard explanation of why continental drift[5] was not taken seriously by geophysicists until the 1960s was that before the development of *plate tectonics,* there was thought to be no plausible mechanism to explain how continents could move, despite that fact that as long ago as 1782 Benjamin Franklin[6] had suggested one that is not so different from the modern theory:

> Such changes in the superficial parts of the globe seemed to me unlikely to happen if the Earth were solid to the centre. I therefore imagined, that the internal parts might be a fluid more dense, and of greater specific gravity than any of the solids we are acquainted with, which, therefore, might swim in or upon that fluid. Thus the surface of the globe would be a shell capable of being broken and disordered by the violent movements of the fluid on which it rested.

In any case, skepticism about Wegener's hypothesis did not prevent the geophysicists of his time from accepting even more fantastical explanations of why geological features on one side of the ocean match those on the other, and why the fauna of Africa and South America resemble one another in some respects. Both there, and in an embarrassingly large number of other parts of the world, land bridges were supposed to have existed in the past, but to have sunk beneath the seas later on. Despite the lack of explanations of how dense matter had once been able to float, or, if you prefer, how buoyant material was later able to sink, land bridges were regarded as more acceptable than drifting continents. Wegener had no recognized expertise in geophysics, and that, no doubt, had something to do with it as well: He was the son of an evangelical minister, he was trained in astronomy, and his practical experience was in meteorology. Nonetheless, he was essentially right.

Aneuploidy

There are some parallels (but also some important differences) between the history of continental drift and that of the causes of cancer. It has been known for more than a century that there is a close association between cancer and a chromosomal abnormality known as *aneuploidy*, but most experts today consider this little more than a coincidence, or that the chromosomal abnormality is more of a consequence than a cause of the cancer.

To understand what is meant by aneuploidy, we need to start from the generalization that most animals and many plants are diploid, as discussed in Chapter 9. This means that in normal individuals, all of the chromosomes apart from the sex chromosomes come in pairs, and even the sex chromosomes

[5]In the first edition of this book, this chapter was entitled *Genomic Drift,* an allusion to continental drift whose relevance will, I hope, become apparent. I have changed it here to avoid confusion with the ideas of neutral mutation as a driving force of evolution, as discussed in Chapter 2.

[6]Benjamin Franklin (1782) Letter to Abbé Jean-Louis Giraud Soulavie.

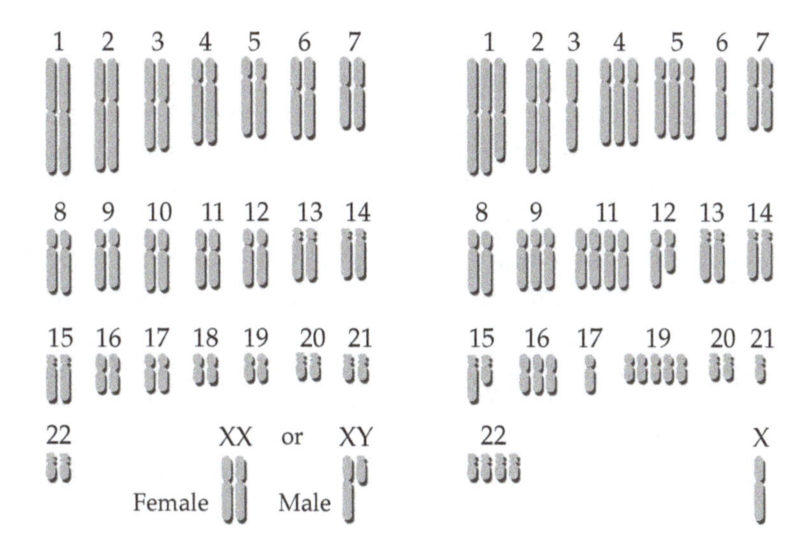

Figure 13.1 Aneuploidy. A normal ("euploid") human cell (*left*) contains 46 chromosomes, 22 pairs of *autosomes* and two sex chromosomes, XX in females, XY in males. In a severely aneuploid cell (*right*), some chromosomes are present in only one copy (3, 6, 17, 21, X), others in more than two copies (1, 4, 5, 9, 11, 16, 19, 22), some are damaged (1, 12, 15), and some are missing (10, 18).

are paired in one of the two sexes. In humans, for example, there are 22 pairs of autosomal (nonsex) chromosomes, together with two sex chromosomes, which form a pair in women (XX) but not in men (XY), making 46 chromosomes in total. As always in biology, it is more complicated than that, because even though women have two X chromosomes, only one of these is expressed in any one cell, on account of *X inactivation* (a high degree of *heterochromatin*).[7] So far as expression is concerned, therefore, only the autosomal chromosomes are paired, and there is no difference between men and women in this respect.

Different species have different numbers of chromosomes, and in Chinese hamsters, which I discuss later on, there are 10 pairs of *autosomes* and one pair of sex chromosomes, making 22 in total. If no mistakes are made during cell division, the counts are preserved after each such division, but this is a highly complicated and still only partially understood process, and mistakes can occur; as a result, a daughter cell may contain one or three examples of an autosome instead of the expected two. When this occurs in the germ line, the result may be an individual with an abnormal number of chromosomes in

[7]This characteristic is not readily seen by casual observation of women, other than in the rare condition Fabry disease, but it is easily seen in cats. Tortoiseshell cats are nearly always female, and have one X chromosome specifying production of one color and the other specifying a different one. The rare exceptions are not true XY males, but sterile XXY intersex individuals, a condition corresponding to Klinefelter syndrome in humans.

every cell, resulting, as noted briefly at the end of Chapter 1, in conditions such as Down syndrome and other trisomies. Errors also occur in somatic cell division, usually with less serious results because only a minority of cells are affected. In either case—in individual cells, or in whole organisms—the occurrence of an abnormal number of chromosomes is called *aneuploidy*.

As long ago as 1890, David von Hansemann noticed that the human cancer cells that he examined were aneuploid (Figure 13.1), and he suggested that the aneuploidy caused the cancer. The general theory of cancer that Theodor Boveri developed from this idea[8] was for many years regarded as a respectable hypothesis. Even though the connection between effect and supposed cause remained obscure, much was known about cancer that was consistent with aneuploidy as its cause. First of all, cancer cells have abnormal rates of growth, hardly surprising if they have a chromosomal composition different from what the cell-division machinery "expects." In popular accounts, this abnormal rate of growth is usually understood as abnormally *fast* growth, but that is an error, because many cancers grow abnormally slowly; what is general is that they do not grow at the same rates as the normal cells that gave rise to them. Second, they show extreme genetic instability, so that in each generation the daughter cells are genetically different from their parent cells, in marked contrast to the genetic identity (or near-identity) that persists through many cell divisions in normal cells. Third, cancers display metabolic defects, with abnormally high metabolic activity, though they may sometimes lack some metabolic pathways completely. All of these and other well-established characteristics of cancer can be understood in terms of Boveri's hypothesis, as we will see later on, but that did not prevent it from being largely discarded and forgotten about in the 1970s after the discovery that certain genes, known as *oncogenes*, are characteristic of certain kinds of cancer.

The first of these stemmed from Peyton Rous' discovery in 1910 that a type of cancer common in chickens was caused by a virus and could be transmitted artificially from one chicken to another. In 1966, more than half a century later, he was awarded the Nobel Prize for this discovery. Not long after this Peter Duesberg discovered the oncogene responsible for the infectious character of this cancer. He found that chickens infected with an RNA virus containing such a gene developed cancer, but did not develop cancer when infected by a virus from which the oncogene had been removed.

This was the first oncogene to be characterized, and it did not establish that *all* cancers are caused by oncogenes, but the idea that oncogenes were needed for cancer proved so appealing that within a few years it became the dominant hypothesis, and largely supplanted the aneuploidy hypothesis in the minds of most cancer researchers. An exception, ironically enough, was Duesberg,

[8]T. Boveri (1914) *Zur Frage der Entstehung maligner Tumoren*, Fischer, Jena, Germany.

who has continued to regard viral oncogenes as a rarity and has accumulated evidence that Boveri was right all along.

All cancer cells are aneuploid

All researchers agree that cancer cells are aneuploid; they could hardly be otherwise, because the evidence is overwhelming. They disagree about whether aneuploidy is the cause or a consequence of the cancer. The standard view is the latter, because it is claimed that no mechanism exists to explain why aneuploidy should result in cancer; in any case, "everyone knows" that genetic mutations cause cancer and that mutagens, or substances in the environment that generate mutations, are also carcinogens, or cancer agents. There are, however, some major exceptions: The carcinogenic effect of asbestos fibers is now so generally accepted that the use of asbestos for insulating buildings is banned over much of the developed world, yet asbestos fibers are chemically inert and efforts to show that they cause mutations have completely failed.[9] What is more, the capacity of asbestos to prevent normal cell division, and hence to cause aneuploidy, is only too clear: Photographs taken with microscopes show cells with embedded asbestos fibers making purely physical barriers to the proper organization of chromosomes necessary for normal cell division. There is no more need to invoke chemistry to explain this than there would be to invoke defective steering wheels to explain the effect on traffic circulation of a tree fallen across a road. This is much like the argument I have used in Chapter 7: Let us reserve our sophisticated explanations for the things that cannot be explained more simply and not waste them on things where a simple explanation accounts for all the facts.

Exceptions like asbestos can perhaps be set on one side. As I have also insisted in Chapter 7, biological phenomena are so extremely complicated that we should not expect too close an agreement between prediction and results, and the sort of perfect correlation that would delight a physicist should arouse suspicion in a biologist. So, even the best of biological theories usually have to contend with one or two inconsistencies that have to be left to future biologists to clarify. There remains the question of timing. If aneuploidy is the consequence of cancer, then we should expect to find some cells that are clearly cancerous but which have not yet become aneuploid and vice versa: If aneuploidy is the cause, we should expect to find noncancerous aneuploid cells. The latter observation is easy to make: After all, any individual with Down syndrome or another trisomy is an entire organism of 10^{14} aneuploid

[9]The building next to the one where I work was remodeled last year, and I was surprised to see wall panels being installed over an insulating material that looks exactly like the asbestos fiber that was used everywhere 40 years ago. Apparently it is *amorphous silica fiber*, but if the carcinogenicity of asbestos is not chemical in origin, who is to say that chemically different substances with similar physical properties will not be carcinogenic? I wonder how thoroughly the asbestos substitutes have been checked for carcinogenicity.

but noncancerous cells. Moreover, although such trisomic individuals do not necessarily develop cancer, they do have a high probability of developing cancer. A person with Down syndrome, for example, has about 30 times the average probability of developing leukemia. Even such a small degree of aneuploidy is also sufficient to produce around 80 characteristic symptoms of the syndrome, implying that it has many biochemical effects.

Does aneuploidy necessarily lead to cancer?

What about the converse observation: Can we find cancerous cells that are not aneuploid? Duesberg and his collaborators studied this question by examining the effects of the carcinogenic chemical dimethylbenzanthracene on Chinese hamster embryo cells. As noted above, a normal diploid cell of this animal has 22 chromosomes, but three weeks after treatment the commonest chromosome number was 23, many cells having other numbers ranging from 20 to 25, and a few having numbers about double those in this range, 44 or 46 (Figure 13.2). In the following week, the counts below 22 or a little above 23 decreased, but many other values between 28 and 46 appeared. All of this, incidentally, occurred months before there were any signs of cancer. In untreated cells used as a control, the majority of cells had 22 chromosomes, though about one-quarter had 23, and other numbers, as high as 49, occurred with detectable frequency.

These observations are sufficiently complicated that it would be rash to propose a simple explanation. Nonetheless, several points appear clear. First of all, aneuploidy is by no means absent from the untreated cells, and more generally, a substantial proportion of cells are aneuploid in individuals who do not develop cancer. Thus, cancer cannot be a necessary consequence of aneuploidy, and the normal organism must be able to tolerate a moderate degree of aneuploidy. The simplest explanation is that although the aneuploid cells in a normal population are defective and likely to die out after a small number of cell divisions, this does not much matter as long as they remain rare, and as long as they can be replaced by normal ("euploid") cells as fast as they die out.

In the pre-cancerous population of Chinese hamster cells, the increase in complication between the third and four weeks can perhaps be interpreted as evidence of the efforts of the population to correct a problem that is beyond correction. Because these efforts can be seen to be making the problem worse, it should not be too surprising if total failure, in the form of cancer, appear some weeks later. Finally, we should note in passing that a count of 22 does not guarantee that a cell is euploid, with a normal set of chromosomes: It could have three or more examples of some and only one or zero of others.

The high proportion of cells in healthy organisms that are not strictly euploid (Figure 13.2) makes it clear that individual aneuploid cells arise all

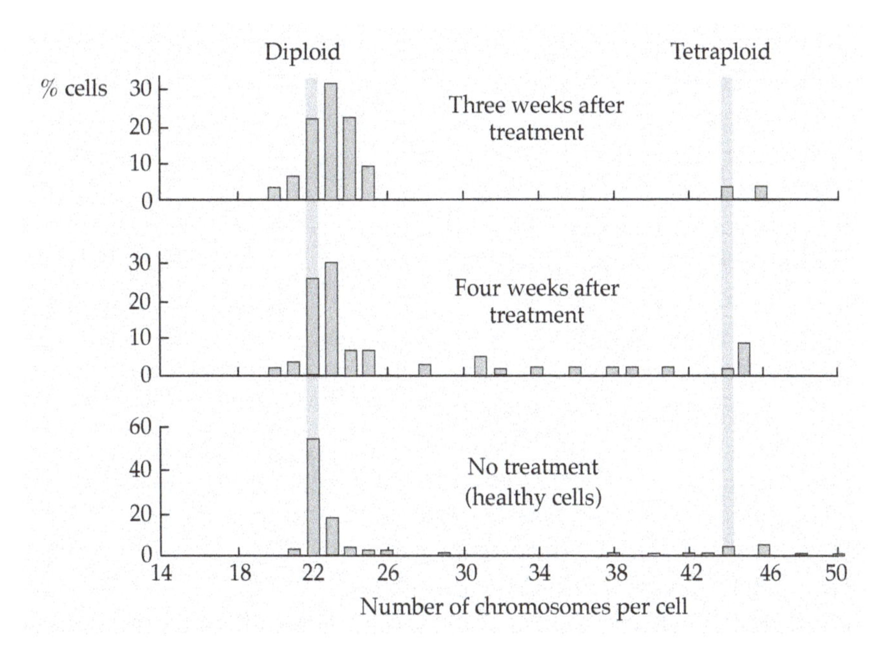

Figure 13.2 Development of aneuploidy in Chinese hamster cells. Three weeks after treatment of cells with a carcinogenic chemical (*top*), only about one-fifth of the cells have the expected number of chromosomes, 22. Many cells have somewhat fewer or somewhat more, and a few have approximately double the expected number, so they are tetraploid rather than diploid. One week later (*middle*), there are numerous cells with chromosome counts intermediate between the diploid and tetraploid values. Aneuploidy is quite common even in normal healthy cells (*bottom*), and fewer than 60% of these cells have the normal diploid count. The illustration is based on experiments of Fabarius, Hehlmann, and Duesberg.[10]

the time and that the appearance of an aneuploid cell is not a rare event. Nonetheless, there may be circumstances that favor this appearance, increasing the likelihood that a particular individual will develop cancer. For example, David Rasnick points out that rodents in general and mice in particular are far more prone to cancer than humans:[11] About one in three laboratory rodents develop cancer in a lifetime of two or three years; about one in three humans develop cancer in a much longer lifetime of 80 or so years. A possible explanation of this large difference is that rodent chromosomes are much more similar to one another in length and organization than human chromosomes, so it is perhaps easier for them to get "mixed up" during meiosis. This might also make them

[10]A. Fabarius, R. Hehlmann, and P. H. Duesberg (2003) "Instability of chromosome structure in cancer cells increases exponentially with degrees of aneuploidy" *Cancer Genetics and Cytogenetics* **143**, 59–72.

[11]D. W. Rasnick (2012) *The Chromosomal Imbalance Theory of Cancer: the Autocatalyzed Progression of Aneuploidy is Cancer*, Science Publishers, St. Helier, Jersey, Channel Islands.

more prone to speciation in appropriate conditions, as mentioned for the mice of Madeira in Chapter 2. However, laboratory rodents are highly inbred, so their tendency to cancer may be misleading in relation to rodents in the wild.

All of this may seem to be taking us rather far from the main themes of this book, but a few years ago Duesberg and Rasnick realized that metabolic control analysis could provide the key to explaining how aneuploidy could lead to cancer. As we have seen in Chapters 8 and 9, increasing or decreasing a gene dose by 50% usually produces no easily observable effect at all. Even complete elimination of a gene often has much less effect than one would guess. This came as a surprise to people who developed techniques for producing mice that lacked selected genes: The characteristics of such *knock-out mice*, as they are called, were supposed to reveal the functions of the genes that had been suppressed. Distressingly often, however, such mice proved to be healthy, with no obvious differences in phenotype. Similar experiments can be done on a much greater scale in organisms like yeast, and it turns out that around 80% of yeast genes are "silent": If any such gene is eliminated,[12] the yeast continues to grow, in many cases at a normal rate. Does this mean that most genes are unnecessary and that the organism could manage perfectly well without them? Well no, it just means that organisms have back-up mechanisms to handle problems that occur when components fail, and that some genes have functions that are needed in extreme conditions and not all the time. A well-fed laboratory mouse experiences a more comfortable environment than a wild mouse in the field; yeast cells growing in a thermostated fermenter encounter fewer stresses than cells living on the walls of a brewery.

PETER H. DUESBERG (1936–) is one of the most reviled scientists active today, on account of his opposition to the view that human immunodeficiency virus (HIV) is the cause of the acquired immunodeficiency syndrome (AIDS). This does not affect his standing as a cancer researcher, however. He was born and educated in Germany, but moved in 1964 to Berkeley, California, where he works at the University of California, principally in cancer research. He was elected to the National Academy of Sciences at the young age of 50.

The early stages of aneuploidy are more likely to involve changes of 50% in gene doses than complete elimination of genes, because any reasonable model of how aneuploidy arises will attribute it to a malfunctioning of cell division that leads to one or three examples of a given chromosome instead of the normal two. As we have seen, the typical effect of changing any one enzyme level by 50% will be no visible effect, but in aneuploidy we are not concerned with "any one enzyme," but with all the enzymes and other proteins (notably transporters, hormones, and regulatory proteins) that are encoded by the affected chromosome, and a large number of individually negligible effects

[12]One at a time: There is no suggestion that one could eliminate all 80% of silent genes, or even a significant proportion of them, without major effects.

can add up to a substantial total. This, after all, is just what the flux summation relationship says: Every one of the flux control coefficients can be individually negligible, but together they account for the whole of the control of the flux.

So, we need to consider the effect on a generalized flux representing the metabolic activity of the cell of changing many of the enzyme concentrations by 50%. If all of them are changed by this amount, then the expected result is simple, any flux will also change by 50%, because no matter how many enzymes there are, their flux control coefficients add up to 1. An error in cell division will typically produce aneuploidy in just one chromosome: In Down syndrome, for example, there are three examples of chromosome 21 instead of two, but chromosome 21 is among the smallest human chromosomes, and the normal pair account for about 1.8% of the human genome, much less than 1/23 (4.3%). Because even the smallest chromosomes carry many genes, however, it is not likely to produce a severe error if we assume that the average flux control coefficient of a gene product of chromosome 21 is approximately the same as the average for all the gene products. So, we can estimate the metabolic effect of Down syndrome as that of increasing the activity of 1.8% of the enzymes by 50%, which comes to about 0.9%.

We have assumed here that flux control coefficients remain constant when the enzyme activities concerned change, but, as noted in Chapter 8, this is too simple: A flux control coefficient normally decreases when the activity of the enzyme increases and vice versa. In their analysis, therefore, Rasnick and Duesberg used a more realistic model that allowed for the changes in flux control coefficients when enzyme activities change, while remaining simplified in some other respects. With this model the metabolic effect in Down syndrome is calculated as 0.6%—smaller than 0.9%, certainly, but not so much smaller that the simple analysis is totally meaningless, especially since 0.9% is small enough to make the point: Overexpressing the whole of chromosome 21 by 50% is expected to produce barely detectable effects on most fluxes. This is consistent with the observation that subjects of Down syndrome have a viable metabolism (though they do have other serious problems), and it emphasizes that even when many enzyme activities are changed, it is difficult to produce large changes in metabolic rates.

In a Chinese hamster, an average pair of chromosomes accounts for one-eleventh of the genes in the genome. But suppose the initial error in cell division affects one of the largest pairs so that, say, 20% of enzyme activities increase by 50%. In this case, the naive calculation ignoring changes in flux control coefficients can be calculated in the head, and is half of 20%, or 10%. The more realistic calculation gives about 7%, which is not negligible, certainly, but is still far from the gross metabolic abnormalities found in cancer. Thus, a single appearance of aneuploidy as a consequence of an error in cell division is not sufficient by itself to produce cancer. However, as we have seen, aneuploidy beyond the normal range of variation is typically detectable in Chinese hamster cells a few weeks after treatment with a carcinogenic chemical, whereas cancer is only detected months after that. Moreover, the changes seen between the third and fourth week after treatment indicate that aneuploid cells become increasingly aneuploid with each cell division, increasingly abnormal from a genetic point of view. Thus, each cell division in the pre-cancerous stage produces a new generation of cells with a greater degree of genetic imbalance, a greater degree of aneuploidy, a greater degree of metabolic abnormality, and a higher risk of aneuploidy-caused cell death, until we arrive at a fully cancerous stage. Rasnick and Duesberg estimate that the problem gets beyond correction when more than about one-third of the cells are aneuploid.

The pursuit of perfection?

What of the pursuit of perfection in all this? Appearance of cancer, followed by death, is hardly a progress toward perfection as we would usually understand it. That is to see the cancer from the point of view of the host, whereas we need to examine it from the point of view of the cancer cells. A cancer is not part of the host, because it has a different genome and is not even of the same species, because it cannot interbreed with the host species. Moreover, it dies with the host, and leaves no descendants. Thus, a cancer is a parasitic species that lives out its entire evolutionary history from speciation to extinction during a fraction of the life span of its host. During that time it passes through various generations and experiences natural selection like any other species, cells that are better able to survive in the peculiar environment of the cancer leaving more descendants than others.

Recognition of aneuploidy as the trigger that leads to cancer does not immediately suggest any particular change in the way cancer should be treated: Once a patient has cancer, it is immaterial whether the original cause was an aneuploid cell division or some effect of an oncogene. On the other hand, the proper treatment of a disease is not the only medically important consideration. There is also the question of prevention, and with any disease understanding the causes helps to prevent its occurrence. So, even though the analysis of metabolic control may not lead to cures for cancer, it may still lead to fewer cancer patients in the first place.

Cancer and the pentose phosphate pathway

This is not the only way of applying the ideas of metabolic regulation to cancer, and improved treatment of cancer patients may follow from an analysis being developed by Marta Cascante in Barcelona. She has long been interested in metabolic regulation, and in collaboration with colleagues in Los Angeles, she studied how current therapies may in some cases be based on faulty notions of metabolic control and, hence, misconceived. Cancer cells typically devote a greater part of their activity than

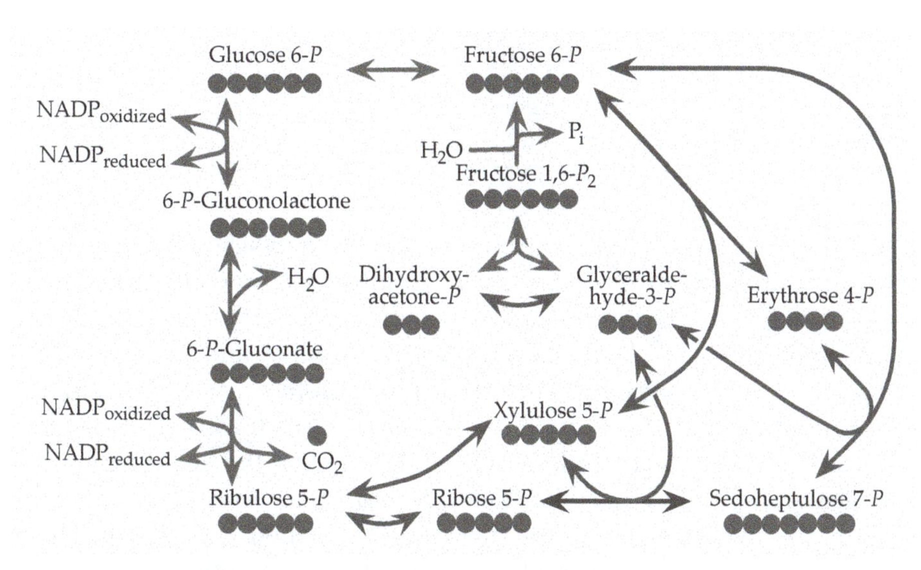

Figure 13.3 The oxidative pentose phosphate pathway. Apart from the change in emphasis in the left-hand part, this is the same as Figure 4.4. Here we are concerned with the oxidation of glucose 6-phosphate to ribulose 5-phosphate and CO_2, with concomitant reduction of oxidized NADP, shown at the left, whereas in Chapters 4 and 5 we ignored this part of the pathway, being concerned only with the non-oxidative part at the right. The oxidized and reduced forms of NADP are illustrated in Figure 13.4.

normal cells do to the production of nucleic acids, and, as we saw in Chapter 4, these include the five-carbon sugar ribose as an essential component. Cancer cells thus typically need more ribose than healthy cells, and, as we also saw in Chapter 4, they can get this by using the pentose phosphate pathway as a route for converting glucose and other six-carbon sugars into ribose.

Normal healthy cells also require a considerable amount of reducing power, which comes in the form of reduced NADP. The non-oxidative part of the pentose phosphate pathway, which we considered in Chapters 4 and 5, cannot provide this, whereas the oxidative part, which we ignored in the earlier discussion, can, as shown in Figure 13.3. In healthy humans, this is the main role of the pathway. Cancer cells are different, because they consume much larger amounts of pentoses than the oxidative pathway can provide, and Cascante and colleagues found that 85% of the ribose molecules isolated from human tumors could be attributed to the non-oxidative pathway.

She therefore considered that the pentose phosphate pathway ought to be a good target for anti-cancer drugs, and accordingly examined which enzymes would be best to try to inhibit. As discussed in Chapter 8, most enzymes in any pathway have only a small influence on the flow of metabolites through the pathway, and as a result, the effect of inhibiting them will be slight unless

Figure 13.4 Oxidized and reduced NADP. Only the parts of the molecules involved in the chemistry are shown; the remainder, accounting for three-quarters of the whole molecule, is shown here as R, but was given in full in Figure 6.1.

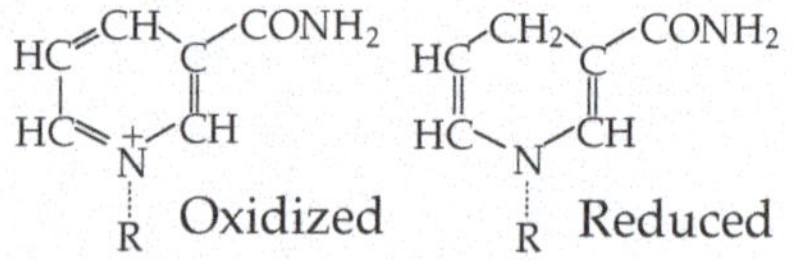

Oxidized Reduced

the inhibition is virtually total. From this sort of analysis Cascante identified the enzyme transketolase as a good target for inhibition, and also noted that the vitamin B_1 deficiency typical of cancer patients is directly related to the same enzyme. In biochemistry, this vitamin is called *thiamine*, and it is necessary for the action of various enzymes, including transketolase. The high activity of the pentose phosphate pathway in cancer cells causes these cells to sequester much of the thiamine that would otherwise be used by the healthy cells.

The usual medical view is that thiamine deficiency in cancer is a problem that needs to be corrected, and the diet of cancer patients is accordingly supplemented with thiamine. Over a long period, thiamine deficiency does create serious medical problems, being responsible for the disease beri beri, which was once widespread. The disease does not develop overnight, however, and in the short term, during the treatment of cancer, the cancer cells may well need the thiamine more than the healthy cells do, so adding it to the diet may do more harm than good.

The study of cancer is relevant to the main themes of this book, not only as an illustration of how the rather abstract ideas of metabolic regulation developed in Chapters 8 to 10 can be brought to bear on a major medical problem, but also as part of the study of what life is. Healthy cells maintain a property known as *organizational invariance*, meaning that they maintain their identities as particular kinds of cells, or, if they are at the stage of development where the different kinds of cells in one individual organism are coming to be different from one another, they change their identities in a pre-programmed and orderly way. Cancer cells do not have this property, because each generation in the development of a cancer is different from the previous one, and within one generation the cells are heterogeneous, with unclear identities. Cancers always die with their hosts (unless artificially maintained as "cell lines" in the laboratory), but, even if they did not, their failure to maintain their organization would condemn them to death.

Most cell lines are difficult to grow in the laboratory, though some rare exceptions, like the famous *HeLa cells*, have continued to thrive many years after they were first isolated in the 1940s from a cancer patient whose names began He... La....[13] These cells can be considered to have achieved a kind

[13]For years there was speculation as to her full name, but it is now known to have been Henrietta Lacks. Her story can be found in *The Immortal Life of Henrietta Lacks*, Rebecca Skloot (2010), Crown Publishers, New York.

of immortality, and have presumably hit upon a structure that allows them organizational invariance. Despite the huge amount of work that has been done on HeLa cells, it is important to remember that they are *not typical* of human cell lines, which remain difficult to culture through more than a few generations.

All of this raises some questions. What do we mean by an organism? What do we mean by saying that an individual is alive? How do we distinguish one individual from another? I will discuss these in the next chapter of this book.

14. The Meaning of Life

> The human body is a machine which winds its own springs. It is the living image of perpetual movement.
>
> Julien Jean Offray de la Mettrie (1748)[1]

> As a subject for serious inquiry, the category "life" has all but vanished from the scientific literature; it is the particulars of life, not its nature, that fill the numberless pages of scientific journals.
>
> Franklin Harold (2001)[2]

> **Vitalism** is the notion that life in living organisms is sustained by a vital principle that cannot be explained in terms of physics and chemistry. This vital principle, often called "the life force," is something quite distinct from the physical body and is responsible for much that happens in health and disease.
>
> Thames Valley University (2008)[3]

> I can epitomize a reductionist approach to organization in general, and to life in particular, as follows: *throw away the organization and keep the underlying matter*. The relational alternative to this says the exact opposite, namely: *throw away the matter and keep the underlying organization.*
>
> Robert Rosen (1991)[4]

IT IS TEMPTING TO INTERPRET the third quotation above as a reference to nineteenth century ideas about life in a course on the history of science, but it is perfectly serious, and illustrates how universities around the world are turning their backs on science as the quest for knowledge in favor of whatever will improve their financial positions. Vitalism as defined in the quotation was a respectable idea in the nineteenth century until it was overthrown by Eduard Buchner's discovery of cell-free fermentation, which I will discuss shortly. This chapter, then, is not so much about the meaning of life, but about the

[1]Le corps humaine est une Machine qui monte elle-même ses ressorts; vivante image du mouvement perpétuel: Julien Jean Offray de la Mettrie (1748) *L'Homme Machine*, Elie Luzac, Leyden. A translation of the whole book (apart from the passages dealing with human sexual attraction and practices, which are replaced with ellipses) made by Gertrude Carman Bussey in 1912 may be found at `http://bactra.org/LaMettrie/Machine/`.

[2]Franklin M. Harold (2001) *The Way of the Cell*, Oxford University Press, Oxford.

[3]Notes for a course at the Plaskett Nutritional Medicine College, Thames Valley University (University of West London): `http://www.dcscience.net/tvu-plaskett-nut-therap-2005.pdf`.

[4]Robert Rosen (1991) *Life Itself: a Comprehensive Inquiry into the Nature, Origin, and Fabrication of Life*, Columbia University Press, New York.

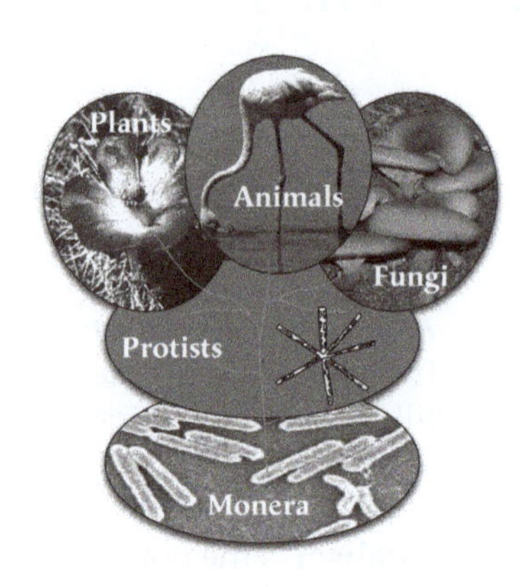

Figure 14.1 The classification of organisms. This is never as straightforward as one might wish. The five-kingdom system shown here appears to make a clean division into animals, plants, fungi, protists, and Monera, but it is now recognized that this last kingdom brings together two domains, the Eubacteria, or classical bacteria, and the Archaea, that differ from one another as much as either of them differs from the other kingdoms. The image of *Asterionella formosa* was provided by Dr. Brigitte Gontero and is used with her permission. The electron micrograph of bacteria was provided by Dr. David Ranava and is used with his permission.

meaning of "life": I will discuss what the word means, and how we recognize whether a system is alive or not, rather than the sort of questions that interest philosophers and theologians: Why are we here? What does it all mean? Is there an afterlife?

Living and nonliving

In 2005 the journal *Science* celebrated 125 years of its existence by asking its editors to assemble a list of "the most compelling puzzles and questions facing scientists today." Many of the questions chosen were questions of physics or astronomy, but many were also questions of biology, and among these one was conspicuously missing: *What is life?* The editors did not explain whether this was missing because they did not think it interesting or important, or because they thought it was already solved, but I will try to indicate in this chapter why I do consider it interesting and important, and why we cannot yet regard it as solved.

Table 14.1 Which are alive?

Living	Nonliving
Fly	Radio
Tree	Automobile
Mule	Virus
Baby	Crystal
Mushroom	The Moon
Ameba	Computer

We all think we can recognize a living organism when we see one, but it is not so easy to *define* "living" in such a way that the definition will include all the entities we consider to be alive, and exclude the ones we do not. With one exception, everyone would agree that the entities in the left-hand column of Table 14.1 are living, whereas those in the right-hand column are not. The exception is the virus: Some experts insist that a virus is a living organism, whereas others, just as expert, insist that it is not. I favor the latter camp,[5] because a virus is not able to maintain its organization, but I will not insist on it, and I mention the example

[5]But my colleagues and friends Chantal Abergel and Jean-Michel Claverie, discoverers of *Megavirus chilensis* and other viruses as large as bacteria, do not.

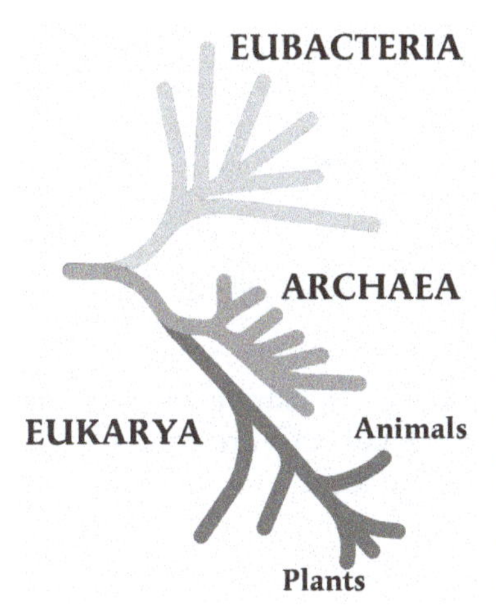

Figure 14.2 Biochemical tree of life. The classification of Figure 14.1, based on traditional ideas, has been made largely obsolete by a classification proposed by Carl Woese. The figure is based on his original analysis, but a tree based on all the data available today would not differ in any important qualitative way, though it would contain vastly more data for single-cell organisms; the plants and animals that loom so large in our parochial notions of what living organisms are would be even more inconspicuous than they already are in Woese's tree.

only to point out that most attempts at generalization in biology can provoke disagreement. Such anomalies have always existed in biology: Whenever anyone has tried to devise an absolute classification with firm lines between classes, difficulties have always appeared, and this will no doubt continue. In Chapter 2, we looked at the difficulties that arise when trying to define species unambiguously, and in Chapter 11, I said that the only definition of *animal* that I know of that includes all animals, as far as it is known, and excludes nearly all organisms that are not animals, is that animals synthesize collagen. Nonetheless, it will not be particularly surprising if an organism that is definitely an animal by all reasonable standards is found that does not contain collagen.

More generally, the classification of organisms into five kingdoms as shown in Figure 14.1 seemed to be solidly based until Carl Woese, in the 1970s, showed—largely on the basis of ribosomal RNA sequences, that is, on biochemical evidence, and thus relevant to the theme of this book—that the kingdom of single-cell organisms known as bacteria (or more formally as Monera) contained two different groups, now known as the Eubacteria, which agree with the classical notions of what bacteria are, and the Archaea, which are often found in extreme environments. One might be tempted to dismiss this as a detail, but it is much more than that, because the Eubacteria and the Archaea differ from one another as much as they differ from the other kingdoms. Indeed, the present view (which may need revision as more information accumulates) is that animals and plants are slightly more closely related to the Archaea than they are to the Eubacteria, as

CARL RICHARD WOESE (1928–2012) was an American microbiologist. Born in Syracuse, New York, he was educated at Amherst College, Massachusetts, where his only training in biology was a single course of biochemistry. He went on to obtain a Ph.D. in biophysics at Yale. In his subsequent career, mainly at the University of Illinois at Urbana, he emphasized the importance of taking account of the huge diversity of single-cell organisms and was responsible for the recognition of the Archaea.

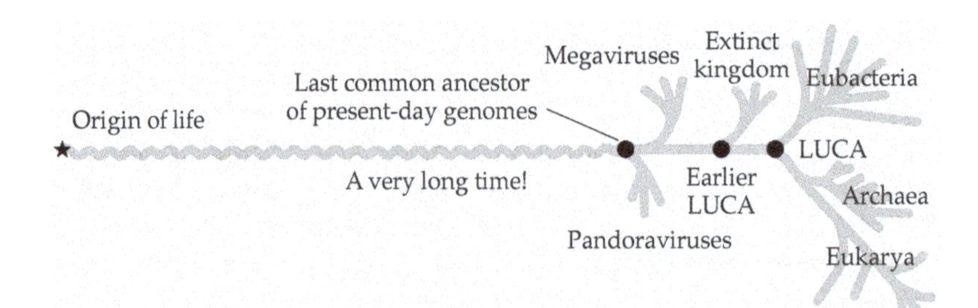

Figure 14.3 Last common ancestor. Figure 14.2 is extended to illustrate the difficulty of defining the *last universal common ancestor*, eliminating details that are not pertinent to the point to be discussed now and adding additional kingdoms.

illustrated in Figure 14.2.

LUCA

Figure 14.2 provides a context to discuss the concept of LUCA, or *cenancestor*, defined as the *last universal common ancestor* of all known organisms, which has come to play an important part of studies of the origins of metabolism. It is redrawn in Figure 14.3 to remove details that we do not need for the immediate discussion and to add others that we do. The ancestor at the point of divergence of the Eubacteria from the Archaea and Eukarya is LUCA. As far as we know, this was the last common ancestor of all present-day organisms, but it was certainly not the first living organism or the last common ancestor of all organisms that have ever lived. If we postulate the existence of a group of organisms that have been completely lost that existed before the divergence of the groups we know today, then LUCA would have been different from LUCA as understood today. Moreover, it is not impossible that some survivors may be found in some obscure habitat, in which case a new definition will be needed. At the other extreme, suppose that all the Eubacteria become extinct in the future but the Archaea and Eukarya survive: This is probably impossible from a practical point of view, because most and maybe all Eukarya could not live without Eubacteria, but it is not impossible conceptually, and if that happens, LUCA will be at the point of divergence of the Archaea and Eukarya.

As I have said, I do not regard viruses as living organisms, but the contrary view may prevail in the future. Some of the giant viruses, as large as bacteria, that have been discovered in the twenty-first century, including the megaviruses and the (very different) pandoraviruses, have genomes that are only remotely related to those of currently known organisms. For example, two-thirds of the genome of the newly discovered *Pithovirus sibericum* appears to be completely different from those of the pandoraviruses or other known genomes.[6] Thus, although LUCA may be the last common ancestor of all current *organisms*, its genome is probably not the last common ancestor of

[6]M. Legendre, J. Bartoli, L. Shmakova, S. Jeudy, K. Labadie, A. Adrait, M. Lescot, O. Poirot, L.

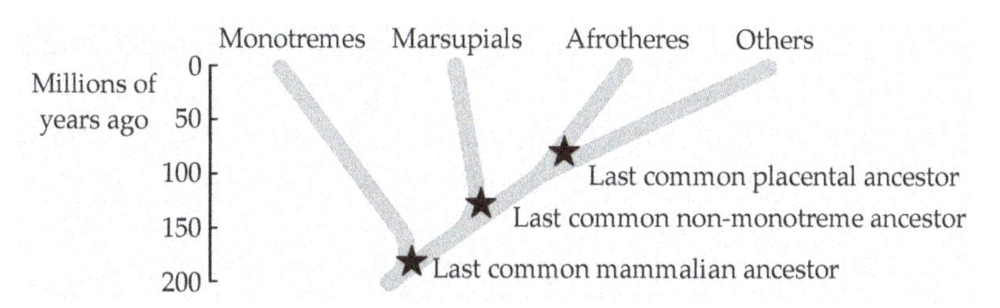

Figure 14.4 Last common mammalian ancestor. The common ancestor of the monotremes (platypuses and echidnas) and the other mammals lived about 180 million years ago. If all the monotremes become extinct, the common ancestor of the surviving mammals will have lived 140 million years ago, and if all the marsupials become extinct, it will have lived 105 million years ago.

all genomes. In this connection it is worth noting that the genomes of the large viruses—known even before the megaviruses and pandoraviruses were discovered—may give a clue to the evolutionary origin of collagen, which, as we saw in Chapter 11, is characteristic of animals and has an amino acid sequence very different from that of nearly all other proteins. Large viral genomes contain what are called *collagen domains*, with repeated Gly-Pro-Pro sequences.[7] These appear to be needed for the transformation of the infected cells, but whether they have any relationship to the collagen of animals remains to be established.

The argument about the special status of LUCA is perhaps easier to follow if we examine organisms that are familiar to everyone, and ask what is the last common ancestor of all mammals (Figure 14.4). In terms of the species that exist today, this is the ancestor at the point of divergence of the monotremes (platypuses and echidnas) from the other mammals and lived about 180 million years ago. The monotremes are few in number, in terms both of species and of individuals, and it is by no means impossible that they will become extinct during the lifetimes of people alive today. If that happens, the common ancestor of the surviving mammals will be the common ancestor of the marsupials and other mammals, and will move forward in time to 140 million years ago, *with almost no effect on non-monotreme mammals*. It is less likely that all the marsupials will become extinct, but if they do, we must move it to the common ancestor of the afrotheres (elephants and their relatives) and the others, and it will have been alive only 105 million years ago. It follows that the last common ancestor of all mammals is not an abolute and fixed entity,

Bertaux, C. Bruley, Y. Couté, E. Rivkina, C. Abergel, and J.-M. Claverie (2014) "Thirty-thousand-year-old distant relative of giant icosahedral DNA viruses with a pandoravirus morphology" *Proceedings of the National Academy of Sciences of the USA* **111**, 4274–4279.

[7]Jan Mrázek and Samuel Karlin (2007) "Distinctive features of large complex virus genomes and proteomes" *Proceedings of the National Academy of Sciences of the USA* **104**, 5127–5132.

but depends on the chance loss of particular lines of descent. By extension, the same applies to LUCA. This whole discussion parallels that of mitochondrial Eve and Y-chromosome Adam in Chapter 2: Just as these have definitions that have changed in the past and will change in the future, so also LUCA is not a fixed entity.

There is often an implication, though less often an unambiguous statement, that the deduced properties of LUCA, especially its metabolism, tell us something about the origin of life. This cannot be true, and it is important to understand that LUCA was already a highly complex organism, containing many proteins and capable of numerous metabolic reactions. Even *Pithovirus sibericum*, with a genome apparently much older than that of LUCA, has a predicted capacity to code for nearly 500 proteins. The first self-organizing systems that gave rise to life must have been much simpler than that, and that is the reason for the long wavy line in Figure 14.3 connecting the origin of life to present organisms.

What does it mean to say that a creature is alive?

Moving now to the main theme of this chapter, how do we decide that all the creatures in Figure 14.2 are alive? What is the common characteristic that distinguishes the living from the nonliving? There is one fundamental way of approaching these questions that I will not discuss, though it is important, which we can call *synthetic biology*: Various research groups are trying to construct a *minimal organism* by using synthetic chemistry to create entities with the properties of living organisms. Instead of this, I will discuss the notion of *self-organization*: What are the requirements that allow an organism to create itself and to maintain itself in the face of changes in its environment and deterioration of its components?

Before discussing self-organization, I should mention a third approach to the definition of life that misses the point completely. Since the rise of molecular biology, which has revolutionized many aspects of biology, it has been common to think that the replication of DNA *is* life, that there is nothing else. This idea was expressed explicitly by Bernardino Fantini, a historian of science at Geneva:

> In 1960 a note to the Academy of Sciences introduced the concept of the operon, a unit of expression coordinated by it. A new dimension of the organization of the genome is introduced.... As a result of profound theoretical innovations and important discontinuities in the traditions of research, ... one sees the appearance of a powerful theoretical edifice, which changes the very foundation of biology and provides new definitions of life and evolution.[8]

[8]En 1960 une note présentée à l'Académie des sciences introduit le concept d'opéron, unité d'expression coordonnée par lui. Une nouvelle dimension de l'organisation du génome est introduite.... À travers des innovations théoriques profondes et des discontinuités importantes

This sort of comment misses the point because it confuses the recipe with the cake. To bake a cake, it is not sufficient to have ingredients available and a set of instructions. It requires a whole organization that can read the instructions and produce a cake. Producing a living organism involves more than just knowing its genome. The wide publicity given to an experiment in recent years made the same mistake. A team led by Craig Venter succeeded in replacing the genome of a bacterium, *Mycoplasma capricolum*, with a completely synthetic copy of the genome of another species, *Mycoplasma genitalium*, and found that the engineered bacterium could replicate.

Creating life

From the technical point of view, this was an impressive achievement, but what did it tell us about life? The press release from the J. Craig Venter Institute described this as the second step in a three-step process to create a synthetic organism (the first step being the sequencing of the genome of *Mycoplasma genitalium*), but was it? An announcement that I had a three-step plan to get me from Marseilles to San Francisco using only the power of my own muscles would probably be greeted with skepticism, but people who believed it would, no doubt, be impressed.

Suppose that I then explained that the first step was to walk to where I had left my bicycle, and that the second was to use it to get to the airport. I could proudly add that both of these had now been crowned with success; the third step still needed to be worked out, and for the moment I would use the services of Air France to take me to San Francisco. My initial claim would then be ridiculed. In making a "synthetic" organism, all of the conceptually difficult part was left to the machinery of an existing organism, *Mycoplasma capricolum*, and all that had been achieved was conceptually straightforward (however difficult it may have been technically), because it did not require significantly new knowledge or understanding.

JULIEN JEAN OFFRAY DE LA METTRIE (1709–1751) was born in Saint-Malo. He studied theology and planned to enter the Church, but switched to medicine. His views on the soul, together with his hedonistic attitude to life and the importance of pleasure, created scandals, first in France and then in the more tolerant society of The Netherlands. After his death in 1751 in Prussia, Frederick the Great said in his eulogy, "All those who are not imposed upon by the insults of the theologians mourn in La Mettrie a good man and a wise physician."

The number of people seriously interested today in the question of what life is has been small, as one might guess from the quotations at the beginning of this chapter, and, more seriously, the different researchers have largely ignored one another's work, neither referring to one another nor trying to create a synthesis of the different theoretical ideas. The inspiration for most of these

dans les traditions de recherche, ... on voit surgir une édifice théorique formidable, qui change les bases mêmes de la biologie et donne de nouvelles définitions de la vie et de l'évolution. Bernardino Fantini, Preface to Jacques Monod (1988) *Pour une éthique de la connaissance*, La Découverte, Paris.

came from the physicist Erwin Schrödinger, but before discussing his ideas I will briefly mention three other authors who reflected on the nature of life much earlier.

The first is Julien de la Mettrie, who argued strongly in his book *L'Homme Machine* (quoted at the beginning of this chapter) for a mechanistic view of life, thereby creating a scandal among theologians with a more traditional idea of the soul. The book also scandalized a wider public on account of its frank discussion of sexual attraction and practices. His rejection of vitalism is accepted by virtually all biologists today. His use of clocks, gears, and shafts as metaphors of life is not now regarded as useful, but his description of the human body as a machine that winds its own springs can be taken as a forerunner of the ideas of self-organization that dominate current thinking about the nature of life.

Cell-free fermentation

In 1897 the German biochemist Eduard Buchner showed that a cell-free extract of yeast could bring about alcoholic fermentation, a process previously thought to require a living organism, especially after the great French physiologist Louis Pasteur had tried and failed to show the same thing earlier in the century.[9]

Although vitalistic ideas did not disappear immediately and completely, Buchner's experiment effectively sounded their death knell and precipitated the explosion of studies of enzymes and metabolism that constitute the heart of biochemistry, because the overthrow of vitalism made it increasingly clear that living organization is based on cellular chemistry and not on a mysterious life force. Many biochemists regard Buchner's work as the birth of biochemistry.

Once the vitalist diversion was out of the way, people could think seriously about what life is and whether it can be reproduced in synthetic chemical systems. Prominent among these was Stéphane Leduc, whose book *La Biologie Synthétique*[10]

STÉPHANE LEDUC (1853–1939) was a French physician and biologist, and Professor in the Medical School of Nantes. He tried to recreate living structures from inorganic materials, but his ideas were not well received, because they were misinterpreted as being tainted with vitalism, and his papers were refused by the Académie des Sciences de Paris. There is a revival of interest in his work today because there are suggestions that structures similar to those he studied may have been present at the origin of life.

[9]The history of vitalism in the nineteenth century is complicated [H. C. Friedmann (1997) "From Friedrich Wöhler's urine to Eduard Buchner's alcohol," pages 67–122 in *New Beer in an Old Bottle*, edited by A. Cornish-Bowden, Universitat de València: `http://tinyurl.com/kyew8q9`]. The argument attracted many of the greatest scientists of the period, some of whom, such as Pasteur, were wrong but had valid reasons for their opinion, whereas others, such as Justus von Liebig, were right for wrong reasons, and used their authority to make scurrilous attacks on their opponents.

[10]S. Leduc (1912) *La Biologie Synthétique*, Poinat, Paris.

Figure 14.5 Osmotic forest. This was created by seeding a solution of sodium silicate in water with small crystals of iron sulfate and copper sulfate. Photograph taken by Ricardo Rojas while an undergraduate student at the University of Chile. © Elsevier. Reproduced with permission from J.-C. Letelier, M. L. Cárdenas, and A. Cornish-Bowden (2011) "From *L'Homme Machine* to metabolic closure: Steps towards understanding life" *Journal of Theoretical Biology* **286**, 100–113.

promoted the idea that *osmotic forests* not only reproduce some aspects of the *appearance* of living structures, but they also capture part of their essence. Today the equipment needed for creating these is often sold as a children's toy: For example, it is sufficient to add a few crystals of colored salts, such as iron sulfate or copper sulfate, to a solution of sodium silicate in water and the result is a spectacular growth that in some cases resembles a living plant or other biological structure, as seen in Figure 14.5. It seems obvious to us today that however much these structures may resemble plants and so on, they tell us nothing about the real structure and organization of living organisms. Although Leduc regarded these as far more than a toy, few people in his time, or since, have been convinced. Nonetheless, active research is proceeding today with the aim of discovering whether inorganic compartments can reproduce some of the conditions at the origin of life that would allow the development of self-organization.[11]

D'ARCY WENTWORTH THOMPSON (1860–1948) was a Scottish biologist and mathematician, as well as a scholar of Greek and Latin. He was Professor of Biology at the University of Dundee for 64 years, starting in 1884, and held the Chair of Natural History at St. Andrews University from 1917. He emphasized the importance of allometric and physical principles in determining the forms of living organisms, and he set out his ideas in detail in his book *On Growth and Form*.

The third of Schrödinger's predecessors who deserves a mention is D'Arcy Thompson, a polymath among polymaths, who wrote a highly influential book, *On Growth and Form*,[12] in which he advanced the idea that many aspects of evolution are constrained by purely mechanical considerations and are thus not accessible to natural selection. We have already encountered the idea

[11] L. M. Barge, I. J. Doloboff, L. M. White, G. D. Stucky, M. J. Russell, and I. Kanik (2011) "Characterization of iron–phosphate–silicate chemical garden structures" *Langmuir* **28**, 3714–3721.

[12] D'Arcy Wentworth Thompson (1945) *On Growth and Form*, 2nd edition, Cambridge University Press, Cambridge.

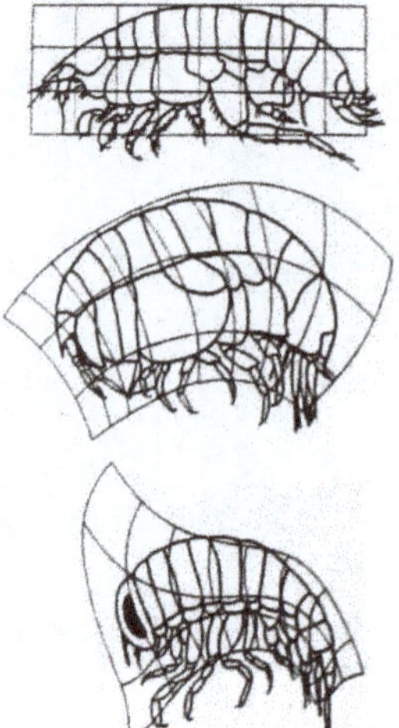

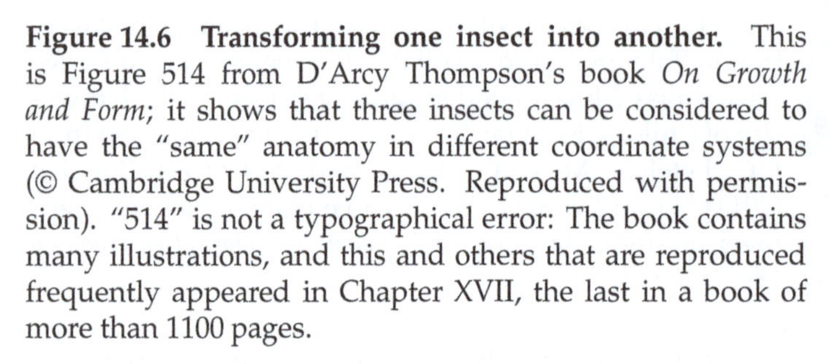

Figure 14.6 Transforming one insect into another. This is Figure 514 from D'Arcy Thompson's book *On Growth and Form*; it shows that three insects can be considered to have the "same" anatomy in different coordinate systems (© Cambridge University Press. Reproduced with permission). "514" is not a typographical error: The book contains many illustrations, and this and others that are reproduced frequently appeared in Chapter XVII, the last in a book of more than 1100 pages.

of allometric constraints in Chapter 11, but D'Arcy Thompson showed that the constraints went far beyond these, and that many aspects of the shapes of organisms needed to be explained in physical terms, especially allometric scaling. Some of the illustrations from his last chapter, such as the one shown in Figure 14.6, are well-known; less well-known is that these come near the end of a long book, and there is a great deal of interesting material before one arrives at them. The main point of this and similar illustrations was made by Denis Diderot a surprisingly long time ago:

> When one considers the animal kingdom, and notices that among the quadrupeds, there is not a single one that does not have the functions and parts—above all the interior parts—entirely similar to those of another quadruped, does not one willingly believe that there was never anything but a single animal prototype of all the animals, of which Nature has only lengthened, shortened, modified, multiplied or obliterated certain organs?[13]

D'Arcy Thompson's book is a wonderful example of scientific writing at its most literary. It is not always easy to read, because he tended to assume that his readers were as erudite as himself and would have no trouble understanding his many quotations (most of them left untranslated) from Latin, Greek, French, German, and Italian (at one point even Provençal). An entire page in French, a long quotation from Buffon,[14] presents an important argument about the hexagonal cells made by honeybees.

In this and other examples, D'Arcy Thompson insists that the result can be understood in terms of physics and mechanics, without needing to invoke

[13]Anonymous, but known to be Denis Diderot (1754) *Pensées sur l'interprétation de la Nature*, no publisher indicated, The Netherlands.

[14]The naturalist George-Louis Leclerc, Comte de Buffon—always known just as Buffon.

"engineering genius" on the part of the bees. This sort of argument has led some readers to regard him as hostile to the whole idea of natural selection, but that is to oversimplify his subtle thinking: He was opposed to regarding natural selection as the unique driving force for evolution and insisted that it needed to be understood in parallel with physical and chemical principles. We have seen other examples in this book: A large part of the evolution of protein sequences is *neutral* (Chapter 2), and the principles of metabolic control derive from mathematical necessity, not from selection (Chapter 8).

Schrödinger's analysis

The starting point for understanding life is none of these, but an influential book by Erwin Schrödinger[15] based on a series of public lectures that he gave in Dublin in 1943. He was the first great physicist to take an interest in the nature of life, and his book stimulated several others to study biological problems. He tried to answer three principal questions about life:

1. How can organisms maintain their organization in the face of a continuous production of entropy as a consequence of the second law of thermodynamics?

2. What is the nature of the hereditary material?

3. Can biology be fully understood (even in principle) in terms of the known laws of physics?

He answered the first question with the statement that "what an organism feeds on is negative entropy." This may seem an unnecessarily poetic way of expressing an idea that is well understood today: The inevitable production of entropy by an organism is compensated for by the ingestion of low-entropy food and the excretion of higher-entropy waste. Although this has sometimes been dismissed as being obvious, it was not well understood at all at the time he was writing, especially by his general audience in Dublin, and his statement undoubtedly cleared away some confusion. Even today his point is not well understood by adherents of creationism (the subject of the next chapter), who often claim that the second law of thermodynamics makes evolution impossible, because they think that increasing the information content of a system is impossible. The point is that continuously increasing entropy is characteristic of a *closed system*, but the

ERWIN RUDOLF JOSEF ALEXANDER SCHRÖDINGER (1887–1961) was an Austrian physicist. His introduction of wave mechanics led to the Nobel Prize for Physics in 1933. After the union of Austria and Germany, he moved to the Dublin Institute of Advanced Studies in Ireland, but he returned to Austria in 1951.

[15]Erwin R. J. A. Schrödinger (1944) *What is Life?* Cambridge University Press, Cambridge. The Canto Classics edition of 2012 includes two other short texts by Schrödinger, *Mind and Matter* and *Autobiographical Sketches*.

branch point is an *open system*, because it continuously receives radiation from the Sun, and organisms are also open systems, for the reasons explained by Schrödinger. In any case, if the creationist argument had any merit, it would apply as much to life as to evolution, but we know that life is possible.

Schrödinger suggested that the hereditary material must be a sort of "aperiodic crystal"—that is, a substance that had a high degree of cyclic regularity, as in a crystal, coupled with nonrepetitive elements whose structures did not interfere with the general regularity, but whose irregularity allowed them to act as what he called a *codescript*. This was an extension of Max Delbrück's argument that the stability of genes in the face of thermal disturbances meant that they must be molecules. After the tremendous increase in knowledge of molecular genetics that has occurred in the half-century that followed Schrödinger's lectures, we can recognize the codescript as a description of DNA, which has a structure that appears completely regular when viewed from a distance, but completely irregular when viewed with enough resolution for the individual bases to be identified.

Schrödinger's last point is that *biology is more general than physics* and that we should look for laws of physics that are not required for physics itself. No such laws have subsequently been identified, and the suggestion has been received with embarrassment, and even ridicule, by his successors. Of the creators of the modern theories of life that I will consider, only Robert Rosen appears to have taken this suggestion seriously. The fragmented nature of modern theories of life, with no one referring to anyone else, at least for most of the time,[16] means that we can discuss them in any order, without expecting to find a common synthesis.

(M, R) systems

I will begin with Rosen's *(M, R) systems*. His ideas are probably the most difficult to understand, because he made no attempt to explain them in ordinary biological terms, or to provide examples. Moreover, he had a tendency to attach private meanings to words that were already in use by biologists with quite different meanings. Although the M in "(M, R) systems" stands for *metabolism* and corresponds with the everyday meaning, the R stands for *repair* and has nothing to do with such things as the repair of damaged DNA. Instead, it refers to the need for organisms to resynthesize enzyme molecules that become inactivated over time, or simply need to be replenished after the dilution that inevitably occurs when an organism grows. It is clearer, therefore, to take the R as standing for

[16]We have tried to move away from this tendency [J.-C. Letelier, M. L. Cárdenas, and A. Cornish-Bowden (2011) "From *L'Homme Machine* to metabolic closure: Steps towards understanding life" *Journal of Theoretical Biology* **286**, 100–113], but there is stll work to be done. Not only did the originators of the various theories ignore one another, but most of their later supporters continue to concentrate their efforts on the theories they like best, paying little or no attention to the others.

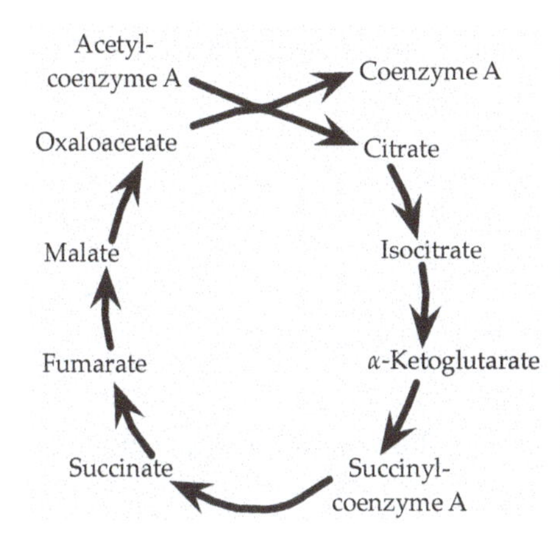

Figure 14.7 Simplified representation of the tricarboxylate cycle. This is redrawn from Figure 11.8 so as to emphasize that although acetyl coenzyme A is consumed by the cycle, and various other substrates and products that are not shown also participate, the cycle intermediates from citrate to oxaloacetate are regenerated with each turn of the cycle, and because they are clearly biological in origin, they fit the definition of an *enzyme* as a "biological catalyst," even though they do not correspond to what we usually think of as enzymes.

replacement.

Rosen's view of life is sometimes summarized in his statement that *organisms are closed to efficient causation*, but most readers are likely to find this too obscure to be understood. It is a reference to Aristotle's classification of four types of cause. The word "cause" does not correspond to what we would call a cause today, and some prefer to leave Aristotle's word αἰτία untranslated, but simply transliterated as "aitia," or, rather than asking what *causes* something, to ask what *makes* something what it is.[17] We can apply the idea both to an example from everyday life, such as what (or who) made the statue *David*, or, as a more biochemical example, what makes glucose 6-phosphate.

ROBERT ROSEN (1934–1998) was an American theoretical biologist. He was born in New York and made his career first at the State University of New York in Buffalo and later at Dalhousie University, Nova Scotia. Earlier he studied with Nicolas Rashevsky at Chicago and regarded his work as an extension of Rashevsky's *relational biology*, in which the relations between entities—that is, their organization, is more fundamental than the entities themselves.

As we see in Figure 14.8, only the efficient cause would usually be called a cause today, and it corresponds to the catalysts of metabolism. Rosen's definition of an organism then begins to make sense if we take it to mean that *all of the catalysts needed by a living organism must be made by the organism itself.* This, in turn, means that the usual distinction between *metabolites* and *enzymes* is less clear cut than we might wish, because if enzymes are made by metabolism, they are themselves metabolites. Moreover, there is a sense in which many of the compounds we call metabolites could be regarded as enzymes, because they satisfy the definition of an enzyme as a biological catalyst: The intermediates in the tricarboxylate cycle (Figure 14.7) are enzymes in this sense.

We thus arrive at the idea that a metabolic system is *closed*: There is no

[17]The word *etiology* (or *aetiology*) derives from this. As a term for the branch of philosophy that deals with causes, it has largely disappeared, but it lives on as a medical term for the study of the causes of diseases.

Statue of David	Aristotle's causes	Glucose 6-phosphate
Marble	*Material cause:* what was it made *from?*	Glucose and ATP
To represent the biblical David	*Formal cause*: what was it made *as*?	A glycolytic intermediate
Jules Cantini	*Efficient cause*: who (what) was it made *by*?	Hexokinase
To beautify the waterfront	*Final cause*: what was it made *for?*	Harnessing energy

Figure 14.8 Aristotle's four "causes" as applied to a statue and a metabolite. The statue of *David* on the waterfront of Marseilles, one of many replicas of Michelangelo's *David* in Florence, was made by Jules Cantini in 1903. Each of Aristotle's *what caused it?* questions is better phrased in modern terms as *what made it?*

beginning and no end, and hence no *hierarchy*. The term *downward causation* sometimes encountered in philosophical accounts of life is thus unsound. It implies that there exists a hierarchy of different levels of regulation such that elements at higher levels control what happens at lower levels, but not vice versa, and that is not possible if there is no hierarchy, and everything affects everything else. Rosen insisted that *an organism is not a machine* and cannot be understood as a machine. This contradicts the title of Mettrie's book, and, more seriously, we will see that other theories of life are less insistent on this point. Figure 14.9 is an attempt to make Rosen's ideas intelligible.

It is important to emphasize that there is no conflict between the idea that organisms are closed to efficient causation and the thermodynamic requirement (Schrödinger's first point) that organisms are open systems, because that is a statement that *organisms are open to material causation*: No one, and

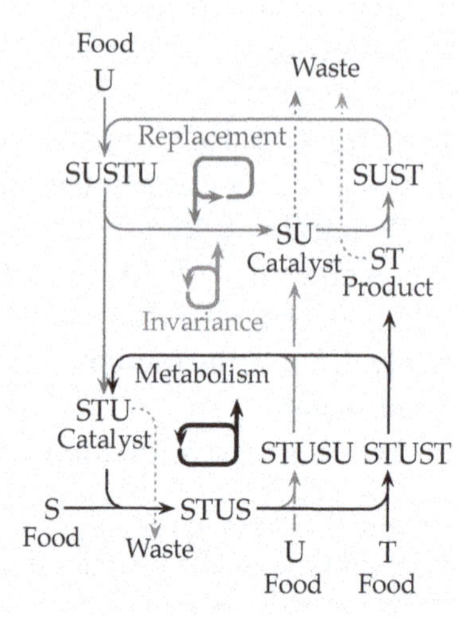

Figure 14.9 An interpretation of an (*M,R*) system. It consists of three interlocking cycles: *metabolism*, in which food molecules S and T are transformed by the action of a catalyst STU into a product ST; *replacement*, in which any STU lost to waste is replaced by reaction of ST with another food molecule U, catalyzed by a different catalyst SU; and *organizational invariance*, in which U is also used to replace SU lost by degradation to waste. This is an attempt to make Rosen's ideas intelligible, but he did not himself present any example resembling this. For a fuller discussion, see A. Cornish-Bowden, M. L. Cárdenas, J.-C. Letelier, and J. Soto-Andrade (2007) "Beyond reductionism: Metabolic circularity as a guiding vision for a real biology of systems" *Proteomics* **7**, 839–845.

Figure 14.10 Structural closure. In a colony of bacteria, here a mixture of *Clostridium acetobutylicum* (straight) and *Desulfovibrio vulgaris* (curved), the individual bacteria are clearly separate from one another: One can readily see where one individual ends or another begins. There is no conflict between this idea and the fact that organisms are open to the flow of matter and closed to efficient causation. Electron micrograph provided by Dr. David Ranava and used with his permission.

certainly not Rosen, is saying that organisms are closed to material causation. Metabolism must always be driven by a source of energy. Some organisms, such as green plants, can use the Sun's light as a source of energy, but animals cannot, and can only obtain energy by converting food into waste.

A major problem with self-organization, most evident in Rosen's approach, but implicit in others, is that a living system must avoid *infinite regress*, if not forever, because organisms eventually die, but at least for a long time. This arises directly from the need to replace catalysts, which is necessary not only because catalysts do not last forever and have lifetimes that are typically orders of magnitude shorter than those of the systems that depend on them. Even if they do not disappear, the catalysts must inevitably become diluted when the systems that use them grow. To maintain its organization, therefore, an organism requires continuous resynthesis of its catalysts, but how does this occur? Resynthesis must also be catalyzed, so other catalysts are also needed, but these catalysts must also be maintained, and so on: We are clearly on the way to an infinite regress in which each attempt to close the cycle opens a new gap. Discussing in detail how this difficulty is surmounted is complicated, and I will not attempt it here. I will simply comment that organizational closure is impossible if specificity is absolute, with every catalyst having just one function, but it becomes possible if at least some (and probably many) are multifunctional. On a limited scale, such multifunctionality is called *moonlighting*, and increasing numbers of examples are being found. For example, the enzyme glyceraldehyde 3-phosphate dehydrogenase, which has been known for many years for its role in metabolism, is now known to fulfill several other apparently unrelated functions, such as a structural role in the crystallin that forms the lens in the vertebrate eye.

Autopoiesis

The next theory of life that I want to discuss is *autopoiesis*, an invented word that literally means "self-construction." It was introduced by the Chilean neuroscientists Humberto Maturana and Francisco Varela. Their initial interest was in trying to understand how the brain functions, and when Maturana started his work the field was dominated (and to some degree still is) by a *positivist* view of the brain as a computer that decodes its input and on that basis selects an appropriate motor action.

One interpretation based on this metaphor was that every percept[18] was coded by a specific neuron. Maturana was strongly influenced by the concepts of cybernetics that were prominent in the 1950s; his difficulty with the positivist view of the brain was that he realized it would not only require specific neurons to detect the percepts, but would also need to be perceived itself by other neurons, which would need.... We are thus on the verge of an infinite regress analogous to the infinite regress in metabolism that Rosen wanted to escape.

Maturana went on with Varela to develop these ideas from neuroscience into a theory of autopoietic systems, which they described in a short book, *De Máquinas y Seres Vivos*,[19] and later a fuller account of autopoiesis.[20] They regarded an autopoietic system as being organized as a network of processes that produces, transforms, and destroys its components, this being done in such a way that it continuously regenerates the network of processes that produce them and constitutes itself as a machine that exists as a concrete unity in space. What does this last requirement mean? More specifically, what is a "concrete unity in space"? What it means is that there must be *structural closure* (Figure 14.10); an organism must be enclosed in a membrane, a cell wall, or a skin, with a definite physical boundary between itself and other organisms. This third type of closure is independent of the closure of organisms to efficient causation and their openness to material causation. Structural closure is necessary to allow one individual to be distinguished from another. It is obviously the case for animals and plants that we can see with our eyes, but it is also true for microscopic organisms, such as bacteria.

HUMBERTO MATURANA ROMESÍN (1928–) is a Chilean neuroscientist and philosopher. He was born in Santiago, studied first medicine and then biology at the University of Chile, and obtained his doctorate at Harvard. His work spans a broad range, but is concerned in particular with cognition.

FRANCISCO JAVIER VARELA GARCÍA GARCÍA (1946–2001) was a Chilean neuroscientist and philosopher. He was born and educated in Santiago, but he spent much of his career outside Chile, in the USA during the period of the military dictatorship, and in France for the last part of his life. He obtained his doctorate at Harvard and went on to develop the theory of autopoiesis with Humberto Maturana.

This requirement for structural closure is an important difference between autopoiesis and (M,R)-systems: It is not clear from Rosen's writings whether he considered the question of structural closure at all, and he certainly did not emphasize it. On the other hand, Maturana and Varela have been vague about

[18] This is not an everyday word. Dictionaries define a *percept* as the mental concept that results from perceiving. When the eye sees something, the brain perceives the input from the eye, and, in the positivist view, constructs a mental image of the object seen.

[19] H. R. Maturana and F. Varela (1973) *De Máquinas y Seres Vivos* ("Of machines and living beings"), Editorial Universitaria, Santiago, Chile.

[20] H. R. Maturana and F. Varela (1980) *Autopoiesis and Cognition: the Realisation of the Living*, D. Reidel Publishing Company, Dordrecht.

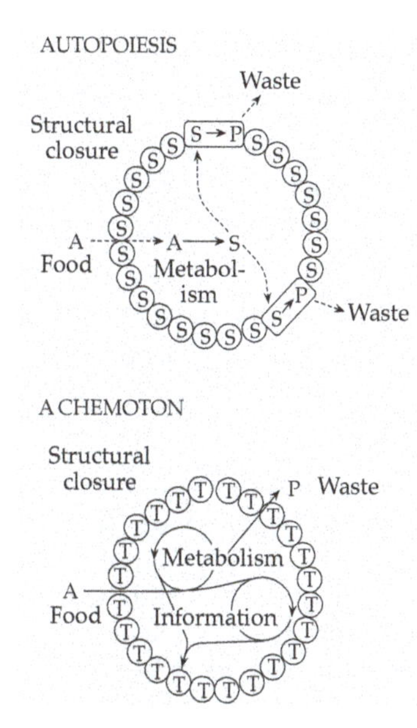

Figure 14.11 Autopoiesis and the chemoton. In **autopoiesis** (*top*) food molecules A enter the system, where they are transformed chemically into molecules S that integrate themselves spontanteously into an enclosing membrane, in which they are gradually transformed (spontaneously) into products P that are excreted as waste. From the thermodynamic point of view, the whole process is driven by the conversion of A into P, assumed to be an irreversible reaction. The **chemoton** (*bottom*) is drawn here in such a way as to make the relationship to autopoiesis as explicit as as possible. The system is driven by a *metabolic* cycle that transforms food into waste, with a series of intermediates (not shown) that are used and regenerated, as in Figure 14.7, and also drives a second *information* cycle that interacts with metabolism to produce the molecules T that bring about *structural closure*.

the need for specific catalysts. In Figure 14.11, we see that an essential function is the use of metabolic reactions to produce an enclosing membranes, but there is no indication in the diagram of the origin of any specificity.

The chemoton

The third theoretical treatment to be considered is the *chemoton* of Tibor Gánti. This is the most firmly rooted in chemistry of all, which is not surprising given Gánti's origins in chemical engineering. Like any engineer, he was more interested in how things can be made to work than in abstract reasoning, and he embodied his ideas in the chemoton model. Nearly all of his work was published in Hungarian and was thus inaccessible to many in the rest of the world until his former student Eörs Szathmáry prepared a translation of his principal work as *The Principles of Life.*[21] This brought together various books and papers, including most of the sixth edition of *As Élet Princípuma* ("The Principle of Life"). It was most valuable, not only for the translation itself, allowing any interested person to read Gánti's work, but also because Szathmáry and the philosopher James Griesemer supplemented it with copious marginal notes and with some additional chapters.

The chemoton resembles autopoiesis, but with the important addition that it explicitly includes an information cycle (lower part of Figure 14.11). Gánti saw the two internal cycles, together with their interaction to produce the molecules that allow structural closure, as the minimum essential features of a living organism. The metabolic cycle resembles the cycles of intermediates like the tricarboxylate cycle (Figure 14.7) found in present-day metabolism in which an overall reaction $A \rightarrow P$ is achieved with the help of a set of

[21]Tibor Gánti (2003) *The Principles of Life*, Oxford University Press, Oxford.

intermediates that are not consumed, so they act as a composite catalyst. It differs from a modern cycle in that in all (or virtually all) known metabolic reactions, there is a specific catalyst to ensure that unwanted side reactions are kept to a minimum, but in Gánti's model there are no specific catalysts: The metabolic cycle by itself is catalytic and that is important, but the individual reactions within it have no specific catalysts, and introducing these would make the model so complicated (because there would have to be additional reactions to produce them) that it would no longer be a minimal model of life. It is not easy to see, therefore, how a chemoton could avoid collapsing into a mass of unwanted reactions.

It could be objected that the model of an (M, R) system illustrated in Figure 14.9 also assumes that the molecules involved just happen to have the property of being able to undergo the specific reactions shown *and no others*. The total number of reactions shown in Figure 14.9 is much smaller than the number implicit in the lower part of Figure 14.11, and it is one thing to assume that a small number of molecules happen to be capable of undergoing a small number of specific reactions, but quite another to suppose the same thing of a larger number of molecules. The whole problem of specificity is a crucial one that needed to be solved by the first organisms, but the way of achieving it remains obscure.

TIBOR GÁNTI (1933–2009) was a chemical engineer who spent his whole life in Hungary (and published almost all of his work in Hungarian). He taught industrial biochemistry and theoretical biology at Eötvös Loránd University and other universities in Hungary, after working first as head of the yeast laboratory of the Yeast Factory of Budapest. He remained closely in touch with industrial chemistry after his academic functions began, because he continued working in the Factory of Industrial Chemicals of Budapest. His aim in developing the *chemoton* model was to arrive at a minimum definition of life.

On the other hand, Figure 14.11 explicitly allows for structural closure, whereas Figure 14.9 (and Rosen's writing in general) does not. Because we know that structural closure is a universal feature of the organisms that we know today, this could be regarded as a major difficulty. However, if life started in natural mineral cavities such as those that exist in Leduc's structures (Figure 14.5), or in spontaneously formed lipid micelles, then we could consider structural closure as something supplied by the environment of the first organisms and not something they needed to make for themselves.

Autocatalytic sets

Stuart Kauffman approached the problem from an entirely different direction.[22] Rather than asking what properties were necessary for a system to be regarded as alive, he asked what sort of conditions might allow self-organization to arise spontaneously from purely chance properties of sets of molecules. Suppose that a particular sequence of units ABC has a small probability of catalyzing a

[22]Stuart A. Kauffman (1993) *The Origins of Order*, Oxford University Press, New York.

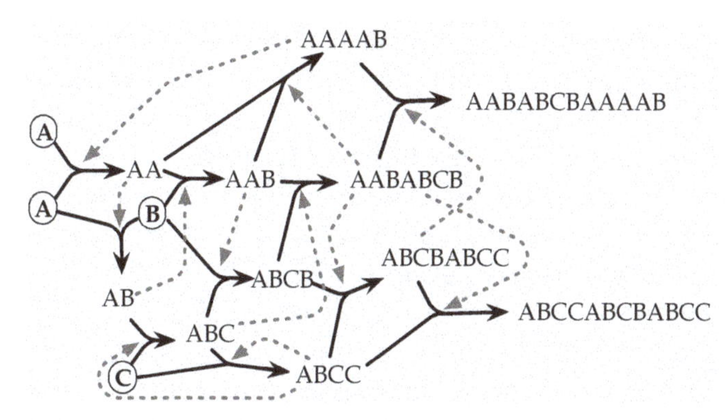

Figure 14.12 An autocatalytic set. All of the members of the set are produced (continuous arrows) by the condensation of other molecules in the set and are catalyzed (broken arrows) by molecules in the set. For example, the reaction AAB + ABCB $\rightarrow$ AABABCB is catalyzed by ABC. The precursor monomer units A, B, and C (framed) are assumed to be available as "food" from the environment. The set is autocatalytic in the sense that the entire system is catalyzed by molecules within itself. There is no requirement that any individual product should catalyze its own formation. That is not forbidden, however, and as the system is drawn, formation of ABCC is autocatalytic in the simpler sense that ABCC catalyzes the reaction C + ABC $\rightarrow$ ABCC. The untidy look of this figure reflects Kauffman's view that self-organization can arise by chance.

reaction of two other similar molecules to generate a longer sequence, AAB + ABCB $\rightarrow$ AABABCB. The individual units can be amino acids, RNA bases, or whatever, it is not important. Because this catalysis is a chance property, the probability that ABC can catalyze that particular reaction is likely to be very small, but as long as it is not zero, Kauffman's idea can function. Let us suppose that the probability that a particular molecule can catalyze a particular reaction is one in a billion, or 10^{-9}. Then we can ask how many different kinds of molecule need to be present before the whole collection can be called an *autocatalytic set*, in which

> Catalytic "closure" must be achieved and maintained. Thus it must be the case that every member of the autocatalytic set has at least one of the possible last steps in its formation catalyzed by some member of the set, and that connected sequences of catalyzed reactions lead from the maintained food set to all members of the autocatalytic set.

We can understand better what this means by examining Figure 14.12. The rather messy look of this figure compared with the more "designed" look of the others we have seen is not accidental. It emphasizes that here we have a type of closure that could arise completely by chance. Every molecule in the diagram (apart from the food molecules) results from a reaction between two of the others that is catalyzed by a member of the set. Kauffman calculated that if the probability is 10^{-9} that any particular molecule can catalyze any particular reaction, then there need to be at least 3×10^8 members in the set for there to be a reasonable probability that the whole system can exist and maintain itself. That is a large number, but it is not impossibly large.

This chapter may appear more incoherent and disconnected than many of the others, but if so that reflects the nature of the subject. Many aspects of the relationship between evolution and biochemistry are now well enough understood to be susceptible to a clear exposition, but that is not yet the case with theories of the nature of life or its origin, which remain fragmentary: There is still work to be done.

STUART ALAN KAUFFMAN (1939–) is an American theoretical biologist with a particular interest in complex systems and the origin of life. He was a Marshall Scholar at the University of Oxford, from which he obtained his BA, after which he studied medicine at the University of California Medical Center in San Francisco. He worked in several different university departments and at the Santa Fe Institute.

15. *The Age of Endarkenment*

Fish have large heads, but these are empty of sense, like the heads of many men.

Julien Jean Offray de la Mettrie (1748)[1]

Darwin gave us a creation story, one in which God was absent and undirected natural processes did all the work. That creation story has held sway for more than a hundred years. It is now on the way out. When it goes, so will all the edifices that have been built on its foundation.

William Dembski (1998)[2]

Now, a mere quarter of a century later, Darwinian evolution is little more than a historical footnote in biology textbooks. Just as students learn that scientists used to believe that the Sun moves around the Earth and maggots are spontaneously generated in rotting meat, so students also learn that scientists used to believe that human beings evolved through random mutations and natural selection. How could a belief that was so influential in 2000 become so obsolete by 2025? Whatever happened to evolutionary theory?

Jonathan Wells (2004)[3]

The Holy Koran, the Prophet's teachings, the majority of Islamic scientists, and the actual facts all prove that the Sun is running in its orbit... and that the Earth is fixed and stable, spread out by God for his mankind.... Anyone who professed otherwise would utter a charge of falsehood toward God, the Koran, and the Prophet.

Abd Al-Aziz ibn Baz (1966)[4]

IN HIS BOOK *The Blind Watchmaker*,[5] Richard Dawkins made no mention of what has become known as "intelligent design,"[6] but the current sense of the term had not been invented in 1986. More surprisingly, though, he discussed classical creationism only briefly, late in the book (apart from a brief mention in the preface), and said nothing about the people who were

[1]Les poissons ont la tête grosse, mais elle est vide de sens, comme celle de bien des hommes. Julien Jean Offray de la Mettrie (1748) *L'Homme Machine* Elie Luzac, Leyden.

[2]W. A. Dembski (1998) "Introduction to mere creation" *Mere Creation* (edited by W. A. Dembski), Intervarsity Press, Downer's Grove, Illinois.

[3]J. Wells (2004) "Whatever happened to evolutionary theory?" *World* **19**, Number 13. The passage quoted is from an article that predicts the state of biology in 2025.

[4]Quoted by T. Dobzhansky (1973) "Nothing in biology makes sense except in the light of evolution" *American Biology Teacher* **35**, 125-129.

[5]R. Dawkins (1986) *The Blind Watchmaker*, Longman Scientific and Technical, London.

[6]In this book I am putting scientific terms in italics at first mention, but I refuse to regard "intelligent design" as a scientific term and will put it in quotation marks instead.

then regarded as its leading lights. Presumably he thought that such matters were not worth bothering with and that no educated person would take them seriously. As late as 2003, when I was writing the first edition of this book, I thought that the threat to biology represented by creationism was mainly a problem in the USA, and that in the rest of the world we could leave it to our American colleagues to worry about. Since then, it has become clear to everyone who cares to look that that attitude is far from confined to the USA, and that creationism is growing all over the western world.

The noun *endarkenment* is not listed in the *Shorter Oxford English Dictionary*, though it does define *endarken* as a verb ("now *rare*, late sixteenth century") meaning to make dark or obscure, but it is a word that we now need, for the retreat from the enlightenment that we thought characterized science and culture from the eighteenth century onwards.

Teaching biology in the USA, Europe, and Turkey

The comforting perception of most scientists is that the trial of John T. Scopes in 1925 was a victory for the rational teaching of biology in the USA. He was convicted of teaching evolution after the state of Tennessee had passed a law against it, but the fundamentalists became a laughing stock (despite winning the case), and only two other states introduced laws forbidding the teaching of evolution in the next half century. This is misleading: Other states did not pass similar laws because *they did not need them,* because there was essentially no teaching of evolution in American schools after 1925. More surprising, perhaps, is that even in other countries, such as the United Kingdom and France, evolution constituted a far smaller component of biological education than most modern biologists realize.

As it happens, I had my first serious exposure to science at Shrewsbury School, the same school that Charles Darwin attended. I was there in 1959, at the time of the 150th anniversary of Darwin's birth, and the 100th anniversary of the publication of *The Origin of Species*. Because he was probably the most famous alumnus of the school (unless you prefer Sir Philip Sidney or Samuel Butler[7]), and his statue can be seen in the center of the town, you might imagine that these two anniversaries would have been the cause of some celebration, but you would be mistaken. My recollection is that they passed essentially unnoticed, except by the Academy of Sciences of the USSR, which sent a commemorative medal to the school.[8] My memory after the lapse of half a century may be at fault, but Peter Hughes, the gifted teacher who taught me chemistry

[7]Samuel Butler (1835–1902), a generation younger than Darwin, and grandson of the Samuel Butler who was headmaster of Shrewsbury when Darwin was there, became one of his fiercest opponents and devoted his later life to increasingly virulent attacks on natural selection. In his successful novel *Erewhon* ("Nowhere"), written before this hostility developed, he also set out his view that machines would eventually become living organisms.

[8]No one seemed to know what to do with this, and it was passed to the teacher of the Russian class that I was taking (in 1959 Russian was the language we were all going to need afterwards).

for three of the five years that I was at Shrewsbury, has confirmed that he had the same recollection. What is clear, anyway, is that evolution formed a far smaller part of the teaching of biology, even at Darwin's own school, than one might have guessed. Theodosius Dobzhansky's famous article[9] was still in the future at that time.

There are many quotations from modern creationists that I could have chosen to prefix this chapter apart from the two I used. I chose them because William Dembski and Jonathan Wells are among the most influential leaders of opinion and are not totally without scientific credentials. Such predictions of the imminent collapse of evolution are not new and have occurred regularly over nearly two centuries. Even from before the publication of *The Origin of Species*, writers were predicting the imminent demise of naturalism in favor of the biblical account of creation.[10] Perhaps unwisely, Wells put a definite date of 2025 on his prediction of when evolution would become little more than a historical footnote in biology textbooks. We have now passed about half of the time between when he wrote that and when it should come to pass, but there is still no sign that a belief that was so influential in 2000 will be obsolete in 2025.

Today creationist propaganda is far from being limited to the USA and is growing everywhere. It still has limited impact in countries like the United Kingdom and France, but strong creationist organizations can be found in Germany, Austria, and the German-speaking part of Switzerland, with "textbooks" intended for use in schools that are produced to high technical standards, and an Education Minister of the German State of Hesse, who said that she believed biblical creation theory should be taught in biology class as a theory, like the theory of evolution. Poland and Russia have strong creationist movements and have had creationist government ministers at times.

ADNAN OKTAR (1956–) is the most visible exponent of Turkish creationism. He has had a major influence on Islamic creationism in western countries and is best known for his book (written under his pseudonym, "Harun Yahya") *Yaratılış Atlası*, translated into English as *Atlas of Creation*, as well as into many other languages. In 1990 he set up an organization called Bilim Araştırma Vakfı ("Science Research Foundation"), a name that does not adequately convey its meaning.

The strongest creationist movement outside the USA is to be found in Turkey, with extensive and expensively produced propaganda that passed largely unnoticed in western Europe[11] until Islamic objections to evolution started to be heard in universities, even in medical schools. The *Atlas of*

[9]T. Dobzhansky (1973) "Nothing in biology makes sense except in the light of evolution" *American Biology Teacher* **35**, 125-129.

[10]See the collection at `http://chem.tufts.edu/AnswersInScience/demise.html`.

[11]Unnoticed by many of us, perhaps, but not by Theodosius Dobzhansky, whose article specifically referred to creationism in the Muslim world.

Creation[12] was produced in Turkey, extensively translated (Arabic, Bosnian, Chinese, Czech, Dutch, English, French, German, Hindi, Indonesian, Italian, Japanese, Persian, Russian, Spanish, Swedish, Urdu ...) and distributed free of charge to educators all over the world. This has made it impossible to continue to ignore the Turkish creationist movement. It differs from the better known movements in that many of the most vocal creationists in the west are *Young-Earth Creationists* who believe that the Earth is only a few thousand years old, whereas the Turkish creationists appear to have no difficulty with a very old Earth, but believe that the biological world has been completely static.

Much of this would be laughable if it were not for the serious threat that it represents for the teaching of biology. After all, as already noted, there have been many predictions that evolutionary theory is on the way out, but there is no sign of this in the writing of biologists. However, in the version known as "intelligent design," creationism is presenting a continuous threat to how biology can be taught at the secondary level in the USA, even if it is so far unsuccessful in attempts to infiltrate teaching in serious universities.

Intelligent design

I will discuss "intelligent design" shortly, but first there should be a mention of the rampant confusion between evolution and the origin of life. In the quotation at the beginning of this chapter, William Dembski claimed that "Darwin gave us a creation story," but Darwin did no such thing. He did not mention the origin of life in his published work, and his only known comment on it is contained in a letter to his friend Joseph Hooker:

> It is often said that all the conditions for the first production of a living organism are now present, which could ever have been present. But if (and Oh! What a big if!) we could conceive in some warm little pond with all sorts of ammonia and phosphoric salts,—light, heat, electricity &c. present, that a protein compound was chemically formed, ready to undergo still more complex changes, at the present day such matter w^{d} be instantly devoured, or absorbed, which would not have been the case before living creatures were formed.

This is often quoted, but we should notice not only that it appeared in a personal letter but also the tentative way in which the suggestion is made: "and Oh! What a big if!" To treat this as a fully developed theory is at best simple-minded and at worst deliberately dishonest.

The origin of life and the theory of evolution are two quite separate areas of research. Few biologists, even among those interested in the origin of life (Chapter 14), would claim that it is a solved problem. On the contrary, everyone accepts that it is fraught with difficulties and uncertainties. But that has nothing to do with evolution, which long ago ceased to be controversial

[12]Harun Yahya (real name Adnan Oktar) (2006) *The Atlas of Creation*, Global Publishing, Istanbul.

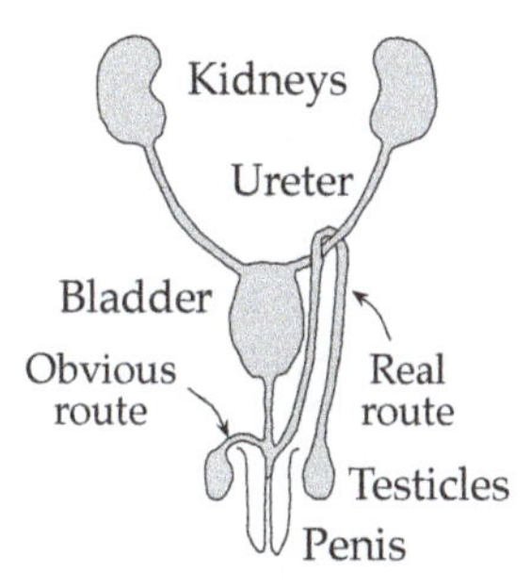

Figure 15.1 Unintelligent design. To get from the testicles to the penis, semen does not take the obvious direct route (*left*), but instead loops around the ureter (*right*). This circuitous route makes no sense in terms of conscious design, but is readily understandable on an evolutionary basis. If we suppose that in the ancestral species the testicles were located inside the body at a much higher level close to the kidneys (as they still are in elephants), then it would have made no difference which side of the ureter the semen passed.

in the minds of virtually all biologists. Such "controversy" as exists is entirely fabricated, mostly by people with no biological credibility: There are engineers, chemists, surgeons, and so on who will argue that schools should "teach the controversy," but biologists? There are none, because there is no controversy. There are arguments about the relative importance of the different mechanisms of evolution, whether, for example, neutral drift is more important than natural selection, as many biochemists would argue, but everyone accepts the reality of evolution itself.

Intelligent design (without the quotation marks!) had an intellectually respectable origin in an argument of William Paley:

> In crossing a heath, suppose I pitched my foot against a *stone*, and were asked how the stone came to be there, I might possibly answer, that, for any thing I knew to the contrary, it had lain there for ever: Nor would it perhaps be very easy to shew the absurdity of this answer. But suppose I had found a *watch* upon the ground, and it should be enquired how the watch happened to be in that place, I should hardly think of the answer which I had before given, that, for any thing I knew, that watch might have always been there.[13]

Paley went on to argue that the existence of an artefact that was clearly designed, like a watch, implied the existence of a designer; because, he thought, biological structures were clearly designed, they must likewise be the work of a designer. In 1805 it would have been difficult to contradict him, because no known process could produce the *appearance* of design without conscious design. This changed completely in 1859 with the publication of *The Origin of Species*, in which Darwin explained that if enough time was available, the appearance of design did not require a conscious designer. The title of Richard Dawkins' book *The Blind Watchmaker* is a deliberate echo of Paley's watch, and in the whole book he explains how the small changes over many millions of years, discarding the majority that are changes for the worse, and retaining only some that result in improvements, can lead to biological structures that have the appearance of conscious design.

[13]W. Paley (1805) *The Evidence of Design from Natural Theology* (9th edition), R. Faulder, London.

WILLIAM PALEY (1743–1805) was an English clergyman and philosopher, whose book *Natural Theology: or, Evidences of the Existence and Attributes of the Deity* introduced the watchmaker analogy to argue that biological structures must be the result of conscious design.

That is only on superficial examination, and closer study of many biological structures shows many design faults that a careful engineer would have avoided: In the vertebrate eye, for example, light must pass through a layer of nerve cells in order to reach the retina;[14] a signal from the brain to the larynx needs to pass around the aorta in order to reach its target, and in a giraffe this represents an unnecessary journey of about 2 meters; in the human, semen needs to loop around the ureter in order to pass from the testicles to the penis (Figure 15.1). Careful examination shows that biology is full of such examples, most of them, like these, not very biochemical, but the regulation of glycine synthesis discussed in Chapter 11 can be regarded as one that makes good sense in terms of the evolutionary origins of large animals, but hard to reconcile with conscious design. A properly designed large animal needs to be able to regulate collagen production independently of its need for C_1 units: Its inability to do this indicates that it has not been designed, but is simply the descendant of smaller animals that needed less collagen. As another biochemical example of poor design, consider the coenzyme NAD, which has a structure that is far from optimal for its function, as mentioned at the beginning of Chapter 6.

Paley's argument is often evoked today, often by people who think it is original and unanswerable, failing to notice that it is much older than the idea of natural selection, and that natural selection provides a convincing response to it. It is the basis of what is called "intelligent design," which was devised not by a scientist but by a retired Professor of Law at the University of California, Phillip Johnson. Although many of its adherents claim that they are real scientists and that "intelligent design" is different from creationism, a remarkable typographical error uncovered by Barbara Forrest, a Professor of Philosophy at Southeastern Louisiana University, tells a different story. In preparation for appearing as an expert witness for a court case in 2005, she managed to obtain various drafts of the book *Of Pandas and People*[15] that was intended to serve as a textbook presenting the creationist case. In particular, she found the following sentences in a draft from late 1987:

> The basic metabolic pathways (reaction chains) of nearly all organisms are the same. Is this because of descent from a common ancestor, or because only these pathways (and their variants) can sustain life? Evolutionists think the former is correct, cdesign proponentsists accept the latter view. Design proponents…

[14]The eye of an octopus is superficially similar, but it avoids this anomaly.

[15]Percival Davis and Dean H. Kenyon (1989) *Of Pandas and People*, Foundation for Thought and Ethics, Richardson, Texas.

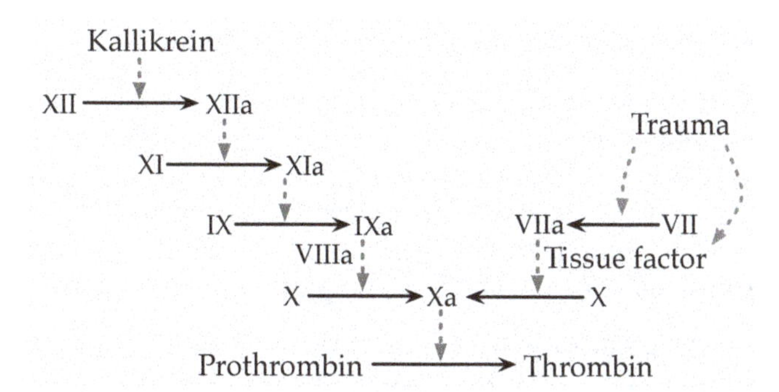

Figure 15.2 Clotting of blood in humans. The effect of trauma and the protein kallikrein is to provoke a *cascade* of reactions between different *factors* (XII, XIIa, and so on) that finally result in the production of thrombin from prothrombin. As discussed in the text, not all animals with a functioning system of clotting have all of the components, so this is not an example of irreducible complexity.

The strange words "cdesign proponentsists" catch the eye. They obviously result from a typographical error, but how did such an error arise? Trying to understand the mechanisms that produce copying errors, and thus, in the long term, evolution, is an important scientific enterprise, and in this instance Forrest's application of the same approach was richly rewarded.

She examined a draft of the book from early 1987 in which "design proponents" appeared throughout as "creationists." The decision of the USA Supreme Court to outlaw the teaching of "creation science" in public schools had led to a hurried search-and-replace operation that had caused the error: It could arise, for example, by selecting "reation" instead of "creation" before replacing it with "design proponent." This sort of example leaves little doubt that "intelligent design" is just a renaming of creationism and not an independent scientific discipline.[16]

The "cdesign proponentsists" perhaps have a parallel with the procollagen cycle (Chapter 11). Collagen is initially synthesized as procollagen, which has only the 20 "standard" amino acids encoded in the DNA. To form the triple helix of collagen, many proline and a few lysine residues in procollagen are transformed into hydroxyproline and hydroxylysine, two amino acids that occur in few other proteins. If these substitutions are made correctly, a viable collagen molecule results, but often they are not done correctly and the collagen molecule is rejected, as, indeed the substitution error discovered by Barbara Forrest did not survive into the published version of *Of Pandas and People*.

Michael Behe has tried to argue that "intelligent design" is different from

[16]These mistakes are easy to make. When I was finalizing the text of this book, I decided that I preferred "faucet" to "tap" in Chapter 8: I made an insufficiently careful search-and-replace operation and was afterwards puzzled why the word "mefaucethor" appeared in Chapter 14.

creationism because it is not Young-Earth Creationism:

> As commonly understood, creationism involves belief in an Earth formed only about 10 thousand years ago, an interpretation of the Bible that is still very popular. For the record, I have no reason to doubt that the Universe is the billions of years old that physicists say it is. Further, I find the idea of common descent (that all organisms share a common ancestor) fairly convincing, and have no particular reason to doubt it.[17]

But that is *not* how creationism is "commonly understood." In the minds of most biologists, there is no yawning chasm between Young-Earth Creationism and the sort of Old-Earth Creationism that Behe accepts, and even if he himself is not a Young-Earth Creationist, many of his supporters who advocate the teaching of "intelligent design" certainly are. The last quoted sentence is interesting, because Behe's grudging support for common descent would be likely to horrify some of his supporters if they recognized that it was there.

Irreducible complexity

Behe's principal argument is that many biochemical systems are "irreducibly complex" and thus could not have arisen step by step from simpler precursors, because if just one component is missing, the entire function fails. One example that he gives is that of the mechanism for blood clotting, which is certainly complex, as illustrated in Figure 15.2. The book *Of Pandas and People*, mentioned earlier, claims that "only when all the components of the [blood-clotting] system are present and in good working order does the system function properly," but this is simply not true. Whales and dolphins lack factor XII, but their blood clots normally; puffer fish lack factor XI as well, but their blood also clots normally.

Another supposed example of "irreducible complexity" is the tricarboxylate cycle, and as we have considered this already in Chapter 11, it will already be familiar. The left-hand side of Figure 15.3 is a drastically simplified version of the cycle (shown in a more complete form in Figure 11.8). It would seem that the cycle of eight reactions requires all eight reactions and would be useless if any were missing: It could not therefore have evolved step by step from a simpler pathway. This cycle includes four oxidative reactions, and thus requires molecular oxygen, which was not present in significant amounts in the primitive atmosphere. Enrique Meléndez-Hevia, whom we have met several times already in this book, and his colleagues[18] noted that the reactions of the left-hand side of the cycle can operate in the direction of reduction, and that if the reaction between succinyl-coenzyme A and succinate is omitted, the

[17]M. J. Behe (2006) *Darwin's Black Box: the Biochemical Challenge to Evolution* (2nd edition), Simon and Schuster, New York.

[18]E. Meléndez-Hevia, T. G. Waddell, and M. Cascante (1996) "The riddle of the Krebs citric acid cycle: assembling the pieces of chemically feasible reactions, and opportunism in the design of metabolic pathways during evolution" *Journal of Molecular Evolution* **43**, 293–303.

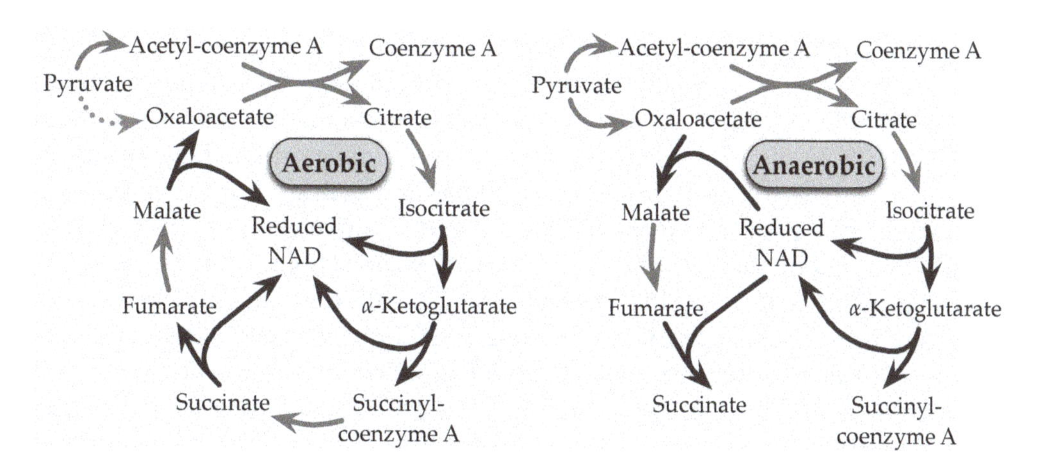

Figure 15.3 Evolution of the tricarboxylate cycle. The cycle as it operates in aerobic organisms today (*left*) is simplified from Figure 11.8. The anaplerotic production of oxaloacetate from pyruvate is not needed when the cycle is just acting as a cycle, but becomes necessary when intermediates are siphoned off for the production of amino acids and other metabolites. The four reactions shown in black are oxidation reactions, producing reduced NAD, and depend on molecular oxygen to reoxidize the NAD. The first organisms cannot have been aerobic, because molecular oxygen was not available in the primitive atmosphere, so the modern cycle would have been impossible. The reactions from oxaloacetate to malate and from fumarate to succinate can operate as reductions, however, and the reaction between malate and fumarate is easily reversible, so the "cycle" can have existed as two linear branches (*right*), with each branch replacing the oxidant or reductant consumed by the other. Because oxaloacetate is a starting material in both branches, its direct production from pyruvate is now necessary.

"cycle" then becomes an ordinary branched pathway with two linear branches, as shown on the right-hand side of Figure 15.3. Each of these branches could evolve step by step like any linear pathway. Addition of just one enzyme to convert succinyl-coenzyme A to succinate would then be sufficient to close the cycle when molecular oxygen appeared in the atmosphere. Thus, the tricarboxylate cycle is not an example of irreducible complexity.

MICHAEL J. BEHE (1952–) is an American biochemist, Professor of Biochemistry at Lehigh University (Pennsylvania). Almost the only real scientist in the "intelligent design" community, he maintains that various biological structures are *irreducibly complex* and could not have been reached by evolution.

The lesson to be drawn from these examples is that it is not sufficient to sit back in one's armchair and assert that a particular feature of biochemistry is too complex to have evolved step by step: We need first to be sure that the "facts" are correct, as the proponents of the examples of blood-clotting and the "impossible" evolution of the tricarboxylate cycle neglected to do, and second, we need to give some hard thought to the question of *how* the feature might have evolved.

Microevolution and macroevolution

Some creationists accept what they call *microevolution* but not *macroevolution*. What they mean by this is the claim that selection, whether natural or artificial, can produce great variety in animals such as dogs, but can never convert one "kind" of animal into another, say a dog into a cat; in more technical terms, they claim that despite the title of Darwin's book, natural selection cannot produce a new species. For Old-Earth Creationists, the problem comes from failing to appreciate the huge amount of time available. Dogs and cats shared a common ancestor more than 50 million years ago, so for natural selection to convert a dog into a cat would require of the order of 100 million years. Artificial selection could certainly proceed much faster, but not fast enough to show perceptible progress in a human lifetime. On the other hand, Young-Earth Creationists, believing the Earth to have existed for only some thousands of years, have logic on their side in this respect (if no other), because a few thousand years would not be sufficient to convert a dog into a cat.

As we have seen in Chapter 2, defining a species is not trivial, even for organisms that reproduce sexually. According to Mayr's definition, the six races of mice found in Madeira could be regarded as six new species that have appeared in the past millennium, though the authors who study them are more cautious and call them six "races." For nonsexual organisms, it is much more difficult, because the capacity to interbreed cannot be used as a criterion. We are forced to use more subjective criteria, such as the opinions of experts, to decide whether two varieties of bacteria constitute different species.[19] In the past, one of the criteria that was accepted to be part of the definition of the gut bacteria *Escherichia coli* as a species was that they could not grow aerobically on citrate as energy and carbon source.

On the other hand, bacteria have the great advantage that we do not need to wait 100 million years to see substantial change, and even during Darwin's lifetime William Dallinger, a minister in the Methodist Church, studied the capacity of microorganisms to adapt to temperatures much higher than those that would cause death in the wild species. A far more elaborate experiment has been in progress for more than 25 years in Richard Lenski's laboratory at Michigan State University, in which cultures of *Escherichia coli* have been followed in detail through more than 55,000 generations. They found a new clone that arose after 31,000 generations and could grow aerobically on citrate as carbon source. By the accepted definition of *Escherichia coli*, therefore, they had found a new species of bacteria that was not *Escherichia coli*.

[19] A full discussion of the questions that need to be considered for deciding which bacteria constitute distinct species is given by M. A. O'Malley (2014) *Philosophy of Microbiology*, Cambridge University Press, Cambridge.

UUU Phe	UCU Ser	UAU Tyr	UGU Cys
UUC Phe	UCC Ser	UAC Tyr	UGC Cys
UUA Leu	UCA Ser	UAA ★	UGA ★
UUG Leu	UCG Ser	UAG ★	UGG Trp
CUU Leu	CCU Pro	CAU His	CGU Arg
CUC Leu	CCC Pro	CAC His	CGC Arg
CUA Leu	CCA Pro	CAA Gln	CGA Arg
CUG Leu	CCG Pro	CAG Gln	CGG Arg
AUU Ile	ACU Thr	AAU Asn	AGU Ser
AUC Ile	ACC Thr	AAC Asn	AGC Ser
AUA Ile	ACA Thr	AAA Lys	AGA Arg
AUG Met	ACG Thr	AAG Lys	AGG Arg
GUU Val	GCU Ala	GAU Asp	GGU Gly
GUC Val	GCC Ala	GAC Asp	GGC Gly
GUA Val	GCA Ala	GAA Glu	GGA Gly
GUG Val	GCG Ala	GAG Glu	GGG Gly

Figure 15.4 The genetic code. This is a conventional representation of the genetic code, as in Figure 1.5. It is included here to facilitate comparison with the representation shown in Figure 15.5, and for the same reason, the three stop codons are shown as ★.

The "Wedge Document" We saw earlier in this chapter that the present state of biology bears little resemblance to what Jonathan Wells predicted would be that of 2025. That is still well in the future, and perhaps we will see big changes in the next 10 years, but we are much closer to another target date proposed by the Discovery Institute, the organization in which Wells is a Senior Fellow. It is dedicated to the dissemination of the ideas of "intelligent design," with the ultimate intention of making it the "dominant perspective in science" within 20 years of the writing of a policy document generally known as the "Wedge Document."[20] This was written in 1998 and we are approaching the end of the target period, so it is not too soon to examine whether design theory is being applied to several specific fields, among which biochemistry is explicitly mentioned. They hoped for "one hundred scientific, academic and technical articles by [their] fellows"—not in 20 years, but in the first five years. The total number is far smaller than that, even in their own flagship journal *Bio-Complexity*. Despite having an Editor-in-Chief, an Managing Editor, and an Editorial Board of 30 members (including, for example, Michael Behe, William Dembski, and Jonathan Wells), this journal published a total of four articles in 2014, a "critical focus," two "critical reviews," and just one research article.[21] This total of four articles per year has never been exceeded since the journal was founded in 2010.

As an example of the sort of research published in *Bio-Complexity*, Jennifer

[20]This was intended to be for internal use, but is now readily available as a PDF file, for example at `http://www.antievolution.org/features/wedge.pdf`.

[21]To put this in perspective, the journal *Biochemistry* published by the Americal Chemical Society has an organization about three times larger and published 776 articles in 2014.

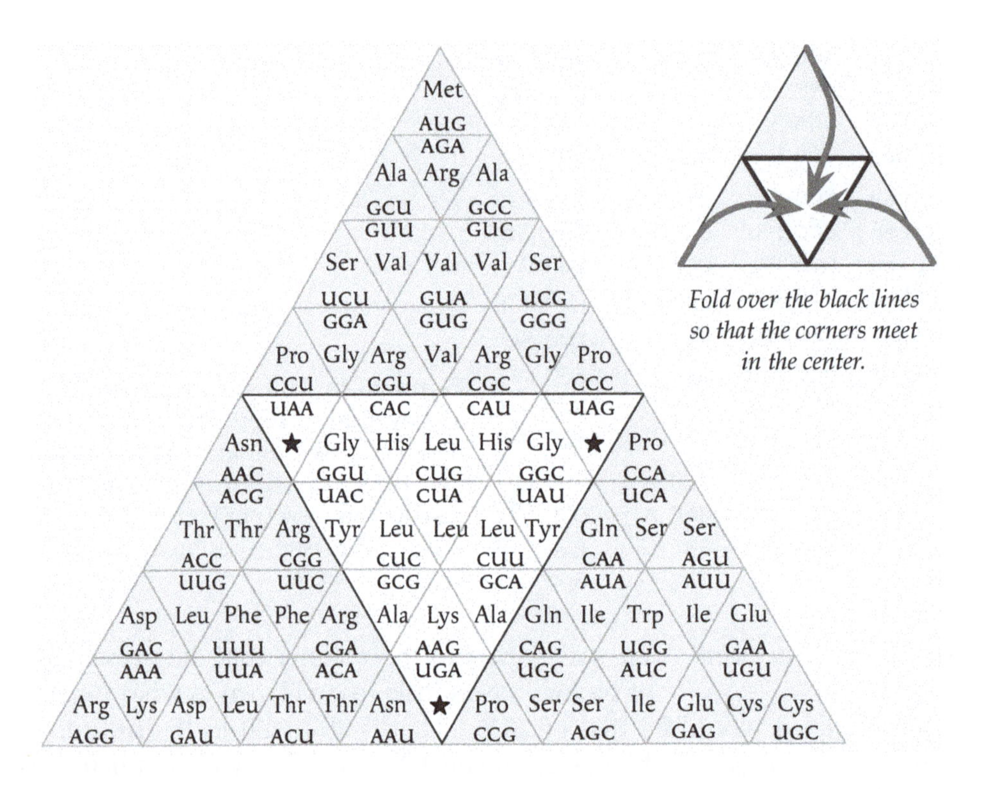

Figure 15.5 Another view of the genetic code. This rearrangement of Figure 15.4 is intended to emphasize aspects of symmetry in the assignment of codons. As presented, it is two-dimensional, but it can be folded along the black lines, as indicated in the small diagram, to produce a tetrahedron, bringing the triangles for methionine, arginine, and cysteine to a point. The three stars ★ represent the three stop codons, placed at the vertices of the central triangle, and the methionine codon AUG, which acts as an initiation codon (see Figure 1.5), at the top of the whole triangle. When the diagram is folded, this codon and the three stop codons are placed close to the vertices of the tetrahedron.

Raff[22] has made a useful analysis of a paper that sets out to reveal symmetry and regularity in the assignments of codons in the genetic code.[23] The main point is to compare the tetrahedral representation shown in Figure 15.5 with the more conventional table in Figure 15.4. The amino acids have been arranged to bring ones of similar properties close together, for example with the hydrophobic amino acids tryptophan, phenylalanine, isoleucine, valine, and leucine all in the central triangles of the faces of the tetrahedron, and so on. This sort of arrangement is subjective and leaves the six arginine codons scattered

[22]`http://tinyurl.com/ko2v8lc`.

[23]F. Castro-Chavez (2012) "A tetrahedral representation of the genetic code emphasizing aspects of symmetry" *Bio-Complexity* **2012** (2), 1–6.

in three pairs that are far from one another.

If we accept this arrangement of amino acids as reasonable, what about the corresponding codons? As can be seen in the conventional table, many amino acids are fully specified by just the first two bases of a codon, and so if the arrangement revealed anything useful about the "design" of the table, one might expect, for example, the four UCX codons to be placed in close proximity, but they are not. GUX, CUX, and UUX are neatly arranged in triangles, but they are the only ones: All of the others are scattered. This analysis may be contrasted with the question discussed in Chapter 3 of why arginine has six codons, whereas the more abundant amino acid glutamate has only two, and Jeffrey Wong suggested the *coevolution hypothesis* that was illustrated in Figure 3.3, whereby aspartate had to lose some of the codons that it may have had in a hypothetical more primitive code in favor of the amino acids that are synthesized from it. We may also contrast it with a different example of how real science is done, as I will now do.

How real science is done

One of the claims sometimes made by people who do not accept the reality of evolution is that it is not testable: Absolutely any observation would be consistent with evolution. This is an argument that can be used with far more justification about the claim that everything we see is the work of a designer, but no matter: How can a scientist test the reality of evolution? David Penny and his colleagues[24] took the question seriously and examined five sets of sequence data for different proteins (cytochome c, two forms of hemoglobin, and two "fibrinopeptides") from 11 organisms (horse, rhesus monkey, human, pig, cattle, sheep, kangaroo, rabbit, dog, rat or mouse, chimpanzee or gorilla). They argued that if an evolutionary relationship between these organisms existed, then analysis of the five sequences would yield similar patterns of relationship, whereas if there was no evolutionary relationship, the results would yield five completely different trees. If you have read as far as this in this book, you will not be surprised to know that they observed the results that evolution predicted.

Another example of how real scientists proceed is provided by the discovery of *Tiktaalik roseae* on Ellesmere Island, in the Canadian Arctic, one of the triumphs of modern evolutionary investigation. Its importance is not so much that it demonstrated the existence of intermediate species between bony fish and four-legged vertebrates—something no biologist doubted—but that knowledge of phylogeny predicted when it should have lived, geology predicted where fossils ought to be located, and exploration confirmed both predictions.[25] Let us look at that in more detail.

Before the discovery of *Tiktaalik roseae*, illustrated in Figure 15.6, the fossil

[24]D. Penny, L. R. Foulds, and M. D. Hendy (1982) "Testing the theory of evolution by comparing

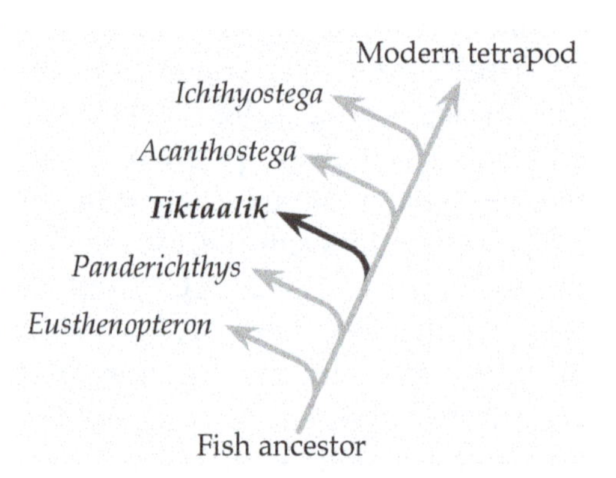

Figure 15.6 *Tiktaalik roseae* and the origin of tetrapods. The modern tetrapods (four-legged vertebrates) are descended from a fishlike ancestor. The fossil record now shows the species illustrated, ranging from ones that are clearly fishlike (*Eusthenopteron*) to ones that are clearly tetrapod-like (*Ichthyostega*). None of the organisms shown should be regarded as ancestral to any of the others. Rather, they represent side branches at different stages of evolution.

record showed clearly fishlike organisms, such as *Eusthenopteron*, and clearly tetrapod-like organisms, such as *Ichthyostega*, with others that were more or less fishlike (*Panderichthys*) or more or less tetrapod-like (*Acanthostega*), but no intermediate between these seemed to exist in the fossil record. Neil Shubin predicted that it should have lived about 375 million years ago, and that one should be able to find fossil evidence in rocks of that age. Such rocks occur in one of the most inhospitable regions of the world, Ellesmere Island in the Canadian Arctic, and an expedition to this region revealed the expected fossils, now known as *Tiktaalik roseae*. It is important to realize that none of the species in Figure 15.6 can be assumed to be ancestors of modern tetrapods, or to one another. It is better to regard them as various side branches from the progression from a fishlike ancestor to modern tetrapods.

Before leaving the topic of creationism, I want to quote some surprising words of Theodosius Dobzhansky that appeared in the same article[26] that I have mentioned already:

> It is wrong to hold creation and evolution as mutually exclusive alternatives. I am a creationist *and* an evolutionist. Evolution is God's, or Nature's, method of Creation. Creation is not an event that happened in 4004 B.C.; it is a process that began 10 billion years ago and is still under way.

Dobzhansky was a practicing member of the Eastern Orthodox Church and was among those who believe that there is no necessary conflict between science and religious faith. R. A. Fisher, several times mentioned already in this book, was a practicing Anglican, and there are also currently active biologists who also see no necessary conflict between science and religion,

phylogenetic trees constructed from five different protein sequences" *Nature* **297**, 197–200.

[25]E. B. Daeschler, N. H. Shubin, and F. A. Jenkins Jr. (2006) "Devonian tetrapod-like fish and the evolution of the tetrapod body plan" *Nature* **440**, 764–771.

[26]T. Dobzhansky (1973) "Nothing in biology makes sense except in the light of evolution" *American Biology Teacher* **35**, 125–129.

such as Dobzhansky's student Francisco Ayala, a Roman Catholic and former Dominican priest; the title alone of his article "The blasphemy of intelligent design"[27] is sufficient to make it clear what he thinks of the blasphemous character of "intelligent design."

I close this chapter, and the book, by quoting the great biologist George Gaylord Simpson, who as long ago as 1961 was concerned about the failure of American schools to teach biology satisfactorily and referred to the argument that one should "teach the controversy," or in other words, to teach both evolution and the arguments against it:[28]

GEORGE GAYLORD SIMPSON (1902–1984) was an American paleontologist whose *Tempo and Mode in Evolution* was a major component of the *Modern Synthesis*, the combination of natural selection with Mendelian genetics that forms the foundation of modern evolutionary theory.

> This was hailed by some teachers as the most "honest" compromise on the problem, but I am afraid I cannot agree. It is less honest—as the student is less able to judge from data in his own hands—than teaching that the Earth is flat and some say it is round. It would be honest only if the teacher pointed out that the authorities who "believe" in evolution ("believe" is a misleading word here, too) are, almost to a man, those who have actually studied the subject in a scientific way and that those who not believe in it are, almost to a man, ignorant of the scientific evidence and swayed by wholly nonscientific consideration.

That was in 1961 (if he had been writing today, he would probably have avoided phrases like "to a man"), but creationism has not disappeared, and has, in fact, become stronger. Nonetheless, the answer to the suggestion to "teach the controversy" is the same as Simpson's: *There is no controversy* among people who know what they are talking about.

[27] F. J. Ayala (2006) "The blasphemy of intelligent design" *History and Philosophy of the Life Sciences*, **28**, 409–421.

[28] G. G. Simpson (1961) "One hundred years without Darwin are enough" *Teachers College Record*, **60**, 617–626. Simpson took the title for his paper from a remark made by the geneticist H. J. Muller 100 years after the publication of *The Origin of Species*.

Index